Electrical Power Systems Technology

Contributors :
**Chengxiong Mao,
Hui Huang,** *et al.*

AURIS REFERENCE LTD.
London, UK

Electrical Power Systems Technology
Contributors : Hui Huang *and* Chengxiong Mao, *et al.*

Auris Reference Ltd., UK

www.aurisreference.com

United Kingdom

Copyright 2016

Printed in 2017 for Sale in the Indian Subcontinent

The information in this book has been obtained from highly regarded resources. The copyrights for individual articles remain with the authors, as indicated. All chapters are distributed under the terms of the Creative Commons Attribution License, which permit unrestricted use, distribution, and reproduction in any medium, provided the original author and source are credited.

Notice

Contributors, whose names have been given on the book cover, are not associated with the Publisher. The editors and the Publisher have attempted to trace the copyright holders of all material reproduced in this publication and apologise to copyright holders if permission has not been obtained. If any copyright holder has not been acknowledged, please write to us so we may rectify.

Reasonable efforts have been made to publish reliable data. The views articulated in the chapters are those of the individual contributors, and not necessarily those of the editors or the Publisher. Editors and/or the Publisher are not responsible for the accuracy of the information in the published chapters or consequences from their use. The Publisher accepts no responsibility for any damage or grievance to individual(s) or property arising out of the use of any material(s), instruction(s), methods or thoughts in the book.

No part of this publication maybe reproduced, stored in a retrieval system or transmitted in any form or by any means, electronic, mechanical, photocopying, recording, scanning or otherwise without prior written permission of the publisher.

Electrical Power Systems Technology

ISBN: 978-1-78154-512-6

British Library Cataloguing in Publication Data
A CIP record for this book is available from the British Library

Exclusively distributed by CBS Publishers & Distributors Pvt. Ltd.

Sales & Distribution Rights only for India, Pakistan, Bangladesh, Sri Lanka, Nepal and Bhutan.This book is not to be sold outside these territories.

Electrical Power Systems Technology

Preface

Electrical power systems includes power generation, electrical power transmission & distribution, control & power system protection, electrical switch-gear, power transformers, control & protection relays, transmission & distribution substations, and all switch yard equipment. All the very basic electrical engineering theories related to this field of technology are also included in this study site. This site is continually improving and in the near future we will add many other functionalities to this website to make it a complete solution for electrical technology. Please give your suggestions to help make the site better. This is a complete reference of electrical engineering and technology.

Engineers involved in electrical system design must also consider safety and legal requirements. When it comes to powering a building, for example, many cities have strict electrical codes in place to protect residents from electrical dangers and fire. National and international organizations also produce standards governing the use of electricity to power everything from cars to televisions. These codes are designed to prevent fires, reduce the risk of electric shock, and protect users from injury or death.

Electric Power Systems comprise the theory and methods for generation, transmission, distribution and consumption of electric energy. Important areas are, among others, power flows, power system dynamics, reliability, new power system components, power quality and analysis of the power market. Electric power has become increasingly important as a way of transmitting and transforming energy in industrial, military and transportation uses. Electric power systems are also at the heart of alternative energy systems, including wind and solar electric, geothermal and small scale hydroelectric generation.

Owing to its lucid style and presentation of advanced topics, the book will be useful to postgraduate students as also to practising engineers.

This page left intentionally blank.

Contents

Preface	*v*
1. ELECTRIC POWER	**1-34**
Introduction	1
Definition	1
Explanation	2
Electricity Generation	7
Battery	19
2. AC AND DC POWER	**35-85**
AC Power	35
Direct Current	51
High-Voltage Direct Current	54
Alternating Current	75
3. POWER MEASUREMENT FUNDAMENTALS	**86-178**
Voltage	86
Electric Current	90
Electrical Resistance and Conductance	112
Electrical Reactance	121
Magnetic Flux	123
Electric Charge	128
Electric Field	137
Electricity Meter	142
Inductance	157
Admittance	177
4. POWER SYSTEM FUNDAMENTALS	**179-212**
History	179
Basics of Electric Power	181

Balancing the Grid　　　　　　　　　　　　　　　　181
　　　Components of Power Systems　　　　　　　　　　　182
　　　Power Systems in Practice　　　　　　　　　　　　　187
　　　Electrical Network　　　　　　　　　　　　　　　　189
　　　Electrical Laws　　　　　　　　　　　　　　　　　　191
　　　Phasor　　　　　　　　　　　　　　　　　　　　　　207

5. ELECTRICAL POWER PRODUCTION SYSTEMS　　　　　213-308
　　　Power Station　　　　　　　　　　　　　　　　　　213

6. ELECTRONIC POWER TRANSFORMER CONTROL STRATEGY IN WIND ENERGY CONVERSION SYSTEMS FOR LOW VOLTAGE RIDE-THROUGH CAPABILITY ENHANCEMENT OF DIRECTLY DRIVEN WIND TURBINES WITH PERMANENT MAGNET SYNCHRONOUS GENERATORS (D-PMSGs)　　　　　　　　　　　　　　　　　　　　309-328

　　　Hui Huang, Chengxiong Mao, Jiming Lu and Dan Wang
　　　Introduction　　　　　　　　　　　　　　　　　　310
　　　System Description and System Model　　　　　　311
　　　Control Strategy of the System　　　　　　　　　　315
　　　Simulation Results　　　　　　　　　　　　　　　　319
　　　Conclusions　　　　　　　　　　　　　　　　　　　324
　　　References　　　　　　　　　　　　　　　　　　　325

List of Contributors

Hui Huang

School of Electrical & Electronic Engineering, Huazhong University of Science and Technology, Wuhan 430074, Hubei, China; E-Mails: dh-1126@163.com (H.H.); cxmao@mail.hust.edu.cn (C.M.); lujiming@mail.hust.edu.cn (J.L.)

Chengxiong Mao

School of Electrical & Electronic Engineering, Huazhong University of Science and Technology, Wuhan 430074, Hubei, China; E-Mails: dh-1126@163.com (H.H.); cxmao@mail.hust.edu.cn (C.M.); lujiming@mail.hust.edu.cn (J.L.)

Jiming Lu

School of Electrical & Electronic Engineering, Huazhong University of Science and Technology, Wuhan 430074, Hubei, China; E-Mails: dh-1126@163.com (H.H.); cxmao@mail.hust.edu.cn (C.M.); lujiming@mail.hust.edu.cn (J.L.)

Dan Wang

School of Electrical & Electronic Engineering, Huazhong University of Science and Technology, Wuhan 430074, Hubei, China; E-Mails: dh-1126@163.com (H.H.); cxmao@mail.hust.edu.cn (C.M.); lujiming@mail.hust.edu.cn (J.L.)

This page left intentionally blank.

Chapter 1

ELECTRIC POWER

INTRODUCTION

Electric power is the rate at which electric energy is transferred by an electric circuit. The SI unit of power is the watt, one joule per second.

Electric power is usually produced by electric generators, but can also be supplied by sources such as electric batteries. Electric power is generally supplied to businesses and homes by the electric power industry. Electric power is usually sold by the kilowatt hour (3.6 MJ) which is the product of power in kilowatts multiplied by running time in hours. Electric utilities measure power using an electricity meter, which keeps a running total of the electric energy delivered to a customer.

DEFINITION

Electric power, like mechanical power, is the rate of doing work, measured in watts, and represented by the letter P. The term *wattage* is used colloquially to mean "electric power in watts." The electric power in watts produced by an electric current I consisting of a charge of Q coulombs every t seconds passing through an electric potential (voltage) difference of V is

$$P = \text{work done per unit time} = \frac{QV}{t} = IV$$

where :
- Q is electric charge in coulombs
- t is time in seconds
- I is electric current in amperes
- V is electric potential or voltage in volts.

EXPLANATION

Electric power is transformed to other forms of power when electric charges move through an electric potential (voltage) difference, which occurs in electrical components in electric circuits. When electric charges move through a potential difference from a high voltage to a low voltage, the energy in the potential is converted to kinetic energy of the charges, which perform work on the device. Devices in which this occurs are called *passive* devices or *loads*; they consume electric power, converting it to other forms such as mechanical work, heat, light, etc. Examples are electrical appliances, such as light bulbs, electric motors, and electric heaters.

If the charges are forced to move by an outside force in the direction from a lower potential to a higher, work is being done *on* the charges, so power is transferred to the electric current from some other type of energy, such as mechanical energy or chemical energy. Devices in which this occurs are called *active* devices or *power sources*; sources of electric current, such as electric generators and batteries.

Passive Sign Convention

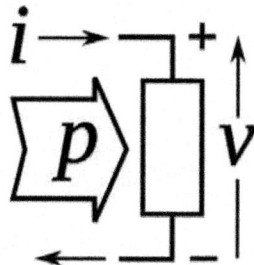

Illustration of the "*reference directions*" of the current (i), voltage (v), and power (p) variables used in the passive sign convention. If positive current is defined as flowing into the terminal which is defined to have positive voltage, then positive power represents electric power flowing into the device (*big arrow*).

In electrical engineering, the **passive sign convention** (**PSC**) is a sign convention or arbitrary standard rule adopted universally by the electrical engineering community for defining the sign of electric power in an electric circuit. The convention defines electric power flowing *into* an electrical component (out of the circuit) as positive, and power flowing *out* of a component as negative. So a passive component which consumes power, such as an appliance or light bulb, will have *positive* power dissipation, while an active component, a source of power such as an electric generator or battery, will have *negative* power dissipation. This is the standard definition of power in electric circuits.

To comply with the convention, the direction of the voltage and current variables used to calculate power and resistance in the component must have a certain relationship : the current variable must be defined so positive current enters the positive voltage terminal of the device. These directions may be different from the directions of the actual current flow and voltage.

The Convention

The passive sign convention states that in components in which the conventional current variable i is defined as entering the device through the terminal which is positive as defined by the voltage variable v, the power p and resistance r are given by

$$p = vi \text{ and } r = v/i$$

In components in which the current i enters the device through the negative voltage terminal, power and resistance are given by

$$p = vi \text{ and } r = -v/i$$

With these definitions, passive components (loads) will have $p > 0$ and $r > 0$, and active components (power sources) will have $p < 0$ and $r < 0$.

Explanation

Active and Passive Components

In electrical engineering, power represents the rate of electrical energy flowing into or out of a given component or control volume. Power is a signed quantity; negative power just represents power flowing in the opposite direction from positive power. From the standpoint of power flow, electrical components in a circuit can be divided into two types :

- In a *load* or *passive* component, such as a light bulb, resistor, or electric motor, electric current (flow of positive charges) moves through the device under the influence of the voltage in the direction of lower electric potential, from the positive terminal to the negative. So work is done *by* the charges *on* the component; potential energy flows out of the charges; and electric power flows from the circuit into the component.
- In a *source* or *active* component, such as a battery or electric generator, current is forced to move through the device in the direction of greater electric potential energy, from the negative to the positive voltage terminal. Work must be done *on* the moving charges by some source of energy in the component, to make them move in this direction. This increases their potential energy, so electric power flows out of the component into the circuit.

Some components can be either a source or a load, depending on the voltage or current through them. For example, a rechargeable battery acts as a source when it is used to produce power, but as a load when it is being recharged.

Since it can flow in either direction, there are two possible ways to define electric power; two possible *reference directions* : either power flowing into the circuit, or power flowing out of the circuit, can be defined as positive. Whichever is defined as positive, the other will be negative. The passive sign convention arbitrarily defines power flowing *out* of the circuit (*into* the component) as positive, so passive components have "positive" power flow.

Reference Directions

The power flow *p* and resistance *r* of an electrical component are related to the voltage *v* and current *i* variables by the defining equation for power and Ohm's law :

$$p = vi \qquad (1)$$
$$r = vi \qquad (2)$$

Like power, voltage and current are signed quantities. The current flow in a wire has two possible directions, so when defining a current variable *i* the direction which represents positive current flow must be indicated, usually by an arrow on the circuit diagram. This is called the *reference direction for current i*. If the actual current is in the opposite direction, the variable *i* will have a negative value. Similarly in defining a voltage variable *v*, the terminal which represents the positive side must be specified, usually with an arrow or plus sign. This is called the *reference direction for voltage v*.

To understand the PSC, it is important to distinguish the reference directions of the variables, *v* and *i*, which can be assigned at will, from the direction of the actual *voltage* and *current*, which is determined by the circuit. The idea of the PSC is that by assigning the reference direction of variables *v* and *i* in a component with the right relationship, the power flow in passive components calculated from Eq. (1) will come out positive, while the power flow in active components will come out negative. It is not necessary to know whether a component produces or consumes power when analyzing the circuit; reference directions can be assigned arbitrarily, directions to currents and polarities to voltages, then the PSC is used to calculate the power in components. If the power comes out positive, the component is a load, converting electric power to some other kind of power. If the power comes out negative, the component is a source, converting some other form of power to electric power.

Sign Conventions

So, choosing the relative direction of the voltage and current variables in a component determines the direction of power flow that is considered positive. The reference directions of the individual variables are not important, only their relation to each other. There are two choices :

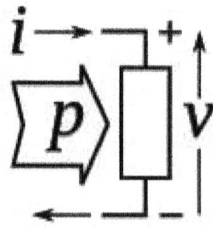

- *Passive sign convention* : Defining the current variable as entering the positive terminal means that if the voltage and current variables have positive values, current flows from the positive to the negative terminal, doing work *on* the component, as occurs in a passive component. So power flowing *into* the component from the line is defined as positive; the power variable represents power *dissipation* in the component. Therefore
    - Active components (power sources) will have negative resistance and negative power flow
    - Passive components (loads) will have positive resistance and positive power flow

    This is the convention normally used.

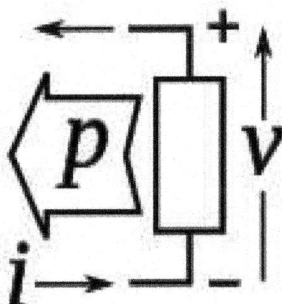

- *Active sign convention* : Defining the current variable as entering the negative terminal means that if the voltage and current variables have positive values, current flows from the negative to the positive terminal, so work is being done *on* the current, and power flows *out* of the component. So power flowing out of the component is defined as positive; the power variable represents power *produced*. Therefore :
    - Active components will have positive resistance and positive power flow
    - Passive components will have negative resistance and negative power flow

    This convention is not used.

In practice it is not necessary to assign the voltage and current variables in a circuit to comply with the PSC. Components in which the variables have a "backward" relationship, in which the current variable enters the negative terminal, can still be made to comply with the PSC by changing the sign of the constitutive relations (1) and (2) used with them. A current entering the negative terminal is equivalent to a negative current entering the positive terminal, so in such a component

$$p = v(-i) = -vi$$

and

$$r = v/(-i) = -v/i$$

Conservation of Energy

One advantage of defining all the variables in a circuit to comply with the PSC is that it makes it easy to express conservation of energy. Since electric energy cannot be created or destroyed, at any given instant every watt of power consumed by a load component must be produced by some source component in the circuit. Therefore the sum of all the power consumed by loads equals the sum of all the power produced by sources. Since with the PSC the power dissipation in sources is negative and power dissipation in loads is positive, the sum of all the power dissipation in all the components in a circuit is always zero

$$\sum_n p_n = \sum_n v_n i_n = 0$$

AC Circuits

Since the sign convention only deals with the directions of the *variables* and not with the direction of the actual *current*, it also applies to alternating current (AC) circuits, in which the direction of the voltage and current periodically reverses. In AC circuits, even though the voltage and current reverse direction, a "formal" direction of current flow and voltage polarity are defined by considering the voltage and current direction in the first half of the cycle "positive". During the second half of the AC cycle, in a resistive AC circuit, both the voltage and current in the device reverse direction, so the sign of the voltage and current reverse. Since the power is the product of voltage and current, the two sign reversals cancel each other, and the sign of the power flow is unchanged.

Alternative Convention in Power Engineering

In practice, the power output of power sources such as batteries and generators is not given in negative numbers, as required by the passive sign convention. No manufacturer sells a "−5 kilowatt generator". The standard practice in electric power circuits is to use positive values for the power and resistance of power sources, as well as loads. This avoids confusion over the meaning of "negative power", and particularly "negative resistance". In order to make the power for both sources and loads come out positive, instead of the PSC, separate sign conventions must be used for sources and loads. These are called the "*generator-load conventions*"

- **Generator convention** : In source components like generators and batteries, the variables V and I are defined according to the *active sign convention*; the current variable is defined as entering the negative terminal of the device.
- **Load convention** : In loads, the variables are defined according to the normal passive sign convention; the current variable is defined as entering the positive terminal.

Using this convention, positive power flow in source components is power *produced*, while positive power flow in load components is power *consumed*. As with the PSC, if the variables in a given component do not conform to the applicable convention, the component can still be made to conform by using negative signs in the constitutive equations (1) and (2)

$$P = -VI \text{ and } R = -V/I$$

This convention may seem preferable to the PSC, since the power P and resistance R always have positive values. However it cannot be used in electronics, because it is not possible to classify some electronic components unambiguously as "sources" or "loads". Some electronic components may act as sources of power with negative resistance in some portions of their operating range, and as absorbers of power with positive resistance in other portions, or even in different portions of the AC cycle. Whether the component acts as a source or load may depend on the current i or voltage v in it, which is not known until the circuit is analyzed. For example, if the voltage across a rechargeable battery's terminals is less than it's open-circuit voltage, it will act as a source, while if the voltage is greater it will act as a load and recharge. Similarly an active device like a transistor can either absorb or produce AC power depending on whether the phase between the voltage and current is less than or greater than 90°. So it is necessary for power and resistance variables to be able to take on both positive and negative values.

Resistive Circuits

In the case of resistive (Ohmic, or linear) loads, Joule's law can be combined with Ohm's law ($V = I \cdot R$) to produce alternative expressions for the dissipated power :

$$P = I^2 R = \frac{V^2}{R},$$

where R is the electrical resistance.

Electromagnetic Fields

Electrical power flows wherever electric and magnetic fields exist together and fluctuate in the same place. In the general case, however, the simple equation $P = IV$ must be replaced by a more complex calculation, the integral of the cross-product of the electrical and magnetic field vectors over a specified area, thus :

$$P = \int_s (E \times H) \cdot dA.$$

The result is a scalar since it is the *surface integral* of the *Poynting vector*.

ELECTRICITY GENERATION

Electricity generation is the process of generating electric power from other sources of primary energy. The fundamental principles of electricity generation were discovered during the 1820s and early 1830s by the British scientist Michael Faraday. His basic method is still used today : electricity is generated by the movement of a loop of wire, or disc of copper between the poles of a magnet. For electric utilities, it is the first process in the delivery of electricity to consumers.

The other processes, electricity transmission, distribution, and electrical power storage and recovery using pumped-storage methods are normally carried out by the electric power industry. Electricity is most often generated at a power station by electromechanical generators, primarily driven by heat engines fueled by chemical combustion or nuclear fission but also by other means such as the kinetic energy of flowing water and wind. Other energy sources include solar photovoltaics and geothermal power.

History

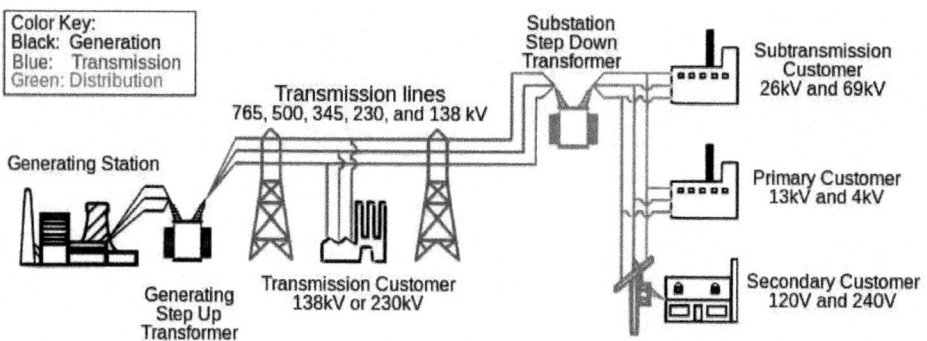

Fig. : Diagram of an electric power system, generation system in black.

Central power stations became economically practical with the development of alternating current power transmission, using power transformers to transmit power at high voltage and with low loss. Electricity has been generated at central stations since 1882. The first power plants were run on water power or coal, and today we rely mainly on coal, nuclear, natural gas, hydroelectric, wind generators, and petroleum, with a small amount from solar energy, tidal power, and geothermal sources. The use of power-lines and power-poles have been significantly important in the distribution of electricity.

Cogeneration

Cogeneration is the practice of using exhaust or extracted steam from a turbine for heating purposes, such as drying paper, distilling petroleum in a refinery or for building heat. Before central power stations were widely introduced it was common for industries, large hotels and commercial buildings to generate their own power and use low pressure exhaust steam for heating. This practice carried on for many years after central stations became common and is still in use in many industries.

Methods of Generating Electricity

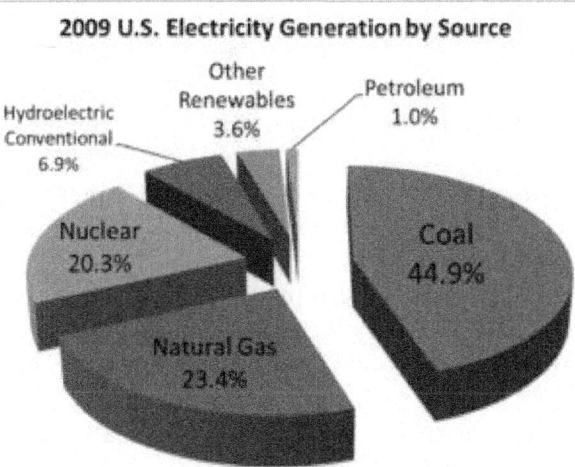

Fig. : Sources of electricity in the U.S. in 2009 fossil fuel generation (mainly coal) was the largest source.

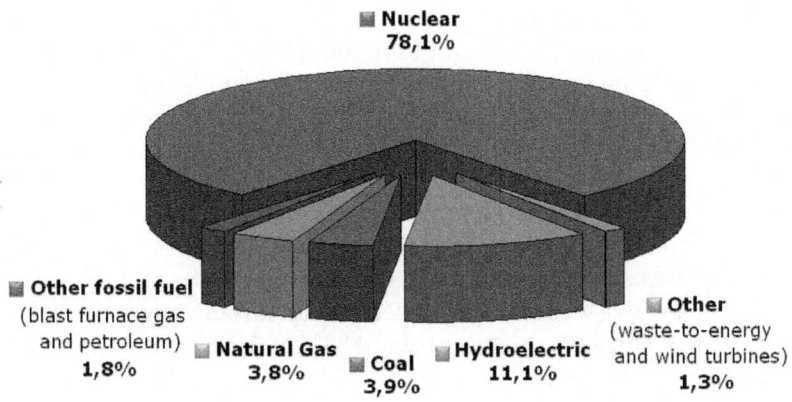

Fig. : Sources of electricity in France in 2006; nuclear power was the main source.

There are seven fundamental methods of directly transforming other forms of energy into electrical energy :
- Static electricity, from the physical separation and transport of charge (examples : triboelectric effect and lightning)
- Electromagnetic induction, where an electrical generator, dynamo or alternator transforms kinetic energy (energy of motion) into electricity. This is the

most used form for generating electricity and is based on Faraday's law. It can be experimented by simply rotating a magnet within closed loops of a conducting material (*e.g.* copper wire)
- Electrochemistry, the direct transformation of chemical energy into electricity, as in a battery, fuel cell or nerve impulse
- Photoelectric effect, the transformation of light into electrical energy, as in solar cells
- Thermoelectric effect, the direct conversion of temperature differences to electricity, as in thermocouples, thermopiles, and thermionic converters.
- Piezoelectric effect, from the mechanical strain of electrically anisotropic molecules or crystals. Researchers at the US Department of Energy's Lawrence Berkeley National Laboratory (Berkeley Lab) have developed a piezoelectric generator sufficient to operate a liquid crystal display using thin films of M13 bacteriophage.
- Nuclear transformation, the creation and acceleration of charged particles (examples : betavoltaics or alpha particle emission)

Static electricity was the first form discovered and investigated, and the electrostatic generator is still used even in modern devices such as the Van de Graaff generator and MHD generators. Charge carriers are separated and physically transported to a position of increased electric potential. Almost all commercial electrical generation is done using electromagnetic induction, in which mechanical energy forces an electrical generator to rotate. There are many different methods of developing the mechanical energy, including heat engines, hydro, wind and tidal power. The direct conversion of nuclear potential energy to electricity by beta decay is used only on a small scale. In a full-size nuclear power plant, the heat of a nuclear reaction is used to run a heat engine. This drives a generator, which converts mechanical energy into electricity by magnetic induction. Most electric generation is driven by heat engines. The combustion of fossil fuels supplies most of the heat to these engines, with a significant fraction from nuclear fission and some from renewable sources. The modern steam turbine (invented by Sir Charles Parsons in 1884) currently generates about 80% of the electric power in the world using a variety of heat sources.

Turbines

All turbines are driven by a fluid acting as an intermediate energy carrier. Many of the heat engines just mentioned are turbines. Other types of turbines can be driven by wind or falling water. Sources include :
- **Steam** - Water is boiled by :
  o Nuclear fission
  o The burning of fossil fuels (coal, natural gas, or petroleum). In hot gas (gas turbine), turbines are driven directly by gases produced by the combustion of natural gas or oil. Combined cycle gas turbine plants are driven by both steam and natural gas. They generate power by burning natural gas in a

gas turbine and use residual heat to generate additional electricity from steam. These plants offer efficiencies of up to 60%.
- o Renewables. The steam is generated by :
  - Biomass
  - Solar thermal energy (the sun as the heat source) : solar parabolic troughs and solar power towers concentrate sunlight to heat a heat transfer fluid, which is then used to produce steam.
  - Geothermal power. Either steam under pressure emerges from the ground and drives a turbine or hot water evaporates a low boiling liquid to create vapour to drive a turbine.
  - Ocean thermal energy conversion (OTEC) : uses the big difference between cooler deep and warmer surface ocean waters to run a heat engine (usually a turbine).
- Other renewable sources :
  - o **Water** (hydroelectric) - Turbine blades are acted upon by flowing water, produced by hydroelectric dams or tidal forces.
  - o **Wind** - Most wind turbines generate electricity from naturally occurring wind. Solar updraft towers use wind that is artificially produced inside the chimney by heating it with sunlight, and are more properly seen as forms of solar thermal energy.

Reciprocating Engines

Small electricity generators are often powered by reciprocating engines burning diesel, biogas or natural gas. Diesel engines are often used for back up generation, usually at low voltages. However most large power grids also use diesel generators, originally provided as emergency back up for a specific facility such as a hospital, to feed power into the grid during certain circumstances. Biogas is often combusted where it is produced, such as a landfill or wastewater treatment plant, with a reciprocating engine or a microturbine, which is a small gas turbine.

Photovoltaic Panels

Photovoltaic panels convert sunlight directly to electricity. Although sunlight is free and abundant, solar electricity is still usually more expensive to produce than large-scale mechanically generated power due to the cost of the panels. Low-efficiency silicon solar cells have been decreasing in cost and multi-junction cells with close to 30% conversion efficiency are now commercially available. Over 40% efficiency has been demonstrated in experimental systems. Until recently, photovoltaics were most commonly used in remote sites where there is no access to a commercial power grid, or as a supplemental electricity source for individual homes and businesses. Recent advances in manufacturing efficiency and photovoltaic technology, combined with subsidies driven by environmental concerns, have dramatically accelerated the deployment of solar panels. Installed capacity is growing by 40% per year led by increases in Germany, Japan, California and New Jersey.

Other Generation Methods

Various other technologies have been studied and developed for power generation. Solid-state generation (without moving parts) is of particular interest in portable applications. This area is largely dominated by thermoelectric (TE) devices, though thermionic (TI) and thermophotovoltaic (TPV) systems have been developed as well. Typically, TE devices are used at lower temperatures than TI and TPV systems. Piezoelectric devices are used for power generation from mechanical strain, particularly in power harvesting. Betavoltaics are another type of solid-state power generator which produces electricity from radioactive decay. Fluid-based magnetohydrodynamic (MHD) power generation has been studied as a method for extracting electrical power from nuclear reactors and also from more conventional fuel combustion systems. Osmotic power finally is another possibility at places where salt and fresh water merges (*e.g.* deltas, ...) Electrochemical electricity generation is also important in portable and mobile applications. Currently, most electrochemical power comes from closed electrochemical cells ("batteries"), which are arguably utilized more as storage systems than generation systems; but open electrochemical systems, known as fuel cells, have been undergoing a great deal of research and development in the last few years. Fuel cells can be used to extract power either from natural fuels or from synthesized fuels (mainly electrolytic hydrogen) and so can be viewed as either generation systems or storage systems depending on their use.

Economics of Generation and Production of Electricity

The selection of electricity production modes and their economic viability varies in accordance with demand and region. The economics vary considerably around the world, resulting in widespread selling prices, Venezuela 3 cents per kWh or Denmark 40 cents per kWh. Hydroelectric plants, nuclear power plants, thermal power plants and renewable sources have their own pros and cons, and selection is based upon the local power requirement and the fluctuations in demand. All power grids have varying loads on them but the daily minimum is the base load, supplied by plants which run continuously. Nuclear, coal, oil and gas plants can supply base load, with the low-carbon option being nuclear. Thermal energy is economical in areas of high industrial density, as the high demand cannot be met by renewable sources. The effect of localized pollution is also minimized as industries are usually located away from residential areas. These plants can also withstand variation in load and consumption by adding more units or temporarily decreasing the production of some units. Nuclear power plants can produce a huge amount of power from a single unit. However, recent disasters in Japan have raised concerns over the safety of nuclear power, and the capital cost of nuclear plants is very high. Hydroelectric power plants are located in areas where the potential energy from falling water can be harnessed for moving turbines and the generation of power. It is not an economically viable source of production where the load varies too much during the annual production cycle and the ability to store the flow of water is limited. Renewable sources other than hydroelectricity (solar power, wind energy, tidal power, etc.) due to advancements in technology, and with mass production, their cost of production has come down and the energy is now in many cases cost-comparative with fossil fuels. Many governments around the world provide subsidies to offset the higher cost of any new power production, and to make the installation of renewable energy systems economically feasible. However, their use is frequently limited by their intermittent nature. If natural gas prices are below $3 per million British thermal units, generating electricity from natural gas is cheaper than generating power by burning coal.

Production

The production of electricity in 2009 was 20,053TWh. Sources of electricity were fossil fuels 67%, renewable energy 16% (mainly hydroelectric, wind, solar and biomass), and nuclear power 13%, and other sources were 3%. The majority of fossil fuel usage for the generation of electricity was coal and gas. Oil was 5.5%, as it is the most expensive common commodity used to produce electrical energy. Ninety-two per cent of renewable energy was hydroelectric followed by wind at 6% and geothermal at 1.8%. Solar photovoltaic was 0.06%, and solar thermal was 0.004%. Data are from OECD 2011-12 Factbook (2009 data).

Source of Electricity (World total year 2008)							
-	Coal	Oil	Natural Gas	Nuclear	Renew-ables	oth-er	Total
Average electric power (TWh/year)	8,263	1,111	4,301	2,731	3,288	568	20,261
Average electric power (GW)	942.6	126.7	490.7	311.6	375.1	64.8	2311.4
Proportion	41%	5%	21%	13%	16%	3%	100%

Data Source IEA/OECD

Total energy consumed at all power plants for the generation of electricity was 4,398,768 ktoe (kilo ton of oil equivalent) which was 36% of the total for primary energy sources (TPES) of 2008. Electricity output (gross) was 1,735,579 ktoe (20,185 TWh), efficiency was 39%, and the balance of 61% was generated heat. A small part (145,141 ktoe, which was 3% of the input total) of the heat was utilized at co-generation heat and power plants. The in-house consumption of electricity and power transmission losses were 289,681 ktoe. The amount supplied to the final consumer was 1,445,285 ktoe (16,430 TWh) which was 33% of the total energy consumed at power plants and heat and power co-generation (CHP) plants.

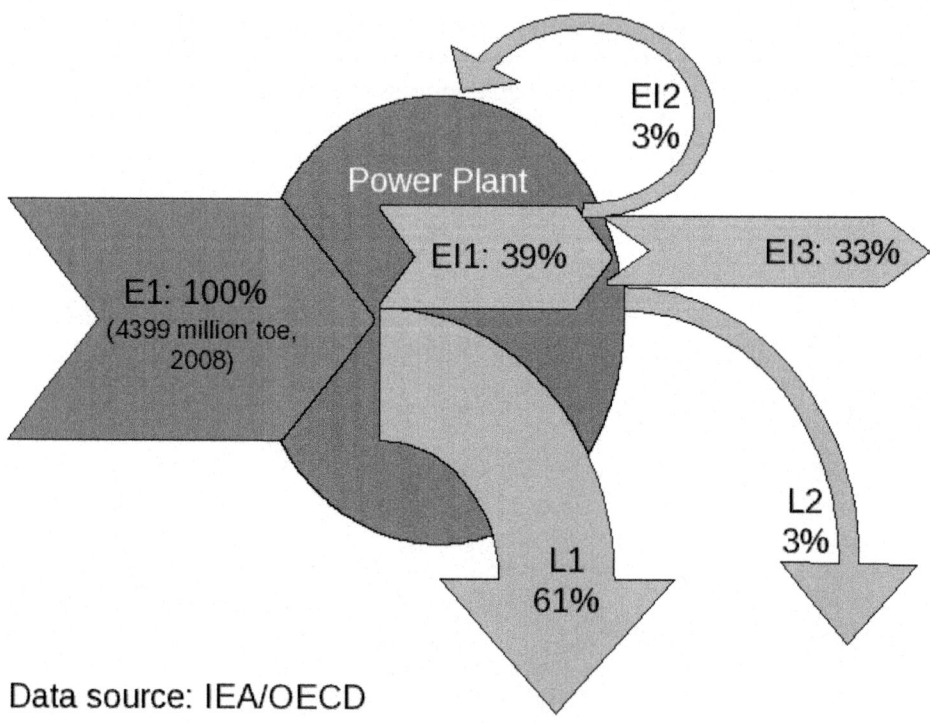

Data source: IEA/OECD

Fig. : Energy Flow of Power Plant.

Historical Results of Production of Electricity

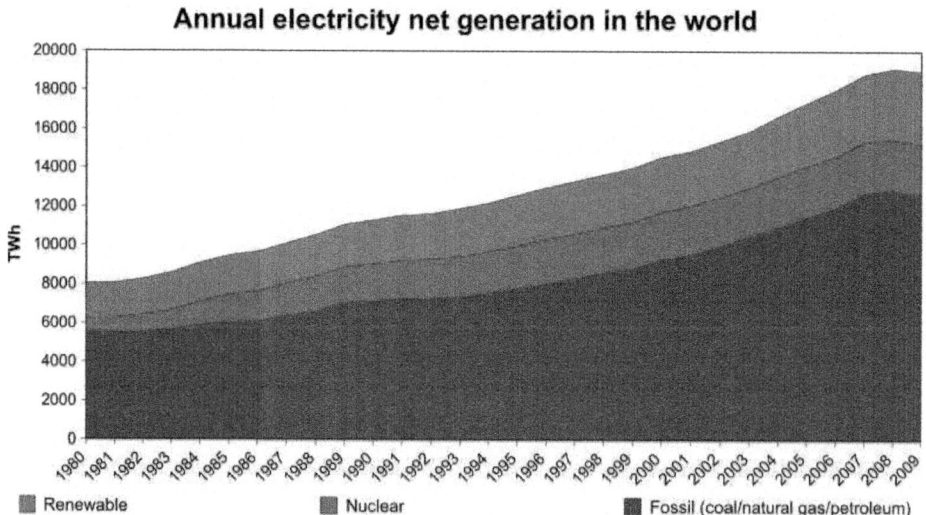

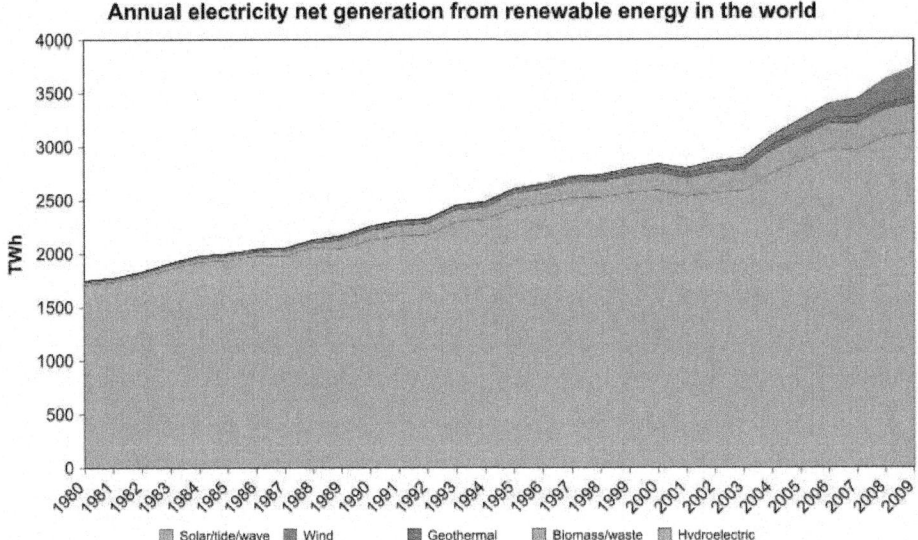

Production By Country

The United States has long been the largest producer and consumer of electricity, with a global share in 2005 of at least 25%, followed by China, Japan, Russia, and India. As of Jan-2010, total electricity generation for the 2 largest generators was as follows : USA : 3992 billion kWh (3992 TWh) and China : 3715 billion kWh (3715 TWh).

List of Countries with Source of Electricity 2008

Data source of values (electric power generated) is IEA/OECD. Listed countries are top 20 by population or top 20 by GDP (PPP) and Saudi Arabia based on CIA World Factbook 2009.

Composition of Electricity by Resource (TWh per year 2008)

Country	Fossil Fuel					Nuclear	rank	Renewable							rank	Bio other*	total	rank
	Coal	Oil	Gas	sub total	rank			Hydroe	Geo Thermal	Solar PV*	Solar Thermal	Wind	Tide	sub total				
World total	8,263	1,111	4,301	13,675	-	2,731	-	3,288	65	12	0.9	219	0.5	3,584	-	271	20,261	-
Proportion	41%	5.5%	21%	67%	-	13%	-	16%	0.3%	0.06%	0.004%	1.1%	0.003%	18%	-	1.3%	100%	-
China	2,733	23	31	2,788	2	68	8	585	-	0.2	-	13	-	598	1	2.4	3,457	2
India	569	34	82	685	5	15	12	114	-	0.02	-	14	-	128.02	6	2.0	830	5
USA	2,133	58	911	3,101	1	838	1	282	17	1.6	0.88	56	-	357	4	73	4,369	1
Indonesia	61	43	25	130	19	-	-	12	8.3	-	-	-	-	20	17	-	149	20
Brazil	13	18	29	59	23	14	13	370	-	-	-	0.6	-	370	3	20	463	9
Pakistan	0.1	32	30	62	22	1.6	16	28	-	-	-	-	-	28	14	-	92	24
Bangladesh	0.6	1.7	31	33	27	-	-	1.5	-	-	-	-	-	1.5	29	-	35	27
Nigeria	-	3.1	12	15	28	-	-	5.7	-	-	-	-	-	5.7	25	-	21	28
Russia	197	16	495	708	4	163	4	167	0.5	-	-	0.01	-	167	5	2.5	1,040	4
Japan	288	139	283	711	3	258	3	83	2.8	2.3	-	2.6	-	91	7	22	1,082	3
Mexico	21	49	131	202	13	9.8	14	39	7.1	0.01	-	0.3	-	47	12	0.8	259	14
Philippines	16	4.9	20	40	26	-	-	9.8	11	0.001	-	0.1	-	21	16	-	61	26
Vietnam	15	1.6	30	47	25	-	-	26	-	-	-	-	-	26	15	-	73	25
Ethiopia	-	0.5	-	0.5	29	-	-	3.3	0.01	-	-	-	-	3.3	28	-	3.8	30
Egypt	-	26	90	115	20	-	-	15	-	-	-	0.9	-	16	20	-	131	22
Germany	291	9.2	88	388	6	148	6	27	0.02	4.4	-	41	-	72	9	29	637	7
Turkey	58	7.5	99	164	16	-	-	33	0.16	-	-	0.85	-	34	13	0.22	198	19
DR Congo	-	0.02	0.03	0.05	30	-	-	7.5	-	-	-	-	-	7.5	22	-	7.5	29

Electric Power

Country	Coal	Oil	Gas	sub total	rank	Nu-clear	rank	Hy-dro	Geo Ther-mal	Solar PV	Solar Ther-mal	Wind	Tide	sub total	rank	Bio other	Total	rank
Iran	0.4	36	173	209	11	-	-	5.0	-	-	-	0.20	-	5.2	26	-	215	17
Thailand	32	1.7	102	135	18	-	-	7.1	0.002	0.003	-	-	-	7.1	23	4.8	147	21
France	27	5.8	22	55	24	439	2	68	-	0.04	-	5.7	0.51	75	8	5.9	575	8
UK	127	6.1	177	310	7	52	10	9.3	-	0.02	-	7.1	-	16	18	11	389	11
Italy	49	31	173	253	9	-	-	47	5.5	0.2	-	4.9	-	58	11	8.6	319	12
South Korea	192	15	81	288	8	151	5	5.6	-	0.3	-	0.4	-	6.3	24	0.7	446	10
Spain	50	18	122	190	14	59	9	26	-	2.6	0.02	32	-	61	10	4.3	314	13
Canada	112	9.8	41	162	17	94	7	383	-	0.03	-	3.8	0.03	386	2	8.5	651	6
Saudi Arabia	-	116	88	204	12	-	-	-	-	-	-	-	-	-	-	-	204	18
Taiwan	125	14	46	186	15	41	11	7.8	-	0.004	-	0.6	-	8.4	21	3.5	238	16
Australia	198	2.8	39	239	10	-	-	12	-	0.2	0.004	3.9	-	16	19	2.2	257	15
Netherlands	27	2.1	63	92	21	4.2	15	0.1	-	0.04	-	4.3	-	4.4	27	6.8	108	23

Solar PV* is Photovoltaics **Bio other*** = 198TWh (Biomass) + 69TWh (Waste) + 4TWh (other)

Environmental Concerns

Variations between countries generating electrical power affect concerns about the environment. In France only 10% of electricity is generated from fossil fuels, the US is higher at 70% and China is at 80%. The cleanliness of electricity is dependent its source. Most scientists agree that emissions of pollutants and greenhouse gases from fossil fuel-based electricity generation account for a significant portion of world greenhouse gas emissions; in the United States, electricity generation accounts for nearly 40% of emissions, the largest of any source. Transportation emissions are close behind, contributing about one-third of U.S. production of carbon dioxide. In the United States, fossil fuel combustion for electric power generation is responsible for 65% of all emissions of sulfur dioxide, the main component of acid rain. Electricity generation is the fourth highest combined source of NOx, carbon monoxide, and particulate matter in the US. In July 2011, the UK parliament tabled a motion that "levels of (carbon) emissions from nuclear power were approximately three times lower per kilowatt hour than those of solar, four times lower than clean coal and 36 times lower than conventional coal". Though Solar PV generation is positioned as environmentally friendly, fabrication of PV cells utilizes substantial amounts of water in addition to toxic chemicals such as phosphorus and arsenic. These are often overlooked when promoting PV. Because of strict environmental regulations in the United States, for example, PV fabrication is often performed in countries with lower standards, such as China, which produces approximately half the world's PV panels.

Lifecycle greenhouse gas emissions by electricity source.		
Technology	Description	50th percentile (g CO_2/kWh$_e$)
Hydroelectric	reservoir	4
Wind	onshore	12
Nuclear	various generation II reactor types	16
Biomass	various	18
Solar thermal	parabolic trough	22
Geothermal	hot dry rock	45
Solar PV	Polycrystaline silicon	46
Natural gas	various combined cycle turbines without scrubbing	469
Coal	various generator types without scrubbing	1001

Water Consumption

Most large scale thermoelectric power stations consume considerable amounts of water for cooling purposes and boiler water make up - 1 L/kWh for once through (*e.g.* river cooling), and 1.7 L/kWh for cooling tower cooling. Water abstraction for cooling water accounts for about 40% of European total water abstraction, although most of this water is returned to its source, albeit slightly warmer. Different cooling systems have different consumption vs. abstraction characteristics. Cooling towers withdraw a small amount of water from the environment and evaporate most of it. Once-through systems withdraw a large amount but return it to the environment immediately, at a higher temperature.

BATTERY

An electric **battery** is a device consisting of one or more electrochemical cells that convert stored chemical energy into electrical energy. Each cell contains a positive terminal, or cathode, and a negative terminal, or anode. Electrolytes allow ions to move between the electrodes and terminals, which allows current to flow out of the battery to perform work.

Primary (single-use or "disposable") batteries are used once and discarded; the electrode materials are irreversibly changed during discharge. Common examples are the alkaline battery used for flashlights and a multitude of portable devices. Secondary (rechargeable batteries) can be discharged and recharged multiple times; the original composition of the electrodes can be restored by reverse current. Examples include the lead-acid batteries used in vehicles and lithium ion batteries used for portable electronics. Batteries come in many shapes and sizes, from miniature cells used to power hearing aids and wrist watches to battery banks the size of rooms that provide standby power for telephone exchanges and computer data centers.

According to a 2005 estimate, the worldwide battery industry generates US$48 billion in sales each year, with 6% annual growth.

Batteries have much lower specific energy (energy per unit mass) than common fuels such as gasoline. This is somewhat mitigated by the fact that batteries deliver their energy as electricity (which can be converted efficiently to mechanical work), whereas using fuels in engines entails a low efficiency of conversion to work.

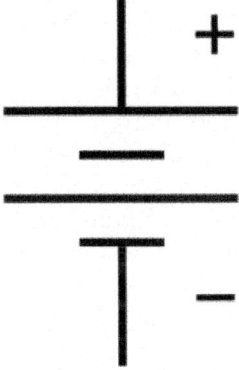

Fig. : The symbol for a battery in a circuit diagram. It originated as a schematic drawing of the earliest type of battery, a voltaic pile.

History

The usage of "battery" to describe a group electrical devices dates to Benjamin Franklin, who in 1748 described multiple Leyden jars by analogy to a battery of cannon (Benjamin Franklin borrowed the term "battery" from the military, which refers to weapons functioning together). Alessandro Volta described the first electrochemical battery, the voltaic pile in 1800. This was a stack of copper and zinc plates, separated by brine soaked paper disks, that could produce a steady current for a considerable length of time. Volta did not appreciate that the voltage was due to chemical reactions. He thought that his cells were an inexhaustible source of energy, and that the associated corrosion effects at the electrodes were a mere nuisance, rather than an unavoidable consequence of their operation, as Michael Faraday showed in 1834.

Although early batteries were of great value for experimental purposes, in practice their voltages fluctuated and they could not provide a large current for a sustained period. The Daniell cell, invented in 1836 by British chemist John Frederic Daniell, was the first practical source of electricity, becoming an industry standard and seeing widespread adoption as a power source for electrical telegraph networks. It consisted of a copper pot filled with a copper sulfate solution, in which was immersed an unglazed earthenware container filled with sulfuric acid and a zinc electrode.

These wet cells used liquid electrolytes, which were prone to leakage and spillage if not handled correctly. Many used glass jars to hold their components, which made them fragile. These characteristics made wet cells unsuitable for portable appliances. Near the end of the nineteenth century, the invention of dry cell batteries, which replaced the liquid electrolyte with a paste, made portable electrical devices practical.

Principle of Operation

Batteries convert chemical energy directly to electrical energy. A battery consists of some number of voltaic cells. Each cell consists of two half-cells connected in series by a conductive electrolyte containing anions and cations. One half-cell includes electrolyte and the negative electrode, the electrode to which anions (negatively charged ions) migrate; the other half-cell includes electrolyte and the positive electrode electrode to which cations (positively charged ions) migrate. Redox reactions power the battery. Cations are reduced (electrons are added) at the cathode during charging, while anions are oxidized (electrons are removed) at the anode during discharge. The electrodes do not touch each other, but are electrically connected by the electrolyte. Some cells use different electrolytes for each half-cell. A separator allows ions to flow between half-cells, but prevents mixing of the electrolytes.

Each half-cell has an electromotive force (or emf), determined by its ability to drive electric current from the interior to the exterior of the cell. The net emf of the cell is the difference between the emfs of its half-cells. Thus, if the electrodes have emfs ε_1 and ε_2, then the net emf is $\varepsilon_2 - \varepsilon_1$ in other words, the net emf is the difference between the reduction potentials of the half-reactions.

The electrical driving force or ΔV across the terminals of a cell is known as the *terminal voltage (difference)* and is measured in volts. The terminal voltage of a cell that is neither charging nor discharging is called the open-circuit voltage and equals the emf of the cell. Because of internal resistance, the terminal voltage of a cell that is discharging is smaller in magnitude than the open-circuit voltage and the terminal voltage of a cell that is charging exceeds the open-circuit voltage. An ideal cell has negligible internal resistance, so it would maintain a constant terminal voltage of ε until exhausted, then dropping to zero. If such a cell maintained 1.5 volts and stored a charge of one coulomb then on complete discharge it would perform 1.5 joules of work. In actual cells, the internal resistance increases under discharge and the open circuit voltage also decreases under discharge. If the voltage and resistance are plotted against time, the resulting graphs typically are a curve; the shape of the curve varies according to the chemistry and internal arrangement employed.

The voltage developed across a cell's terminals depends on the energy release of the chemical reactions of its electrodes and electrolyte. Alkaline and zinc-carbon cells have different chemistries, but approximately the same emf of 1.5 volts; likewise NiCd and NiMH cells have different chemistries, but approximately the same emf of 1.2 volts. The high electrochemical potential changes in the reactions of lithium compounds give lithium cells emfs of 3 volts or more.

Categories and Types of Batteries

Batteries are classified into primary and secondary forms.

- *Primary* batteries irreversibly transform chemical energy to electrical energy. When the supply of reactants is exhausted, energy cannot be readily restored to the battery.
- *Secondary* batteries can be recharged; that is, they can have their chemical reactions reversed by supplying electrical energy to the cell, approximately restoring their original composition.

Some types of primary batteries used, for example, for telegraph circuits, were restored to operation by replacing the electrodes. Secondary batteries are not indefinitely rechargeable due to dissipation of the active materials, loss of electrolyte and internal corrosion.

Primary Batteries

Primary batteries, or primary cells, can produce current immediately on assembly. These are most commonly used in portable devices that have low current drain, are used only intermittently, or are used well away from an alternative power source, such as in alarm and communication circuits where other electric power is only intermittently available. Disposable primary cells cannot be reliably recharged, since the chemical reactions are not easily reversible and active materials may not return to their original forms. Battery manufacturers recommend against attempting to recharge primary cells.

In general, these have higher energy densities than rechargeable batteries, but disposable batteries do not fare well under high-drain applications with loads under 75 ohms (75 Ω).

Common types of disposable batteries include zinc–carbon batteries and alkaline batteries.

Secondary Batteries

Secondary batteries, also known as *secondary cells*, or *rechargeable batteries*, must be charged before first use; they are usually assembled with active materials in the discharged state. Rechargeable batteries are (re)charged by applying electric current, which reverses the chemical reactions that occur during discharge/use. Devices to supply the appropriate current are called chargers.

The oldest form of rechargeable battery is the lead–acid battery. This technology contains liquid electrolyte in an unsealed container, requiring that the battery be kept upright and the area be well ventilated to ensure safe dispersal of the hydrogen gas it produces during overcharging. The lead–acid battery is relatively heavy for the amount of electrical energy it can supply. Its low manufacturing cost and its high surge current levels make it common where its capacity (over approximately 10 Ah) is more important than weight and handling issues. A common application is the modern car battery, which can, in general, deliver a peak current of 450 amperes.

The sealed valve regulated lead–acid battery (VRLA battery) is popular in the automotive industry as a replacement for the lead–acid wet cell. The VRLA battery uses an immobilized sulfuric acid electrolyte, reducing the chance of leakage and extending shelf life. VRLA batteries immobilize the electrolyte. The two types are :

- *Gel batteries* (or "gel cell") use a semi-solid electrolyte.
- *Absorbed Glass Mat* (AGM) batteries absorb the electrolyte in a special fiberglass matting.

Other portable rechargeable batteries include several sealed "dry cell" types, that are useful in applications such as mobile phones and laptop computers. Cells of this type (in order of increasing power density and cost) include nickel–cadmium (NiCd), nickel–zinc (NiZn), nickel metal hydride (NiMH), and lithium-ion (Li-ion) cells. Li-ion has by far the highest share of the dry cell rechargeable market. NiMH has replaced NiCd in most applications due to its higher capacity, but NiCd remains in use in power tools, two-way radios, and medical equipment.

Recent developments include batteries with embedded electronics such as USBCELL, which allows charging an AA battery through a USB connector, and smart battery packs with state-of-charge monitors and battery protection circuits that prevent damage on over-discharge. Low self-discharge (LSD) allows secondary cells to be charged prior to shipping.

Battery Cell Types

Many types of electrochemical cells have been produced, with varying chemical processes and designs, including galvanic cells, electrolytic cells, fuel cells, flow cells and voltaic piles.

Wet Cell

A *wet cell* battery has a liquid electrolyte. Other names are *flooded cell*, since the liquid covers all internal parts, or *vented cell*, since gases produced during operation can escape to the air. Wet cells were a precursor to dry cells and are commonly used as a learning tool for electrochemistry. They can be built with common laboratory supplies, such as beakers, for demonstrations of how electrochemical cells work. A particular type of wet cell known as a concentration cell is important in understanding corrosion. Wet cells may be primary cells (non-rechargeable) or secondary cells (rechargeable). Originally, all practical primary batteries such as the Daniell cell were built as open-top glass jar wet cells. Other primary wet cells are the Leclanche cell, Grove cell, Bunsen cell, Chromic acid cell, Clark cell, and Weston cell. The Leclanche cell chemistry was adapted to the first dry cells. Wet cells are still used in automobile batteries and in industry for standby power for switchgear, telecommunication or large uninterruptible power supplies, but in many places batteries with gel cells have been used instead. These applications commonly use lead–acid or nickel–cadmium cells.

Dry Cell

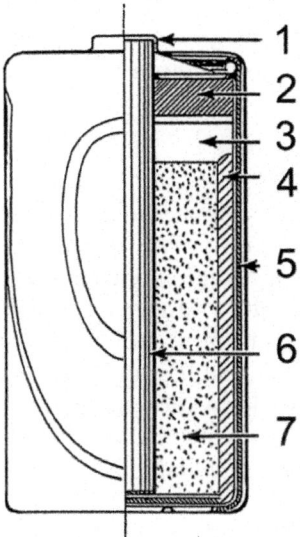

Line art drawing of a dry cell :
1. Brass cap,

2. Plastic seal,
3. Expansion space,
4. Porous cardboard,
5. Zinc can,
6. Carbon rod,
7. Chemical mixture.

A *dry cell* uses a paste electrolyte, with only enough moisture to allow current to flow. Unlike a wet cell, a dry cell can operate in any orientation without spilling, as it contains no free liquid, making it suitable for portable equipment. By comparison, the first wet cells were typically fragile glass containers with lead rods hanging from the open top and needed careful handling to avoid spillage. Lead-acid batteries did not achieve the safety and portability of the dry cell until the development of the gel battery.

A common dry cell is the zinc-carbon battery, sometimes called the dry Leclanché cell, with a nominal voltage of 1.5 volts, the same as the alkaline battery (since both use the same zinc-manganese dioxide combination).

A standard dry cell comprises a zinc anode, usually in the form of a cylindrical pot, with a carbon cathode in the form of a central rod. The electrolyte is ammonium chloride in the form of a paste next to the zinc anode. The remaining space between the electrolyte and carbon cathode is taken up by a second paste consisting of ammonium chloride and manganese dioxide, the latter acting as a depolariser. In some designs, the ammonium chloride is replaced by zinc chloride.

Molten Salt

Molten salt batteries are primary or secondary batteries that use a molten salt as electrolyte. They operate at high temperatures and must be well insulated to retain heat.

Reserve

A reserve battery can be stored unassembled (unactivated and supplying no power) for a long period (perhaps years). When the battery is needed, then it is assembled (*e.g.*, by adding electrolyte); once assembled, the battery is charged and ready to work. For example, a battery for an electronic artillery fuze might be activated by the impact of firing a gun : The acceleration breaks a capsule of electrolyte that activates the battery and powers the fuze's circuits. Reserve batteries are usually designed for a short service life (seconds or minutes) after long storage (years). A water-activated battery for oceanographic instruments or military applications becomes activated on immersion in water.

Battery Cell Performance

A battery's characteristics may vary over load cycle, over charge cycle, and over lifetime due to many factors including internal chemistry, current drain, and temperature.

Capacity and Discharge

A battery's *capacity* is the amount of electric charge it can deliver at the rated voltage. The more electrode material contained in the cell the greater its capacity. A small cell has less capacity than a larger cell with the same chemistry, although they develop the same open-circuit voltage. Capacity is measured in units such as amp-hour (A·h).

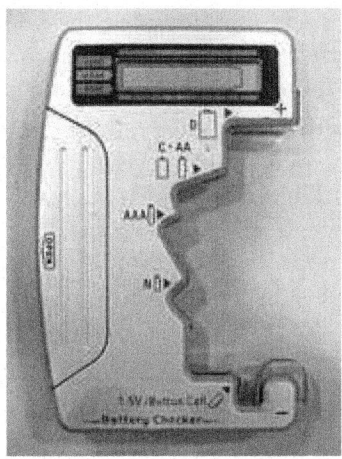

Fig. : A device to check battery voltage.

The rated capacity of a battery is usually expressed as the product of 20 hours multiplied by the current that a new battery can consistently supply for 20 hours at 68 °F (20 °C), while remaining above a specified terminal voltage per cell. For example, a battery rated at 100 A·h can deliver 5 A over a 20-hour period at room temperature.

The fraction of the stored charge that a battery can deliver depends on multiple factors, including battery chemistry, the rate at which the charge is delivered (current), the required terminal voltage, the storage period, ambient temperature and other factors.

The higher the discharge rate, the lower the capacity. The relationship between current, discharge time and capacity for a lead acid battery is approximated (over a typical range of current values) by Peukert's law :

$$t = \frac{Q_p}{I^k}$$

where :

Q_p is the capacity when discharged at a rate of 1 amp.
I is the current drawn from battery (A).
t is the amount of time (in hours) that a battery can sustain.
k is a constant around 1.3.

Batteries that are stored for a long period or that are discharged at a small fraction of the capacity lose capacity due to the presence of generally irreversible *side reactions* that consume charge carriers without producing current. This phenomenon is known as internal self-discharge. Further, when batteries are recharged, additional side reactions can occur, reducing capacity for subsequent discharges. After enough recharges, in essence all capacity is lost and the battery stops producing power.

Internal energy losses and limitations on the rate that ions pass through the electrolyte cause battery efficiency to vary. Above a minimum threshold, discharging at a low rate delivers more of the battery's capacity than at a higher rate.

Installing batteries with varying A·h ratings does not affect device operation (although it may affect the operation interval) rated for a specific voltage unless load limits are exceeded. High-drain loads such as digital cameras can reduce total capacity, as happens with alkaline batteries. For example, a battery rated at 2000 mAh for a 10- or 20-hour discharge would not sustain a current of 1 A for a full two hours as its stated capacity implies.

C Rate

The C-rate is the multiple of the current over the current that a battery can sustain for one hour. A rate of 1 C means that an entire 1.6Ah battery would be discharged in one hour at a discharge current of 1.6 A. A 2C rate would mean a discharge current of 3.2 A, over one half-hour.

Fast-Charging, Large and Light Batteries

As of 2012 Lithium iron phosphate (LiFePO$_4$) battery technology was the fastest-charging/discharging, fully discharging in 10–20 seconds.

As of 2013, the world's largest battery was in Hebei Province, China. It stored 36 megawatt-hours of electricity at a cost of $500 million. Another large battery, composed of Ni–Cd cells, was in Fairbanks, Alaska. It covers 2,000 square metres (22,000 sq ft) — bigger than a football pitch — and weighs 1,300 tonnes, It was manufactured by ABB to provide backup power in the event of a blackout. The battery can provide 40 megawatts of power for up to seven minutes. Sodium–sulfur batteries have been used to store wind power. A 4.4 megawatt-hour battery system that can deliver 11 megawatts for 25 minutes stabilizes the output of the Auwahi wind farm in Hawaii. Lithium–sulfur batteries were used on the longest and highest solar-powered flight. The recharging speed of lithium-ion batteries can be increased by manufacturing changes.

Battery Lifetime

Available capacity of all batteries drops with decreasing temperature. In contrast to most of today's batteries, the Zamboni pile, invented in 1812, offers a very long service life without refurbishment or recharge, although it supplies current only in the nanoamp range. The Oxford Electric Bell has been ringing almost continuously since 1840 on its original pair of batteries, thought to be Zamboni piles.

Self-Discharge

Disposable batteries typically lose 8 to 20 per cent of their original charge per year when stored at room temperature (20°–30°C). This is known as the "self-discharge" rate, and is due to non-current-producing "side" chemical reactions that occur within the cell even when no load is applied. The rate of side reactions is reduced for batteries are stored at lower temperatures, although some can be damaged by freezing.

Old rechargeable batteries self-discharge more rapidly than disposable alkaline batteries, especially nickel-based batteries; a freshly charged nickel cadmium (NiCd) battery loses 10% of its charge in the first 24 hours, and thereafter discharges at a rate of about 10% a month. However, newer low self-discharge nickel metal hydride (NiMH) batteries and modern lithium designs display a lower self-discharge rate (but still higher than for primary batteries).

Corrosion

Internal parts may corrode and fail, or the active materials may be slowly converted to inactive forms.

Physical Component Changes

The active material on the battery plates changes chemical composition on each charge and discharge cycle, active material may be lost due to physical changes of volume; further limiting the number of times the battery can be recharged.

Most nickel-based batteries are partially discharged when purchased, and must be charged before first use. Newer NiMH batteries are ready to be used when purchased, and have only 15% discharge in a year.

Some deterioration occurs on each charge–discharge cycle. Degradation usually occurs because electrolyte migrates away from the electrodes or because active material detaches from the electrodes.

Low-capacity NiMH batteries (1700–2000 mA·h) can be charged some 1,000 times, whereas high-capacity NiMH batteries (above 2500 mA·h) last about 500 cycles. NiCd batteries tend to be rated for 1,000 cycles before their internal resistance permanently increases beyond usable values.

Charge/Discharge Speed

Fast charging increases component changes, shortening battery lifespan.

Overcharging

If a charger cannot detect when the battery is fully charged then overcharging is likely, damaging it.

Memory Effect

NiCd cells, if used in a particular repetitive manner, may show a decrease in capacity called "memory effect". The effect can be avoided with simple practices. NiMH cells, although similar in chemistry, suffer less from memory effect.

Environmental Conditions

Automotive lead-acid rechargeable batteries must endure stress due to vibration, shock, and temperature range. Because of these stresses and sulfation of their lead plates, few automotive batteries last beyond six years of regular use. Automotive starting (SLI : *Starting, Lighting, Ignition*) batteries have many thin plates to maximize current. In general, the thicker the plates the longer the life. They are typically discharged only slightly before recharge.

"Deep-cycle" lead-acid batteries such as those used in electric golf carts have much thicker plates to extend longevity. The main benefit of the lead-acid battery is its low cost; its main drawbacks are large size and weight for a given capacity and voltage. Lead-acid batteries should never be discharged to below 20% of their capacity, because internal resistance will cause heat and damage when they are recharged. Deep-cycle lead-acid systems often use a low-charge warning light or a low-charge power cut-off switch to prevent the type of damage that will shorten the battery's life.

Storage

Battery life can be extended by storing the batteries at a low temperature, as in a refrigerator or freezer, which slows the side reactions. Such storage can extend the life of alkaline batteries by about 5%; rechargeable batteries can hold their charge much longer, depending upon type. To reach their maximum voltage, batteries must be returned to room temperature; discharging an alkaline battery at 250 mA at 0°C is only half as efficient as at 20°C. Alkaline battery manufacturers such as Duracell do not recommend refrigerating batteries.

Battery Sizes

Primary batteries readily available to consumers range from tiny button cells used for electric watches, to the No. 6 cell used for signal circuits or other long duration applications. Secondary cells are made in very large sizes; very large batteries can power a submarine or stabilize an electrical grid and help level out peak loads.

Hazards

Explosion

A battery explosion is caused by misuse or malfunction, such as attempting to recharge a primary (non-rechargeable) battery, or a short circuit. Car batteries are most likely to explode when a short-circuit generates very large currents. Car batteries produce hydrogen, which is very explosive, when they are overcharged (because of electrolysis of the water in the electrolyte). The amount of overcharging is usually very small and generates little hydrogen, which dissipates quickly. However, when "jumping" a car battery, the high current can cause the rapid release of large volumes of hydrogen, which can be ignited explosively by a nearby spark, for example, when disconnecting a jumper cable.

When a battery is recharged at an excessive rate, an explosive gas mixture of hydrogen and oxygen may be produced faster than it can escape from within the battery, leading to pressure build-up and eventual bursting of the battery case. In extreme cases, battery acid may spray violently from the casing and cause injury. Overcharging — that is, attempting to charge a battery beyond its electrical capacity — can also lead to a battery explosion, in addition to leakage or irreversible damage. It may also cause damage to the charger or device in which the overcharged battery is later used. In addition, disposing of a battery via incineration may cause an explosion as steam builds up within the sealed case.

Leakage

Many battery chemicals are corrosive, poisonous or both. If leakage occurs, either spontaneously or through accident, the chemicals released may be dangerous.

For example, disposable batteries often use a zinc "can" both as a reactant and as the container to hold the other reagents. If this kind of battery is overdischarged, the reagents can emerge through the cardboard and plastic that form the remainder of the container. The active chemical leakage can then damage the equipment that the batteries power. For this reason, many electronic device manufacturers recommend removing the batteries from devices that will not be used for extended periods of time.

Toxic Materials

Many types of batteries employ toxic materials such as lead, mercury, and cadmium as an electrode or electrolyte. When each battery reaches end of life it must be disposed of to prevent environmental damage. Battery are one form of electronic waste (e-waste).

E-waste recycling services recover toxic substances, which can then be used for new batteries.

Of the nearly three billion batteries purchased annually in the United States, about 179,000 tons end up in landfills across the country.

In the United States, the Mercury-Containing and Rechargeable Battery Management Act of 1996 banned the sale of mercury-containing batteries, enacted uniform labelling requirements for rechargeable batteries and required that rechargeable batteries be easily removable. California and New York City prohibit the disposal of rechargeable batteries in solid waste, and along with Maine require recycling of cell phones. The rechargeable battery industry operates nationwide recycling programs in the United States and Canada, with dropoff points at local retailers.

The Battery Directive of the European Union has similar requirements, in addition to requiring increased recycling of batteries and promoting research on improved battery recycling methods.

In accordance with this directive all batteries to be sold within the EU must be marked with the "collection symbol" (A crossed-out wheeled bin). This must cover at least 3% of the surface of prismatic batteries and 1.5% of the surface of cylindrical batteries. All packaging must be marked likewise.

Ingestion

Batteries may be harmful or fatal if swallowed.

Small button cells can be swallowed, in particular by young children. While in the digestive tract, the battery's electrical discharge may lead to tissue damage; such damage is occasionally serious and can lead to death. Ingested disk batteries do not usually cause problems unless they become lodged in the gastrointestinal tract. The most common place for disk batteries to become lodged is the esophagus, resulting in clinical sequelae. Batteries that successfully traverse the esophagus are unlikely to lodge elsewhere. The likelihood that a disk battery will lodge in the esophagus is a function of the patient's age and battery size. Disk batteries of 16 mm have become lodged in the esophagi of 2 children younger than 1 year. Older children do not have problems with batteries smaller than 21–23 mm. Liquefaction necrosis may occur because sodium hydroxide is generated by the current produced by the battery (usually at the anode). Perforation has occurred as rapidly as 6 hours after ingestion.

Battery Chemistry

Primary Batteries and Their Characteristics

Chemistry	Anode (-)	Cathode (+)	Maximum Voltage (Theoretical) (V)	Working Voltage (Practical) (V)	Specific energy [MJ/kg]	Elaboration	Shelf Life At 25°C (80% Capacity) (Months)
Zinc–carbon	Zn	MnO_2	1.6	1.2	0.13	Inexpensive.	18
Zinc–chloride			1.5			Also known as "heavy-duty", inexpensive.	

Chemistry	Anode (-)	Cathode (+)	Maximum Voltage (Theoretical) (V)	Working Voltage (Practical) (V)	Specific energy [MJ/kg]	Elaboration	Shelf Life At 25°C (80% Capacity) (Months)
Alkaline (zinc–manganese dioxide)	Zn	MnO$_2$	1.5	1.15	0.4–0.59	Moderate energy density. Good for high- and low-drain uses.	30
Nickel oxyhydroxide (zinc–manganese dioxide/nickel oxyhydroxide)			1.7			Moderate energy density. Good for high drain uses.	
Lithium (lithium–copper oxide) Li–CuO			1.7			No longer manufactured. Replaced by silver oxide (IEC-type "SR") batteries.	
Lithium (lithium–iron disulfide) LiFeS2			1.5			Expensive. Used in 'plus' or 'extra' batteries.	
Lithium (lithium–manganese dioxide) LiMnO$_2$			3.0		0.83–1.01	Expensive. Used only in high-drain devices or for long shelf-life due to very low rate of self-discharge. 'Lithium' alone usually refers to this type of chemistry.	
Lithium (lithium–carbon fluoride) Li–(CF)$_n$	Li	(CF)n	3.6	3.0			120

Chemistry	Anode (-)	Cathode (+)	Maximum Voltage (Theoretical) (V)	Working Voltage (Practical) (V)	Specific energy [MJ/kg]	Elaboration	Shelf Life At 25°C (80% Capacity) (Months)
Lithium (lithium-chromium oxide) Li–CrO$_2$	Li	CrO$_2$	3.8	3.0			108
Mercury oxide	Zn	HgO	1.34	1.2		High-drain and constant voltage. Banned in most countries because of health concerns.	36
Zinc–air	Zn	O$_2$	1.6	1.1	1.59	Used mostly in hearing aids.	
Silver-oxide (silver–zinc)	Zn	Ag$_2$O	1.85	1.5	0.47	Very expensive. Used only commercially in 'button' cells.	30
Magnesium	Mg	MnO$_2$	2.0	1.5			40

Secondary (rechargeable) batteries and their characteristics

Chemistry	Cell Voltage	Specific energy [MJ/kg]	Comments
NiCd	1.2	0.14	Inexpensive. High-/low-drain, moderate energy density. Can withstand very high discharge rates with virtually no loss of capacity. Moderate rate of self-discharge. Environmental hazard due to Cadmium – use now virtually prohibited in Europe.
Lead–acid	2.1	0.14	Moderately expensive. Moderate energy density. Moderate rate of self-discharge. Higher discharge rates result in considerable loss of capacity. Environmental hazard due to Lead. Common use – Automobile batteries

Chemistry	Cell Volt-age	Specific energy [MJ/kg]	Comments
NiMH	1.2	0.36	Inexpensive. Performs better than alkaline batteries in higher drain devices. Traditional chemistry has high energy density, but also a high rate of self-discharge. Newer chemistry has low self-discharge rate, but also a ~25% lower energy density. Used in some cars.
NiZn	1.6	0.36	Moderately inexpensive. High drain device suitable. Low self-discharge rate. Voltage closer to alkaline primary cells than other secondary cells. No toxic components. Newly introduced to the market (2009). Has not yet established a track record. Limited size availability.
AgZn	1.86 1.5	0.46	Smaller volume than equivalent Li-ion. Extremely expensive due to silver. Very high energy density. Very high drain capable. For many years considered obsolete due to high silver prices. Cell suffers from oxidation if unused. Reactions are not fully understood. Terminal voltage very stable but suddenly drops to 1.5 volts at 70-80% charge (believed to be due to presence of both argentous and argentic oxide in positive plate – one is consumed first). Has been used in lieu of primary battery (moon buggy). Is being developed once again as a replacement for Li-ion.
Lithium ion	3.6	0.46	Very expensive. Very high energy density. Not usually available in "common" battery sizes. Very common in laptop computers, moderate to high-end digital cameras, camcorders, and cellphones. Very low rate of self-discharge. Terminal voltage unstable (varies from 4.2 to 3.0 volts during discharge). Volatile : Chance of explosion if short-circuited, allowed to overheat, or not manufactured with rigorous quality standards.

Homemade Cells

Almost any liquid or moist object that has enough ions to be electrically conductive can serve as the electrolyte for a cell. As a novelty or science demonstration, it is possible to insert two electrodes made of different metals into a lemon, potato, etc. and generate small amounts of electricity. "Two-potato clocks" are also widely available in hobby and toy stores; they consist of a pair of cells, each consisting of a potato (lemon, et. cetera) with two electrodes inserted into it, wired in series to form a battery with enough voltage to power a digital clock. Homemade cells of this kind are of no practical use.

A voltaic pile can be made from two coins (such as a nickel and a penny) and a piece of paper towel dipped in salt water. Such a pile generates a very low voltage but, when many are stacked in series, they can replace normal batteries for a short time.

Sony has developed a biological battery that generates electricity from sugar in a way that is similar to the processes observed in living organisms. The battery generates electricity through the use of enzymes that break down carbohydrates.

Lead acid cells can easily be manufactured at home, but a tedious charge/discharge cycle is needed to 'form' the plates. This is a process in which lead sulfate forms on the plates, and during charge is converted to lead dioxide (positive plate) and pure lead (negative plate). Repeating this process results in a microscopically rough surface, increasing the surface area. This increases the current the cell can deliver.

Daniell cells are easy to make at home. Aluminium–air batteries can be produced with high-purity aluminium. Aluminium foil batteries will produce some electricity, but are not efficient, in part because a significant amount of (combustible) hydrogen gas is produced.

Chapter 2

AC AND DC POWER

AC POWER

Power in an electric circuit is the rate of flow of energy past a given point of the circuit. In alternating current circuits, energy storage elements such as inductors and capacitors may result in periodic reversals of the direction of energy flow. The portion of power that, averaged over a complete cycle of the AC waveform, results in net transfer of energy in one direction is known as real power. The portion of power due to stored energy, which returns to the source in each cycle, is known as reactive power.

Real, Reactive, and Apparent Power

In a simple alternating current (AC) circuit consisting of a source and a linear load, both the current and voltage are sinusoidal. If the load is purely resistive, the two quantities reverse their polarity at the same time. At every instant the product of voltage and current is positive, indicating that the direction of energy flow does not reverse. In this case, only real power is transferred.

If the loads are purely *reactive*, then the voltage and current are 90 degrees out of phase. For half of each cycle, the product of voltage and current is positive, but on the other half of the cycle, the product is negative, indicating that on average, exactly as much energy flows toward the load as flows back. There is no net energy flow over one cycle. In this case, only reactive energy flows—there is no net transfer of energy to the load.

Practical loads have resistance, inductance, and capacitance, so both real and reactive power will flow to real loads. Power engineers measure apparent power as the magnitude of the vector sum of real and reactive power. Apparent power is the product of the root-mean-square of voltage and current.

Engineers care about apparent power, because even though the current associated with reactive power does no work at the load, it heats the wires, wasting energy. Conductors, transformers and generators must be sized to carry the total current, not just the current that does useful work.

Another consequence is that adding the apparent power for two loads will not accurately give the total apparent power unless they have the same displacement between current and voltage (the same power factor).

Conventionally, capacitors are considered to generate reactive power and inductors to consume it. If a capacitor and an inductor are placed in parallel, then the currents flowing through the inductor and the capacitor tend to cancel rather than add. This is the fundamental mechanism for controlling the power factor in electric power transmission; capacitors (or inductors) are inserted in a circuit to partially cancel reactive power 'consumed' by the load.

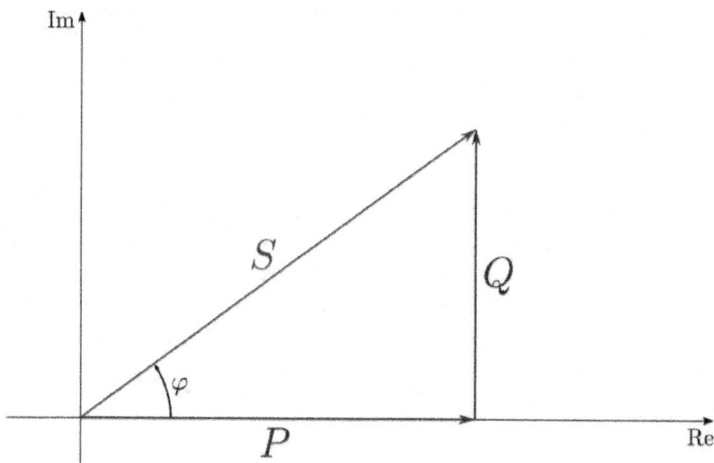

Fig. : The complex power is the vector sum of real and reactive power. The apparent power is the magnitude of the complex power.

Real power, P

Reactive power, Q

Complex power, S

Apparent power, $|S|$

Phase of current, φ

Engineers use the following terms to describe energy flow in a system (and assign each of them a different unit to differentiate between them) :

- **Real power**, P, or **active power** : watt (W)
- **Reactive power**, Q : volt-ampere reactive (var)
- **Complex power**, S : volt-ampere (VA)
- **Apparent power**, $|S|$: the magnitude of complex power S : volt-ampere (VA)
- **Phase of voltage relative to current**, φ : the angle of difference (in degrees) between voltage and current; current lagging voltage (quadrant I vector), current leading voltage (quadrant IV vector).

AC and DC Power

In the diagram, P is the real power, Q is the reactive power (in this case positive), S is the complex power and the length of S is the apparent power. Reactive power does not do any work, so it is represented as the **imaginary axis** of the vector diagram. Real power does do work, so it is the real axis.

The unit for all forms of power is the watt (symbol : W), but this unit is generally reserved for real power. Apparent power is conventionally expressed in volt-amperes (VA) since it is the product of rms voltage and rms current. The unit for reactive power is expressed as var, which stands for volt-ampere reactive. Since reactive power transfers no net energy to the load, it is sometimes called "wattless" power. It does, however, serve an important function in electrical grids and its lack has been cited as a significant factor in the Northeast Blackout of 2003.

Understanding the relationship among these three quantities lies at the heart of understanding power engineering. The mathematical relationship among them can be represented by vectors or expressed using complex numbers, $S = P + jQ$ (where j is the imaginary unit).

Reactive Power

Reactive power flow is needed in an alternating-current transmission system to support the transfer of real power over the network. In alternating current circuits, energy is stored temporarily in inductive and capacitive elements, which can result in the periodic reversal of the direction of energy flow. The portion of power flow remaining, after being averaged over a complete AC waveform, is the real power; that is, energy that can be used to do work (for example, overcome friction in a motor, or heat an element). On the other hand, the portion of power flow that is temporarily stored in the form of magnetic or electric fields, due to inductive and capacitive network elements, and then returned to source, is known as *reactive power*.

AC connected devices that store energy in the form of a magnetic field include devices called inductors, which consist of a large coil of wire. When a voltage is initially placed across the coil, a magnetic field builds up, and it takes a period of time for the current to reach full value. This causes the current to lag behind the voltage in phase; hence, these devices are said to *absorb* reactive power.

A capacitor is an AC device that stores energy in the form of an electric field. When current is driven through the capacitor, it takes a period of time for a charge to build up to produce the full voltage difference. On an AC network, the voltage across a capacitor is constantly changing – the capacitor will oppose this change, causing the voltage to lag behind the current. In other words, the current leads the voltage in phase; hence, these devices are said to *generate* reactive power.

Energy stored in capacitive or inductive elements of the network give rise to reactive power flow. Reactive power flow strongly influences the voltage levels across the network. Voltage levels and reactive power flow must be carefully controlled to allow a power system to be operated within acceptable limits.

Reactive Power Control

Transmission connected generators are generally required to support reactive power flow. For example, on the United Kingdom transmission system generators are required by the Grid Code Requirements to supply their rated power between the limits of 0.85 power factor lagging and 0.90 power factor leading at the designated terminals. The system operator will perform switching actions to maintain a secure and economical voltage profile while maintaining a reactive power balance equation:

Generator_MVARs + System_gain + Shunt_capacitors = MVAR_Demand + Reactive_losses + Shunt_reactors.

The 'System gain' is an important source of reactive power in the above power balance equation, which is generated by the capacitive nature of the transmission network itself. By making decisive switching actions in the early morning before the demand increases, the system gain can be maximized early on, helping to secure the system for the whole day.

To balance the equation some pre-fault reactive generator use will be required. Other sources of reactive power that will also be used include shunt capacitors, shunt reactors, Static VAR Compensators and voltage control circuits.

Unbalanced Polyphase Systems

While real power and reactive power are well defined in any system, the definition of apparent power for unbalanced polyphase systems is considered to be one of the most controversial topics in power engineering. Originally, apparent power arose merely as a figure of merit. Major delineations of the concept are attributed to Stanley's *Phenomena of Retardation in the Induction Coil* (1888) and Steinmetz's *Theoretical Elements of Engineering*. However, with the development of three phase power distribution, it became clear that the definition of apparent power and the power factor could not be applied to unbalanced polyphase systems. In 1920, a "Special Joint Committee of the AIEE and the National Electric Light Association" met to resolve the issue. They considered two definitions:

$$pf = \frac{P_a + P_b + P_c}{|S_a| + |S_b| + |S_c|}$$

that is, the quotient of the sums of the real powers for each phase over the sum of the apparent power for each phase.

$$pf = \frac{P_a + P_b + P_c}{|P_a + P_b + P_c + j(Q_a + Q_b + Q_c)|}$$

that is, the quotient of the sums of the real powers for each phase over the magnitude of the sum of the complex powers for each phase.

The 1920 committee found no consensus and the topic continued to dominate discussions. In 1930 another committee formed and once again failed to resolve the question. The transcripts of their discussions are the lengthiest and most con-

AC and DC Power

troversial ever published by the AIEE (Emanuel, 1993). Further resolution of this debate did not come until the late 1990s.

Basic Calculations Using Real Numbers

A perfect resistor stores no energy, so current and voltage are in phase. Therefore there is no reactive power and $P = S$. Therefore for a perfect resistor

$$P = S = V_{RMS} I_{RMS} = I^2_{RMS} R = \frac{V^2_{RMS}}{R}$$

For a perfect capacitor or inductor there is no net power transfer, so all power is reactive. Therefore for a perfect capacitor or inductor :

$$P = 0$$

$$Q = |S| = V_{RMS} I_{RMS} = I^2_{RMS} |X| = \frac{V^2_{RMS}}{|X|}$$

Where X is the reactance of the capacitor or inductor.

If X is defined as being positive for an inductor and negative for a capacitor then we can remove the modulus signs from S and X and get

$$Q = I^2_{RMS} X = \frac{V^2_{RMS}}{X}$$

Instantaneous power is defined as :

$$P_{inst}(t) = V(t) I(t)$$

where $v(t)$ and $i(t)$ are the time varying voltage and current waveforms.

This definition is useful because it applies to all waveforms, whether they are sinusoidal or not. This is particularly useful in power electronics, where non-sinusoidal waveforms are common.

In general, we are interested in the real power averaged over a period of time, whether it is a low frequency line cycle or a high frequency power converter switching period. The simplest way to get that result is to take the integral of the instantaneous calculation over the desired period.

$$P_{avg} = \frac{1}{t_2 - t_1} \int_{t_1}^{t_2} V(t) I(t) \, dt$$

This method of calculating the average power gives the real power regardless of harmonic content of the waveform. In practical applications, this would be done in the digital domain, where the calculation becomes trivial when compared to the use of rms and phase to determine real power.

$$P_{avg} = \frac{1}{n} \sum_{k=1}^{n} V[k] I[k]$$

Multiple Frequency Systems

Since an RMS value can be calculated for any waveform, apparent power can be calculated from this.

For real power it would at first appear that we would have to calculate loads of product terms and average all of them. However if we look at one of these product terms in more detail we come to a very interesting result.

$A \cos(\omega_1 t + k_1) \cos(\omega_2 t + k_2)$

$= \dfrac{A}{2} \cos[(\omega_1 t + k_1) + (\omega_2 t + k_2)] + \dfrac{A}{2} \cos[(\omega_1 t + k_1) - (\omega_2 t + k_2)]$

$= \dfrac{A}{2} \cos[(\omega_1 + \omega_2)t + k_1 + k_2] + \dfrac{A}{2} \cos[(\omega_1 - \omega_2)t + k_1 - k_2]$

however the time average of a function of the form $\cos(\omega t + k)$ is zero provided that ω is non-zero. Therefore the only product terms that have a non-zero average are those where the frequency of voltage and current match. In other words it is possible to calculate real (average) power by simply treating each frequency separately and adding up the answers.

Furthermore, if we assume the voltage of the mains supply is a single frequency (which it usually is), this shows that harmonic currents are a bad thing. They will increase the rms current (since there will be non-zero terms added) and therefore apparent power, but they will have no effect on the real power transferred. Hence, harmonic currents will reduce the power factor.

Harmonic currents can be reduced by a filter placed at the input of the device. Typically this will consist of either just a capacitor (relying on parasitic resistance and inductance in the supply) or a capacitor-inductor network. An active power factor correction circuit at the input would generally reduce the harmonic currents further and maintain the power factor closer to unity.

Power Factor

The **power factor** of an AC electrical power system is defined as the ratio of the real power flowing to the load, to the apparent power in the circuit, and is a dimensionless number between -1 and 1. Real power is the capacity of the circuit for performing work in a particular time. Apparent power is the product of the current and voltage of the circuit. Due to energy stored in the load and returned to the source, or due to a non-linear load that distorts the wave shape of the current drawn from the source, the apparent power will be greater than the real power. A negative power factor occurs when the device which is normally the load generates power which then flows back towards the device which is normally considered the generator.

In an electric power system, a load with a low power factor draws more current than a load with a high power factor for the same amount of useful power transferred. The higher currents increase the energy lost in the distribution sys-

AC AND DC POWER

tem, and require larger wires and other equipment. Because of the costs of larger equipment and wasted energy, electrical utilities will usually charge a higher cost to industrial or commercial customers where there is a low power factor.

Linear loads with low power factor (such as induction motors) can be corrected with a passive network of capacitors or inductors. Non-linear loads, such as rectifiers, distort the current drawn from the system. In such cases, active or passive power factor correction may be used to counteract the distortion and raise the power factor. The devices for correction of the power factor may be at a central substation, spread out over a distribution system, or built into power-consuming equipment.

Linear Circuits

Instantaneous and average power calculated from AC voltage and current with a zero power factor ($\varphi = 90$, $\cos \varphi = 0$). The blue line shows all the power is stored temporarily in the load during the first quarter cycle and returned to the grid during the second quarter cycle, so no real power is consumed.

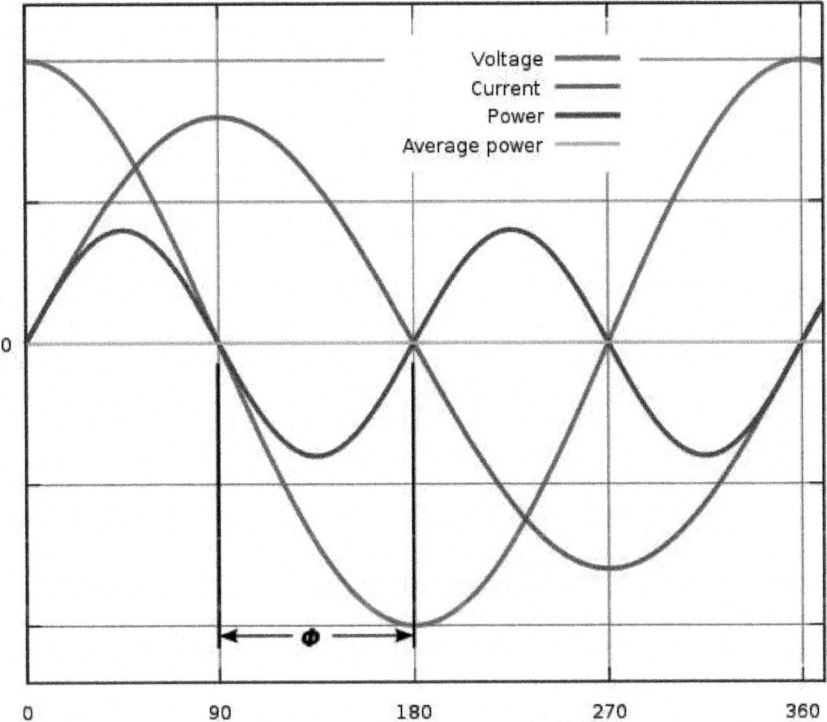

Fig. : Instantaneous and average power calculated from AC voltage and current with a lagging power factor ($\varphi = 45$, $\cos \varphi = 0.71$). The blue line shows some of the power is returned to the grid during the part of the cycle labelled φ.

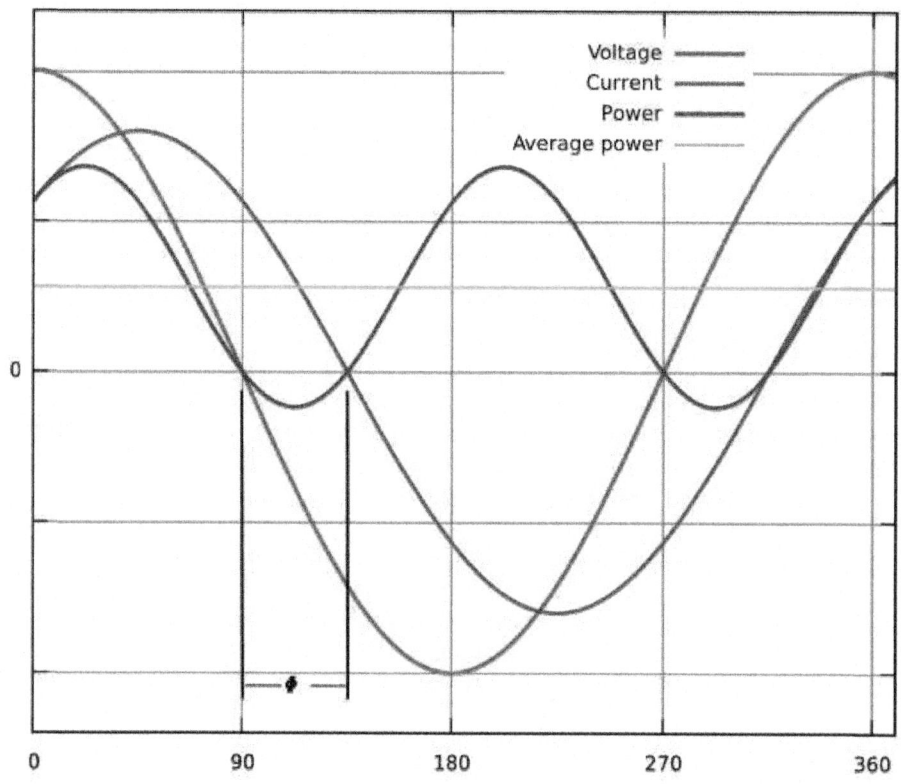

In a purely resistive AC circuit, voltage and current waveforms are in step (or in phase), changing polarity at the same instant in each cycle. All the power entering the load is consumed (or dissipated). Where reactive loads are present, such as with capacitors or inductors, energy storage in the loads results in a time difference between the current and voltage waveforms. During each cycle of the AC voltage, extra energy, in addition to any energy consumed in the load, is temporarily stored in the load in electric or magnetic fields, and then returned to the power grid a fraction of a second later in the cycle. The "ebb and flow" of this non-productive power increases the current in the line. Thus, a circuit with a low power factor will use higher currents to transfer a given quantity of real power than a circuit with a high power factor. A linear load does not change the shape of the waveform of the current, but may change the relative timing (phase) between voltage and current.

Circuits containing purely resistive heating elements (filament lamps, cooking stoves, etc.) have a power factor of 1.0. Circuits containing inductive or capacitive elements (electric motors, solenoid valves, lamp ballasts, and others) often have a power factor below 1.0.

AC AND DC POWER

Definition and Calculation

AC power flow has the three components: real power (also known as active power) (P), measured in watts (W); apparent power (S), measured in volt-amperes (VA); and reactive power (Q), measured in reactive volt-amperes (var).

The power factor is defined as:

$$= \frac{P}{S}$$

In the case of a perfectly sinusoidal waveform, P, Q and S can be expressed as vectors that form a vector triangle such that:

$$S^2 = P^2 + Q^2$$

If φ is the phase angle between the current and voltage, then the power factor is equal to the cosine of the angle, cos φ, and :

$$|P| = |S| \cos \varphi$$

Since the units are consistent, the power factor is by definition a dimensionless number between −1 and 1. When power factor is equal to 0, the energy flow is entirely reactive, and stored energy in the load returns to the source on each cycle. When the power factor is 1, all the energy supplied by the source is consumed by the load. Power factors are usually stated as "leading" or "lagging" to show the sign of the phase angle. Capacitive loads are leading (current leads voltage), and inductive loads are lagging (current lags voltage).

If a purely resistive load is connected to a power supply, current and voltage will change polarity in step, the power factor will be unity (1), and the electrical energy flows in a single direction across the network in each cycle. Inductive loads such as transformers and motors (any type of wound coil) consume reactive power with current waveform lagging the voltage. Capacitive loads such as capacitor banks or buried cable generate reactive power with current phase leading the voltage. Both types of loads will absorb energy during part of the AC cycle, which is stored in the device's magnetic or electric field, only to return this energy back to the source during the rest of the cycle.

For example, to get 1 kW of real power, if the power factor is unity, 1/kVA of apparent power needs to be transferred (1 kW ÷ 1 = 1/kVA). At low values of power factor, more apparent power needs to be transferred to get the same real power. To get 1 kW of real power at 0.2 power factor, 5/kVA of apparent power needs to be transferred (1 kW ÷ 0.2 = 5/kVA). This apparent power must be produced and transmitted to the load in the conventional fashion, and is subject to the usual distributed losses in the production and transmission processes.

Electrical loads consuming alternating current power consume both real power and reactive power. The vector sum of real and reactive power is the apparent power. The presence of reactive power causes the real power to be less than the apparent power, and so, the electric load has a power factor of less than 1.

Power Factor Correction of Linear Loads

A high power factor is generally desirable in a transmission system to reduce transmission losses and improve voltage regulation at the load. It is often desirable to adjust the power factor of a system to near 1.0. When reactive elements supply or absorb reactive power near the load, the apparent power is reduced. Power factor correction may be applied by an electric power transmission utility to improve the stability and efficiency of the transmission network. Individual electrical customers who are charged by their utility for low power factor may install correction equipment to reduce those costs.

Power factor correction brings the power factor of an AC power circuit closer to 1 by supplying reactive power of opposite sign, adding capacitors or inductors that act to cancel the inductive or capacitive effects of the load, respectively. For example, the inductive effect of motor loads may be offset by locally connected capacitors. If a load had a capacitive value, inductors (also known as *reactors* in this context) are connected to correct the power factor. In the electricity industry, inductors are said to *consume* reactive power and capacitors are said to *supply* it, even though the energy is just moving back and forth on each AC cycle.

The reactive elements can create voltage fluctuations and harmonic noise when switched on or off. They will supply or sink reactive power regardless of whether there is a corresponding load operating nearby, increasing the system's no-load losses. In the worst case, reactive elements can interact with the system and with each other to create resonant conditions, resulting in system instability and severe overvoltage fluctuations. As such, reactive elements cannot simply be applied without engineering analysis.

An **automatic power factor correction unit** consists of a number of capacitors that are switched by means of contactors. These contactors are controlled by a regulator that measures power factor in an electrical network. Depending on the load and power factor of the network, the power factor controller will switch the necessary blocks of capacitors in steps to make sure the power factor stays above a selected value.

Instead of using a set of switched capacitors, an unloaded synchronous motor can supply reactive power. The reactive power drawn by the synchronous motor is a function of its field excitation. This is referred to as a *synchronous condenser*. It is started and connected to the electrical network. It operates at a leading power factor and puts vars onto the network as required to support a system's voltage or to maintain the system power factor at a specified level.

The condenser's installation and operation are identical to large electric motors. Its principal advantage is the ease with which the amount of correction can be adjusted; it behaves like an electrically variable capacitor. Unlike capacitors, the amount of reactive power supplied is proportional to voltage, not the square of voltage; this improves voltage stability on large networks. Synchronous condensors are often used in connection with high-voltage direct-current transmission projects or in large industrial plants such as steel mills.

AC and DC Power

For power factor correction of high-voltage power systems or large, fluctuating industrial loads, power electronic devices such as the Static VAR compensator or STATCOM are increasingly used. These systems are able to compensate sudden changes of power factor much more rapidly than contactor-switched capacitor banks, and being solid-state require less maintenance than synchronous condensers.

Non-Linear Loads

A non-linear load on a power system is typically a rectifier (such as used in a power supply), or some kind of arc discharge device such as a fluorescent lamp, electric welding machine, or arc furnace. Because current in these systems is interrupted by a switching action, the current contains frequency components that are multiples of the power system frequency. Distortion power factor is a measure of how much the harmonic distortion of a load current decreases the average power transferred to the load.

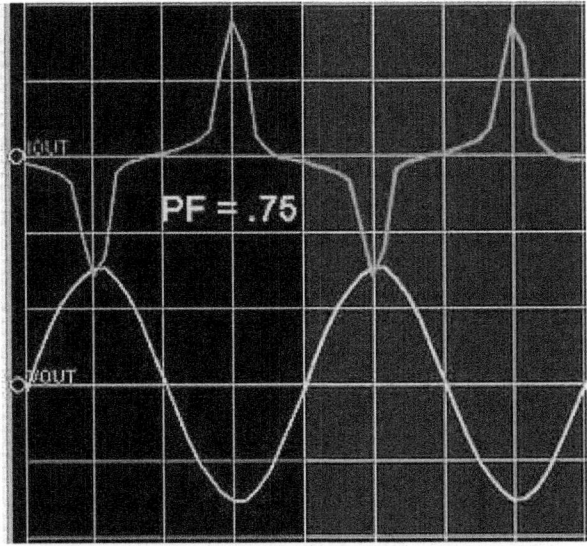

Fig. : Sinusoidal voltage and non-sinusoidal current give a distortion power factor of 0.75 for this computer power supply load.

Non-Sinusoidal Components

Non-linear loads change the shape of the current waveform from a sine wave to some other form. Non-linear loads create harmonic currents in addition to the original (fundamental frequency) AC current. Filters consisting of linear capacitors and inductors can prevent harmonic currents from entering the supplying system.

In linear circuits having only sinusoidal currents and voltages of one frequency, the power factor arises only from the difference in phase between the current and voltage. This is "displacement power factor". The concept can be generalized to a total, distortion, or true power factor where the apparent power includes all harmonic components. This is of importance in practical power systems that contain non-linear loads such as rectifiers, some forms of electric lighting, electric arc furnaces, welding equipment, switched-mode power supplies and other devices.

A typical multi-meter will give incorrect results when attempting to measure the AC current drawn by a non-sinusoidal load; the instruments sense the average value of a rectified waveform. The average response is then calibrated to the effective, RMS value. An RMS sensing multi-meter must be used to measure the actual RMS currents and voltages (and therefore apparent power). To measure the real power or reactive power, a watt meter designed to work properly with non-sinusoidal currents must be used.

Distortion Power Factor

The *distortion power factor* describes how the harmonic distortion of a load current decreases the average power transferred to the load.

$$\text{distortion power factor} = \frac{1}{\sqrt{1+\text{THD}_i^2}} = \frac{I_{1,}\text{rms}}{I\,\text{rms}}$$

THD_i is the total harmonic distortion of the load current. $I_{1,}$ rms is the fundamental component of the current and I_{rms} is the total current – both are root mean square-values (distortion power factor can also be used to describe individual order harmonics, using the corresponding current in place of total current). This definition with respect to total harmonic distortion assumes that the voltage stays undistorted (sinusoidal, without harmonics). This simplification is often a good approximation for stiff voltage sources (not being affected by changes in load downstream in the distribution network). Total harmonic distortion of typical generators from current distortion in the network is on the order of 1–2%, which can have larger scale implications but can be ignored in common practice.

The result when multiplied with the displacement power factor (DPF) is the overall, true power factor or just power factor (PF) :

$$PF = \text{DPF}\,\frac{I_{1,}\text{rms}}{I_{rms}}$$

Distortion in Three-Phase Networks

In practice, the local effects of distortion current on devices in a three-phase distribution network rely on the magnitude of certain order harmonics rather than the total harmonic distortion.

For example, the triplen, or zero-sequence, harmonics (3rd, 9th, 15th, etc.) have the property of being in-phase when compared line-to-line. In a delta-wye transformer, these harmonics can result in circulating currents in the delta windings and result in greater resistive heating. In a wye-configuration of a transformer, triplen harmonics will not create these currents, but they will result in a non-zero current in the neutral wire. This could overload the neutral wire in some cases and create error in kilowatt-hour metering systems and billing revenue. The presence of current harmonics in a transformer also result in larger eddy currents in the magnetic core of the transformer. Eddy current losses generally increase as the square of the frequency, lowering the transformer's efficiency, dissipating additional heat, and reducing its service life.

Negative-sequence harmonics (5th, 11th, 17th, etc.) combine 120 degrees out of phase, similarly to the fundamental harmonic but in a reversed sequence. In generators and motors, these currents produce magnetic fields which oppose the rotation of the shaft and sometimes result in damaging mechanical vibrations.

Switched-Mode Power Supplies

A particularly important class of non-linear loads is the millions of personal computers that typically incorporate switched-mode power supplies (SMPS) with rated output power ranging from a few watts to more than 1 kW. Historically, these very-low-cost power supplies incorporated a simple full-wave rectifier that conducted only when the mains instantaneous voltage exceeded the voltage on the input capacitors. This leads to very high ratios of peak-to-average input current, which also lead to a low distortion power factor and potentially serious phase and neutral loading concerns.

A typical switched-mode power supply first makes a DC bus, using a bridge rectifier or similar circuit. The output voltage is then derived from this DC bus. The problem with this is that the rectifier is a non-linear device, so the input current is highly non-linear. That means that the input current has energy at harmonics of the frequency of the voltage.

This presents a particular problem for the power companies, because they cannot compensate for the harmonic current by adding simple capacitors or inductors, as they could for the reactive power drawn by a linear load. Many jurisdictions are beginning to legally require power factor correction for all power supplies above a certain power level.

Regulatory agencies such as the EU have set harmonic limits as a method of improving power factor. Declining component cost has hastened implementation of two different methods. To comply with current EU standard EN61000-3-2, all switched-mode power supplies with output power more than 75 W must include passive power factor correction, at least. 80 Plus power supply certification requires a power factor of 0.9 or more.

Power Factor Correction in Non-Linear Loads

Passive PFC

The simplest way to control the harmonic current is to use a filter : it is possible to design a filter that passes current only at line frequency (50 or 60 Hz). This filter reduces the harmonic current, which means that the non-linear device now looks like a linear load. At this point the power factor can be brought to near unity, using capacitors or inductors as required. This filter requires large-value high-current inductors, however, which are bulky and expensive.

A passive PFC requires an inductor larger than the inductor in an active PFC, but costs less.

This is a simple way of correcting the non-linearity of a load by using capacitor banks. It is not as effective as active PFC. One example of this is a valley-fill circuit.

Active PFC

An *active power factor corrector* (active PFC) is a power electronic system that changes the waveshape of current drawn by a load to improve the power factor. The purpose is to make the load circuitry that is power factor corrected appear purely resistive (apparent power equal to real power). In this case, the voltage and current are in phase and the reactive power consumption is zero. This enables the most efficient delivery of electrical power from the power company to the consumer.

```
610W Continuous @ 40C (670W Peak)
Up to 90% (10dB) Less Noise per Watt
EPS12V / NVIDIA® SLI™ Certified
High Efficiency (83%), .99 Active PFC
+12VDC @ 49A (Large Single Rail)
24-pin, 8-pin, 4-pin M/B Connectors
2 PCI-E and 15 Drive Connectors
Automatic Fan Speed Control Circuit
Black Finish (Copper on request)
5-Year Warranty and Tech Support
```

Fig. : Specifications taken from the packaging of a 610 W PC power supply showing Active PFC rating.

Some types of active PFC are :
- Boost
- Buck
- Buck-boost.

Active power factor correctors can be single-stage or multi-stage.

In the case of a switched-mode power supply, a boost converter is inserted between the bridge rectifier and the main input capacitors. The boost converter attempts to maintain a constant DC bus voltage on its output while drawing a current that is always in phase with and at the same frequency as the line voltage. Another switchmode converter inside the power supply produces the desired output voltage from the DC bus. This approach requires additional semi-conductor switches and control electronics, but permits cheaper and smaller passive components. It is frequently used in practice. For example, SMPS with passive PFC can achieve power factor of about 0.7–0.75, SMPS with active PFC, up to 0.99 power factor, while a SMPS without any power factor correction has a power factor of only about 0.55–0.65.

Due to their very wide input voltage range, many power supplies with active PFC can automatically adjust to operate on AC power from about 100 V (Japan) to 230 V (Europe). That feature is particularly welcome in power supplies for laptops.

Dynamic PFC

Dynamic power factor correction (DPFC), sometimes referred to as "real-time power factor correction," is used for electrical stabilization instances of rapid load changes. When electrical networks experience rapid load changes, especially with the presence of non-linear loads, standard power factor correction is unable to adjust with the constantly changing, *i.e.* dynamic, electrical network, causing over or under correction. DPFC has the ability with semi-conductors to connect capacitors or inductors to electrical networks without disturbing the electrical network and causing unnecessary stress to electrical components, such as fuses and capacitors. Implementation of DPFC improves power quality by reducing current, especially reactive current, bringing stability to the electricity.

For companies, especially manufacturers, poor power quality leads to higher electrical bills due to power quality penalties. While PFC is available, few companies use it because PFC tends to be expensive without being cost effective. Today, DPFC devices range from sampling that takes place once per wave cycle (50 Hz/60 Hz) to over 8000 times per wave cycle.

Importance of Power Factor in Distribution Systems

Power factors below 1.0 require a utility to generate more than the minimum volt-amperes necessary to supply the real power (watts). This increases generation and transmission costs. For example, if the load power factor were as low as 0.7, the apparent power would be 1.4 times the real power used by the load. Line current in the circuit would also be 1.4 times the current required at 1.0 power factor, so the losses in the circuit would be doubled (since they are proportional to the square of the current). Alternatively all components of the system such as generators, conductors, transformers, and switchgear would be increased in size (and cost) to carry the extra current.

Utilities typically charge additional costs to commercial customers who have a power factor below some limit, which is typically 0.9 to 0.95. Engineers are often interested in the power factor of a load as one of the factors that affect the efficiency of power transmission.

With the rising cost of energy and concerns over the efficient delivery of power, active PFC has become more common in consumer electronics. Current Energy Star guidelines for computers call for a power factor of ≥ 0.9 at 100% of rated output in the PC's power supply. According to a white paper authored by Intel and the U.S. Environmental Protection Agency, PCs with internal power supplies will require the use of active power factor correction to meet the ENERGY STAR 5.0 Program Requirements for Computers.

In Europe, IEC 555-2 requires power factor correction be incorporated into consumer products.

Measuring the Power Factor

The power factor in a single-phase circuit (or balanced three-phase circuit) can be measured with the wattmeter-ammeter-voltmeter method, where the power in watts is divided by the product of measured voltage and current. The power factor of a balanced polyphase circuit is the same as that of any phase. The power factor of an unbalanced polyphase circuit is not uniquely defined.

A direct reading power factor meter can be made with a moving coil meter of the electrodynamic type, carrying two perpendicular coils on the moving part of the instrument. The field of the instrument is energized by the circuit current flow. The two moving coils, A and B, are connected in parallel with the circuit load. One coil, A, will be connected through a resistor and the second coil, B, through an inductor, so that the current in coil B is delayed with respect to current in A. At unity power factor, the current in A is in phase with the circuit current, and coil A provides maximum torque, driving the instrument pointer toward the 1.0 mark on the scale. At zero power factor, the current in coil B is in phase with circuit current, and coil B provides torque to drive the pointer towards 0. At intermediate values of power factor, the torques provided by the two coils add and the pointer takes up intermediate positions.

Another electromechanical instrument is the polarized-vane type. In this instrument a stationary field coil produces a rotating magnetic field, just like a polyphase motor. The field coils are connected either directly to polyphase voltage sources or to a phase-shifting reactor if a single-phase application. A second stationary field coil, perpendicular to the voltage coils, carries a current proportional to current in one phase of the circuit. The moving system of the instrument consists of two vanes that are magnetized by the current coil. In operation the moving vanes take up a physical angle equivalent to the electrical angle between the voltage source and the current source. This type of instrument can be made to register for currents in both directions, giving a four-quadrant display of power factor or phase angle.

Digital instruments can be made that either directly measure the time lag between voltage and current waveforms and so calculate the power factor, or by measuring both true and apparent power in the circuit and calculating the quotient. The first method is only accurate if voltage and current are sinusoidal; loads such as rectifiers distort the waveforms from the sinusoidal shape.

Mnemonics

English-language power engineering students are advised to remember : "ELI the ICE man" or "ELI on ICE" – the voltage E leads the current I in an inductor L; the current leads the voltage in a capacitor C.

AC AND DC POWER

Another common mnemonic is CIVIL – in a capacitor (C) the current (I) leads voltage (V), voltage (V) leads current (I) in an inductor (L).

DIRECT CURRENT

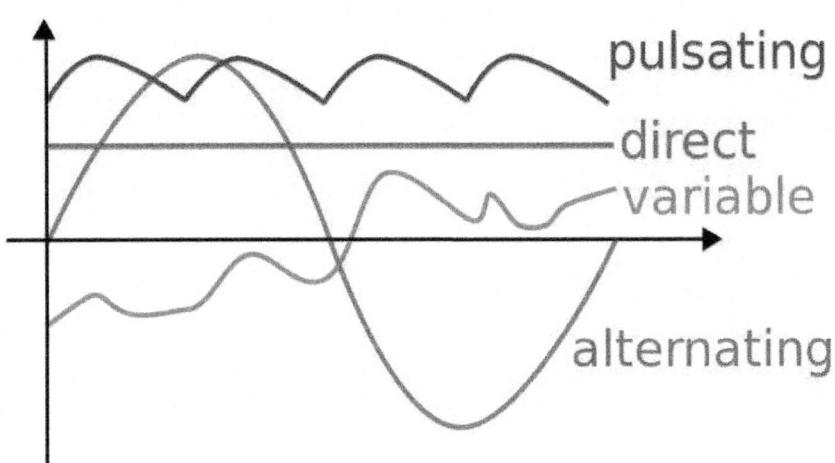

Fig. : Direct Current (red curve). The horizontal axis measures time; the vertical, current or voltage.

Direct current (DC) is the unidirectional flow of electric charge. Direct current is produced by sources such as batteries, thermocouples, solar cells, and commutator-type electric machines of the dynamo type. Direct current may flow in a conductor such as a wire, but can also flow through semi-conductors, insulators, or even through a vacuum as in electron or ion beams. The electric current flows in a constant direction, distinguishing it from alternating current (AC). A term formerly used for *direct current* was **galvanic current**.

The abbreviations *AC* and *DC* are often used to mean simply *alternating* and *direct*, as when they modify *current* or *voltage*.

Direct current may be obtained from an alternating current supply by use of a current-switching arrangement called a rectifier, which contains electronic elements (usually) or electromechanical elements (historically) that allow current to flow only in one direction. Direct current may be made into alternating current with an inverter or a motor-generator set.

The first commercial electric power transmission (developed by Thomas Edison in the late nineteenth century) used direct current. Because of the significant advantages of alternating current over direct current in transforming and transmission, electric power distribution is nearly all alternating current today. In the mid-1950s, high-voltage direct current transmission was developed, and is now an option instead of long-distance high voltage alternating current systems. For long distance underseas cables (*e.g.* between countries, such as NorNed), this is the only technically feasible option. For applications requiring direct current, such as third rail power systems, alternating current is distributed to a substation, which utilizes a rectifier to convert the power to direct current.

Direct current is used to charge batteries, and in nearly all electronic systems, as the power supply. Very large quantities of direct-current power are used in production of aluminum and other electrochemical processes. Direct current is used for some railway propulsion, especially in urban areas. High-voltage direct current is used to transmit large amounts of power from remote generation sites or to interconnect alternating current power grids.

Various Definitions

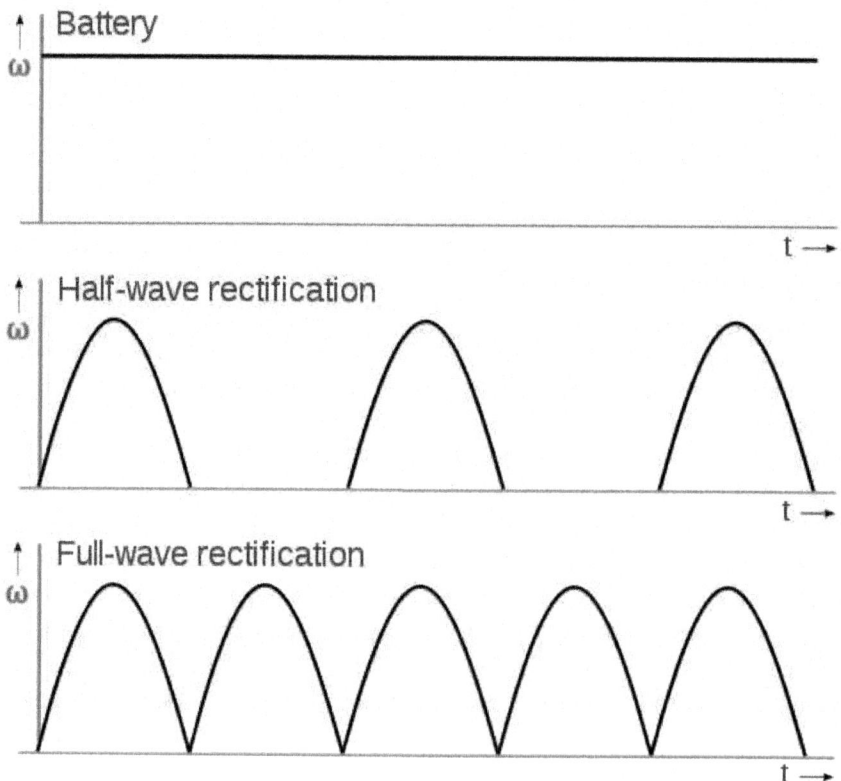

Fig. : Types of direct current.

The term DC is used to refer to power systems that use only one polarity of voltage or current, and to refer to the constant, zero-frequency, or slowly varying local mean value of a voltage or current. For example, the voltage across a DC voltage source is constant as is the current through a DC current source. The DC solution of an electric circuit is the solution where all voltages and currents are constant. It can be shown that any stationary voltage or current waveform can be decomposed into a sum of a DC component and a zero-mean time-varying component; the DC component is defined to be the expected value, or the average value of the voltage or current over all time.

Although DC stands for "direct current", DC often refers to "constant polarity". Under this definition, DC voltages can vary in time, as seen in the raw output of a rectifier or the fluctuating voice signal on a telephone line.

Some forms of DC (such as that produced by a voltage regulator) have almost no variations in voltage, but may still have variations in output power and current.

Circuits

A direct current circuit is an electrical circuit that consists of any combination of constant voltage sources, constant current sources, and resistors. In this case, the circuit voltages and currents are independent of time. A particular circuit voltage or current does not depend on the past value of any circuit voltage or current. This implies that the system of equations that represent a DC circuit do not involve integrals or derivatives with respect to time.

If a capacitor or inductor is added to a DC circuit, the resulting circuit is not, strictly speaking, a DC circuit. However, most such circuits have a DC solution. This solution gives the circuit voltages and currents when the circuit is in DC steady state. Such a circuit is represented by a system of differential equations. The solution to these equations usually contain a time varying or transient part as well as constant or steady state part. It is this steady state part that is the DC solution. There are some circuits that do not have a DC solution. Two simple examples are a constant current source connected to a capacitor and a constant voltage source connected to an inductor.

In electronics, it is common to refer to a circuit that is powered by a DC voltage source such as a battery or the output of a DC power supply as a DC circuit even though what is meant is that the circuit is DC powered.

Applications

Direct-current installations usually have different types of sockets, connectors, switches, and fixtures, mostly due to the low voltages used, from those suitable for alternating current. It is usually important with a direct-current appliance not to reverse polarity unless the device has a diode bridge to correct for this (most battery-powered devices do not).

The Unicode code point for the direct current symbol, found in the Miscel-laneous Technical block, is U+2393

Fig. : This symbol is found on many electronic devices that either require or produce direct current.

DC is commonly found in many extra-low voltage applications and some low-voltage applications, especially where these are powered by batteries, which can produce only DC, or solar power systems, since solar cells can produce only DC. Most automotive applications use DC, although the alternator is an AC device which uses a rectifier to produce DC. Most electronic circuits require a DC power supply. Applications using fuel cells (mixing hydrogen and oxygen together with a catalyst to produce electricity and water as byproducts) also produce only DC.

The vast majority of automotive applications use "12-volt" DC power; a few have a 6 V or a 42 V electrical system.

Light aircraft electrical systems are typically 12 V or 28 V.

Through the use of a DC-DC converter, high DC voltages such as 48 V to 72 V DC can be stepped down to 36 V, 24 V, 18 V, 12 V or 5 V to supply different loads. In a telecommunications system operating at 48 V DC, it is generally more efficient to step voltage down to 12 V to 24 V DC with a DC-DC converter and power equipment loads directly at their native DC input voltages versus operating a 48 V DC to 120 V AC inverter to provide power to equipment.

Many telephones connect to a twisted pair of wires, and use a bias tee to internally separate the AC component of the voltage between the two wires (the audio signal) from the DC component of the voltage between the two wires (used to power the phone).

Telephone exchange communication equipment, such as DSLAM, uses standard −48 V DC power supply. The negative polarity is achieved by grounding the positive terminal of power supply system and the battery bank. This is done to prevent electrolysis depositions.

HIGH-VOLTAGE DIRECT CURRENT

A **high-voltage, direct current** (**HVDC**) electric power transmission system uses direct current for the bulk transmission of electrical power, in contrast with the more common alternating current (AC) systems. For long-distance transmission, HVDC systems may be less expensive and suffer lower electrical losses. For underwater power cables, HVDC avoids the heavy currents required to charge and discharge the cable capacitance each cycle. For shorter distances, the higher cost of DC conversion equipment compared to an AC system may still be warranted, due to other benefits of direct current links.

HVDC allows power transmission between unsynchronized AC transmission systems. Since the power flow through an HVDC link can be controlled independently of the phase angle between source and load, it can stabilize a network against disturbances due to rapid changes in power. HVDC also allows transfer of power between grid systems running at different frequencies, such as 50 Hz and 60 Hz. This improves the stability and economy of each grid, by allowing exchange of power between incompatible networks.

The modern form of HVDC transmission uses technology developed extensively in the 1930s in Sweden (ASEA) and in Germany. Early commercial installations included one in the Soviet Union in 1951 between Moscow and Kashira, and a 100 kV, 20 MW system between Gotland and mainland Sweden in 1954. The longest HVDC link in the world is currently the Xiangjiaba–Shanghai 2,071 km (1,287 mi), ±800 kV, 6400 MW link connecting the Xiangjiaba Dam to Shanghai, in the People's Republic of China. Early in 2013, the longest HVDC link will be the Rio Madeira link in Brazil, which consists of two bipoles of ±600 kV, 3150 MW each, connecting Porto Velho in the state of Rondônia to the São Paulo area, where the length of the DC line is 2,375 km (1,476 mi).

High Voltage Transmission

High voltage is used for electric power transmission to reduce the energy lost in the resistance of the wires. For a given quantity of power transmitted, doubling the voltage will deliver the same power at only half the current. Since the power lost as heat in the wires is proportional to the square of the current for a given conductor size, but does not depend on the voltage, doubling the voltage reduces the line losses per unit of electrical power delivered by a factor of 4. While power lost in transmission can also be reduced by increasing the conductor size, larger conductors are heavier and more expensive.

High voltage cannot readily be used for lighting or motors, so transmission-level voltages must be reduced for end-use equipment. Transformers are used to change the voltage levels in alternating current (AC) transmission circuits. AC became dominant after the War of Currents competition between the direct current (DC) system of Thomas Edison and the AC system of George Westinghouse because transformers made voltage changes practical, and AC generators were more efficient than those using DC.

Practical conversion of power between AC and DC became possible with the development of power electronics devices such as mercury-arc valves and, starting in the 1970s, semi-conductor devices as thyristors, integrated gate-commutated thyristors (IGCTs), MOS-controlled thyristors (MCTs) and insulated-gate bipolar transistors (IGBT).

History of HVDC Technology

Electromechanical (Thury) Systems

The first long-distance transmission of electric power was demonstrated using direct current in 1882 at Miesbach-Munich Power Transmission, but only 1.5 kW was transmitted. An early method of high-voltage DC transmission was developed by the Swiss engineer René Thury and his method was put into practice by 1889 in Italy by the *Acquedotto De Ferrari-Galliera* company. This system used series-connected motor-generator sets to increase the voltage. Each set was insulated from electrical ground and driven by insulated shafts from a prime mover. The transmission line was operated in a 'constant current' mode, with up to 5,000 volts across each machine, some machines having double commutators to reduce the voltage on each commutator. This system transmitted 630 kW at 14 kV DC over a distance of 120 km. The Moutiers–Lyon system transmitted 8,600 kW of hydroelectric power a distance of 200 km, including 10 km of underground cable. This system used eight series-connected generators with dual commutators for a total voltage of 150,000 volts between the positive and negative poles, and operated from c.1906 until 1936. Fifteen Thury systems were in operation by 1913. Other Thury systems operating at up to 100 kV DC worked into the 1930s, but the rotating machinery required high maintenance and had high energy loss. Various other electromechanical devices were tested during the first half of the 20th century with little commercial success.

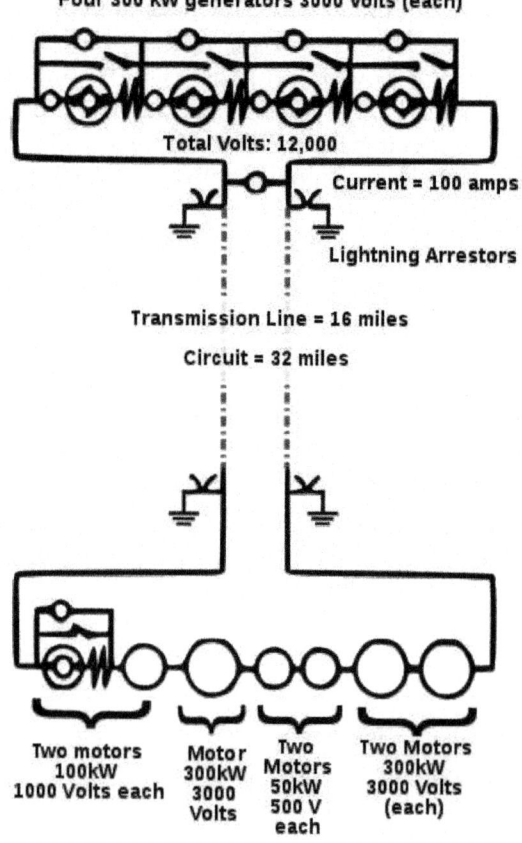

Fig.: Schematic diagram of a Thury HVDC transmission system.

One technique attempted for conversion of direct current from a high transmission voltage to lower utilization voltage was to charge series-connected batteries, then reconnect the batteries in parallel to serve distribution loads. While at least two commercial installations were tried around the turn of the 20th century, the technique was not generally useful owing to the limited capacity of batteries, difficulties in switching between series and parallel connections, and the inherent energy inefficiency of a battery charge/discharge cycle.

Mercury Arc Valves

First proposed in 1914, the grid controlled mercury-arc valve became available for power transmission during the period 1920 to 1940. Starting in 1932, General Electric tested mercury-vapour valves and a 12 kV DC transmission line, which also served to convert 40 Hz generation to serve 60 Hz loads, at Mechanicville, New York. In 1941, a 60 MW, ±200 kV, 115 km buried cable link was designed for the city of Berlin using mercury arc valves (Elbe-Project), but owing to the collapse of the German government in 1945 the project was never completed. The nominal justification for the project was that, during war-time, a buried cable would be less conspicuous as a bombing target. The equipment was moved to the Soviet Union and was put into service there as the Moscow–Kashira HVDC system. The Moscow–Kashira system and the 1954 connection by ASEA between the mainland of Sweden and the island of Gotland marked the beginning of the modern era of HVDC transmission.

Mercury arc valves require an external circuit to force the current to zero and thus turn off the valve. In HVDC applications, the AC power system itself provides the means of *commutating* the current to another valve in the converter. Consequently, converters built with mercury arc valves are known as line-commutated converters (LCC). LCCs require rotating synchronous machines in the AC systems to which they are connected, making power transmission into a passive load impossible.

Mercury arc valves were common in systems designed up to 1972, the last mercury arc HVDC system (the Nelson River Bipole 1 system in Manitoba, Canada) having been put into service in stages between 1972 and 1977. Since then, all mercury arc systems have been either shut down or converted to use solid state devices. The last HVDC system to use mercury arc valves was the Inter-Island HVDC link between the North and South Islands of New Zealand, which used them on one of its two poles. The mercury arc valves were decommissioned on 1 August, 2012, ahead of commissioning of replacement thyristor converters.

Thyristor Valves

Since 1977, new HVDC systems have used only solid-state devices, in most cases thyristor valves. Like mercury arc valves, thyristors require connection to an external AC circuit in HVDC applications to turn them on and off. HVDC using thyristor valves is also known as line-commutated converter (LCC) HVDC.

Development of thyristor valves for HVDC began in the late 1960s. The first complete HVDC scheme based on thyristor valves was the Eel River scheme in Canada, which was built by General Electric and went into service in 1972.

On March 15, 1979, a 1920 MW thyristor based direct current connection between Cabora Bassa and Johannesburg (1,410 km) was energised. The conversion equipment was built in 1974 by AEG, and BBC (Brown Boveri Company) and Siemens were partners in the project, the late completion date a result of the civil war. The transmission voltage of ±533 kV was the highest in the world at the time.

Capacitor-Commutated Converters (CCC)

Line-commutated converters have some limitations in their use for HVDC systems. This results from requiring the AC circuit to turn off the thyristor current and the need for a short period of 'reverse' voltage to effect the turn-off (turn-off time). An attempt to address these limitations is the *Capacitor-Commutated Converter (CCC)* which has been used in a small number of HVDC systems. The CCC differs from a conventional HVDC system in that it has series capacitors inserted into the AC line connections, either on the primary or secondary side of the converter transformer. The series capacitors partially offset the *commutating inductance* of the converter and help to reduce fault currents. This also allows a smaller *extinction angle* to be used with a converter/inverter, reducing the need for reactive power support. However, CCC has remained only a niche application because of the advent of voltage-source converters (VSC) which completely eliminate the need for an extinction (turn-off) time.

Voltage-Source Converters (VSC)

Widely used in motor drives since the 1980s, voltage-source converters started to appear in HVDC in 1997 with the experimental Hellsjön–Grängesberg project in Sweden. By the end of 2011, this technology had captured a significant proportion of the HVDC market.

The development of higher rated insulated-gate bipolar transistors (IGBTs), gate turn-off thyristors (GTOs) and integrated gate-commutated thyristors (IGCTs), has made smaller HVDC systems economical. The manufacturer ABB Group calls this concept *HVDC Light*, while Siemens calls a similar concept *HVDC PLUS* (*Power Link Universal System*) and Alstom call their product based upon this technology *HVDC MaxSine*. They have extended the use of HVDC down to blocks as small as a few tens of megawatts and lines as short as a few score kilometres of overhead line. There are several different variants of VSC technology : most installations built until 2012 use pulse width modulation in a circuit that is effectively an ultra-high-voltage motor drive. Current installations, including HVDC PLUS and HVDC MaxSine, are based on variants of a converter called a *Modular Multi-Level Converter* (MMC).

Multi-level converters have the advantage that they allow harmonic filtering equipments to be reduced or eliminated altogether. By way of comparison, AC harmonic filters of typical line-commutated converter stations cover nearly half of the converter station area.

With time, voltage-source converter systems will probably replace all installed simple thyristor-based systems, including the highest DC power transmission applications.

Advantages of Hvdc Over AC Transmission

The most common reason for choosing HVDC over AC transmission is that HVDC is more economic than AC for transmitting large amounts of power point-to-point over long distances. A long distance, high power HVDC transmission scheme generally has lower capital costs and lower losses than an AC transmission link.

Even though HVDC conversion equipment at the terminal stations is costly, overall savings in capital cost may arise because of significantly reduced transmission line costs over long distance routes. HVDC needs fewer conductors than an AC line, as there is no need to support three phases. Also, thinner conductors can be used since HVDC does not suffer from the skin effect. These factors can lead to large reductions in transmission line cost for a long distance HVDC scheme.

Depending on voltage level and construction details, HVDC transmission losses are quoted as about 3.5% per 1,000 km, which is less than typical losses in an AC transmission system.

HVDC transmission may also be selected because of other technical benefits that it provides for the power system. HVDC schemes can transfer power between separate AC networks. HVDC powerflow between separate AC systems can be automatically controlled to provide support for either network during transient conditions, but without the risk that a major power system collapse in one network will lead to a collapse in the second.

The combined economic and technical benefits of HVDC transmission can make it a suitable choice for connecting energy sources that are located far away from the main load centers.

Specific applications where HVDC transmission technology provides benefits include :
- Undersea cables transmission schemes (*e.g.*, 250 km Baltic Cable between Sweden and Germany, the 580 km NorNed cable between Norway and the Netherlands, and 290 km Basslink between the Australian mainland and Tasmania).
- Endpoint-to-endpoint long-haul bulk power transmission without intermediate 'taps', usually to connect a remote generating plant to the main grid, for example the Nelson River DC Transmission System in Canada.
- Increasing the capacity of an existing power grid in situations where additional wires are difficult or expensive to install.
- Power transmission and stabilization between unsynchronised AC networks, with the extreme example being an ability to transfer power between countries

that use AC at different frequencies. Since such transfer can occur in either direction, it increases the stability of both networks by allowing them to draw on each other in emergencies and failures.
- Stabilizing a predominantly AC power-grid, without increasing fault levels (prospective short circuit current).

Cable Systems

Long undersea / underground high voltage cables have a high electrical capacitance compared with overhead transmission lines, since the live conductors within the cable are surrounded by a relatively thin layer of insulation (the dielectric), and a metal sheath. The geometry is that of a long co-axial capacitor. The total capacitance increases with the length of the cable. This capacitance is in a parallel circuit with the load. Where alternating current is used for cable transmission, additional current must flow in the cable to charge this cable capacitance. This extra current flow causes added energy loss via dissipation of heat in the conductors of the cable, raising its temperature. Additional energy losses also occur as a result of dielectric losses in the cable insulation.

However, if direct current is used, the cable capacitance is charged only when the cable is first energized or if the voltage level changes; there is no additional current required. For a long AC powered undersea cable, the entire current-carrying ability of the conductor would be needed to supply the charging current alone. This cable capacitance issue limits the length and power carrying ability of AC powered cables. DC powered cables are only limited by their temperature rise and Ohm's Law. Although some leakage current flows *through* the dielectric insulator, this is small compared to the cable's rated current.

Overhead Line Systems

The capacitive effect of long underground or undersea cables in AC transmission applications also applies to AC overhead lines, although to a much lesser extent. Nevertheless, for a long AC overhead transmission line, the current flowing just to charge the line capacitance can be significant, and this reduces the capability of the line to carry useful current to the load at the remote end. Another factor that reduces the useful current carrying ability of AC lines is the skin effect, which causes a non-uniform distribution of current over the cross-sectional area of the conductor. Transmission line conductors operating with HVDC current do not suffer from either of these constraints. Therefore, for the same conductor losses (or heating effect), a given conductor can carry more current to the load when operating with HVDC than AC.

Finally, depending upon the environmental conditions and the performance of overhead line insulation operating with HVDC, it may be possible for a given transmission line to operate with a constant HVDC voltage that is approximately the same as the peak AC voltage for which it is designed and insulated. The power delivered in an AC system is defined by the root mean square (RMS) of an AC voltage, but RMS is only about 71% of the peak voltage. Therefore, if the HVDC line can operate continuously with an HVDC voltage that is the same as the peak voltage of the AC equivalent line, then for a given current (where HVDC current is the same as the RMS current in the AC line), the power transmission capability when operating with HVDC is approximately 40% higher than the capability when operating with AC.

Asynchronous Connections

Because HVDC allows power transmission between unsynchronized AC distribution systems, it can help increase system stability, by preventing cascading failures from propagating from one part of a wider power transmission grid to another. Changes in load that would cause portions of an AC network to become unsynchronized and to separate, would not similarly affect a DC link, and the power flow through the DC link would tend to stabilize the AC network. The magnitude and direction of power flow through a DC link can be directly controlled, and changed as needed to support the AC networks at either end of the DC link. This has caused many power system operators to contemplate wider use of HVDC technology for its stability benefits alone.

Disadvantages

The disadvantages of HVDC are in conversion, switching, control, availability and maintenance.

HVDC is less reliable and has lower availability than alternating current (AC) systems, mainly due to the extra conversion equipment. Single-pole systems have availability of about 98.5%, with about a third of the downtime unscheduled due to faults. Fault-tolerant bipole systems provide high availability for 50% of the link capacity, but availability of the full capacity is about 97% to 98%.

The required converter stations are expensive and have limited overload capacity. At smaller transmission distances, the losses in the converter stations may be bigger than in an AC transmission line for the same distance. The cost of the converters may not be offset by reductions in line construction cost and lower line loss.

Operating a HVDC scheme requires many spare parts to be kept, often exclusively for one system, as HVDC systems are less standardized than AC systems and technology changes faster.

In contrast to AC systems, realizing multi-terminal systems is complex (especially with line commutated converters), as is expanding existing schemes to multi-terminal systems. Controlling power flow in a multi-terminal DC system requires good communication between all the terminals; power flow must be actively regulated by the converter control system instead of the inherent impedance and phase angle properties of the transmission line. Multi-terminal systems are rare. As of 2012 only two are in service : the Hydro Québec – New England transmission between Radisson, Sandy Pond and Nicolet and the Sardinia–mainland Italy link which was modified in 1989 to also provide power to the island of Corsica.

HVDC circuit breakers are difficult to build because some mechanism must be included in the circuit breaker to force current to zero, otherwise arcing and contact wear would be too great to allow reliable switching. In November 2012, ABB announced development of the world's first HVDC circuit breaker.

The ABB breaker contains four switching elements, two mechanical (one high-speed and one low-speed) and two semi-conductor (one high-voltage and one low-voltage). Normally, power flows through the low-speed mechanical switch, the high-speed mechanical switch and the low-voltage semi-conductor switch. The last two switches are paralleled by the high-voltage semi-conductor switch.

Initially, all switches are closed (on). Because the high-voltage semi-conductor switch has much greater resistance than the mechanical switch plus the low-voltage semi-conductor switch, current flow through it is low. To disconnect, first the low-voltage semi-conductor switch opens. This diverts the current through the high-voltage semi-conductor switch. Because of its relatively high resistance, it begins heating very rapidly. Then the high-speed mechanical switch is opened. Unlike the low-voltage semi-conductor switch, which is only capable of standing off the voltage drop of the closed high-voltage semi-conductor switch, this is capable of standing off the full voltage. Because no current is flowing through this switch when it opens, it is not damaged by arcing. Then, the high-voltage semi-conductor switch is opened. This actually cuts the power. However, it only cuts power to a very low level; it is not quite 100% off. A final low-speed mechanical switch disconnects the residual current.

Costs of High Voltage DC Transmission

Generally, providers of HVDC systems, such as Alstom, Siemens and ABB, do not specify cost details of particular projects. It may be considered a commercial matter between the provider and the client.

Costs vary widely depending on the specifics of the project (such as power rating, circuit length, overhead vs. cabled route, land costs, and AC network improvements required at either terminal). A detailed comparison of DC vs. AC transmission costs may be required in situations where there is no clear technical advantage to DC alone, and economical reasoning drives the selection.

However, some practitioners have provided some information :

For an 8 GW 40 km link laid under the English Channel, the following are approximate primary equipment costs for a 2000 MW 500 kV bipolar conventional HVDC link (exclude way-leaving, on-shore reinforcement works, consenting, engineering, insurance, etc.)

- Converter stations ~£110M (~173.7M USD)
- Subsea cable + installation ~£1M/km (~1.6M USD/km)

AC and DC Power

So for an 8 GW capacity between England and France in four links, little is left over from £750M for the installed works. Add another £200–300M for the other works depending on additional onshore works required.

An April 2010 announcement for a 2,000 MW, 64 km line between Spain and France is estimated at EUR 700 million. This includes the cost of a tunnel through the Pyrenees.

The Conversion Process

Converter

At the heart of an HVDC converter station, the equipment which performs the conversion between AC and DC is referred to as the *converter*. Almost all HVDC converters are inherently capable of converting from AC to DC (*rectification*) or from DC to AC (*inversion*), although in many HVDC systems, the system as a whole is optimised for power flow in only one direction. Irrespective of how the converter itself is designed, the station which is operating (at a given time) with power flow from AC to DC is referred to as the *rectifier* and the station which is operating with power flow from DC to AC is referred to as the *inverter*.

Early HVDC systems used electromechanical conversion (the Thury system) but all HVDC systems built since the 1940s have used electronic (static) converters. Electronic converters for HVDC are divided into two main categories :

- Line-commutated converters (LCC)
- Voltage-sourced converters, or current-source converters.

Line-commutated converters

Most of the HVDC systems in operation today are based on line-commutated converters.

The basic LCC configuration uses a three-phase bridge rectifier or *six-pulse bridge*, containing six electronic switches, each connecting one of the three phases to one of the two DC rails. A complete switching element is usually referred to as a *valve*, irrespective of its construction. However, with a phase change only every 60°, considerable harmonic distortion is produced at both the DC and AC terminals when this arrangement is used.

An enhancement of this arrangement uses 12 valves in a *twelve-pulse bridge*. The AC is split into two separate three phase supplies before transformation. One of the sets of supplies is then configured to have a star (wye) secondary, the other a delta secondary, establishing a 30° phase difference between the two sets of three phases. With twelve valves connecting each of the two sets of three phases to the two DC rails, there is a phase change every 30°, and harmonics are considerably reduced. For this reason the twelve-pulse system has become standard on most line-commutated converter HVDC systems built since the 1970s.

With line commutated converters, the converter has only one degree of freedom – the *firing angle*, which represents the time delay between the voltage across a valve becoming positive (at which point the valve would start to conduct if it were made from diodes) and the thyristors being turned on. The DC output voltage of the converter steadily becomes less positive as the firing angle is increased : firing angles of up to 90° correspond to rectification and result in positive DC voltages, while firing angles above 90° correspond to inversion and result in negative DC voltages. The practical upper limit for the firing angle is about 150–160° because above this, the valve would have insufficient *turn-off time*.

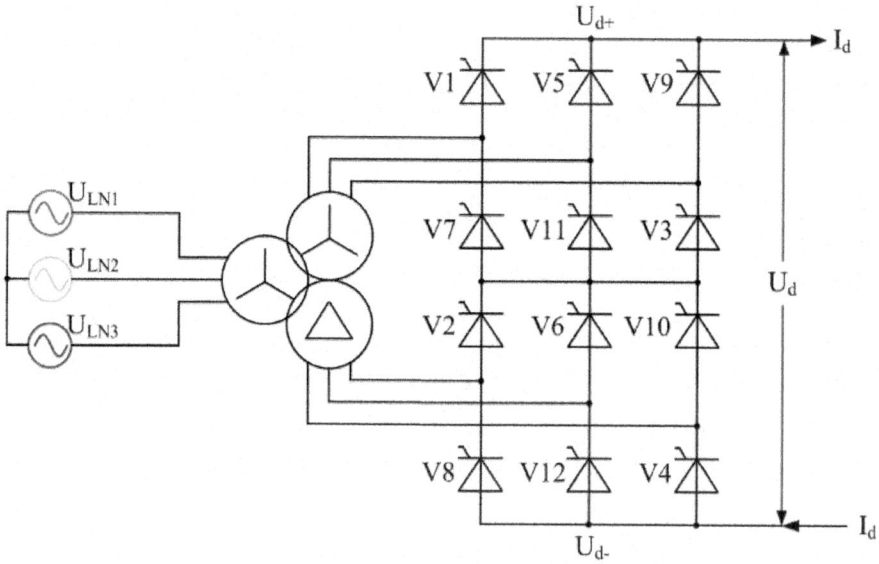

Fig. : A twelve-pulse bridge rectifier.

Early LCC systems used mercury-arc valves, which were rugged but required high maintenance. Because of this, many mercury-arc HVDC systems were built with bypass switchgear across each six-pulse bridge so that the HVDC scheme could be operated in six-pulse mode for short periods of maintenance. The last mercury arc system was shut down in 2012.

The thyristor valve was first used in HVDC systems in 1972. The thyristor is a solid-state semi-conductor device similar to the diode, but with an extra-control terminal that is used to switch the device on at a particular instant during the AC cycle. Because the voltages in HVDC systems, up to 800 kV in some cases, far exceed the breakdown voltages of the thyristors used, HVDC thyristor valves are built using large numbers of thyristors in series. Additional passive components such as grading capacitors and resistors need to be connected in parallel with each thyristor in order to ensure that the voltage across the valve is evenly shared between the thyristors. The thyristor plus its grading circuits and other auxiliary equipment is known as a *thyristor level*.

Each thyristor valve will typically contain tens or hundreds of thyristor levels, each operating at a different (high) potential with respect to earth. The command information to turn on the thyristors therefore cannot simply be sent using a wire connection – it needs to be isolated. The isolation method can be magnetic but is usually optical. Two optical methods are used : indirect and direct optical triggering. In the indirect optical triggering method, low-voltage control electronics send light pulses along optical fibres to the *high-side* control electronics, which derives its power from the voltage across each thyristor. The alternative direct optical triggering method dispenses with most of the high-side electronics, instead using light pulses from the control electronics to switch light-triggered thyristors (LTTs), although a small monitoring electronics unit may still be required for protection of the valve.

In a line-commutated converter, the DC current (usually) cannot change direction; it flows through a large inductance and can be considered almost constant. On the AC side, the converter behaves approximately as a current source, injecting both grid-frequency and harmonic currents into the AC network. For this reason, a line commutated converter for HVDC is also considered as a *current-source inverter*.

Voltage-Sourced Converters

Because thyristors can only be turned on (not off) by control action, the control system only has one degree of freedom – when to turn on the thyristor. This is an important limitation in some circumstances.

With some other types of semi-conductor device such as the insulated-gate bipolar transistor (IGBT), both turn-on and turn-off can be controlled, giving a second degree of freedom. As a result, they can be used to make *self-commutated converters*. In such converters, the polarity of DC voltage is usually fixed and the DC voltage, being smoothed by a large capacitance, can be considered constant. For this reason, an HVDC converter using IGBTs is usually referred to as a *voltage sourced converter*. The additional controllability gives many advantages, notably the ability to switch the IGBTs on and off many times per cycle in order to improve the harmonic performance. Being self-commutated, the converter no longer relies on synchronous machines in the AC system for its operation. A voltage sourced converter can therefore feed power to an AC network consisting only of passive loads, something which is impossible with LCC HVDC.

HVDC systems based on voltage sourced converters normally use the six-pulse connection because the converter produces much less harmonic distortion than a comparable LCC and the twelve-pulse connection is unnecessary.

Most of the VSC HVDC systems built until 2012 were based on the *two level converter*, which can be thought of as a six pulse bridge in which the thyristors have been replaced by IGBTs with inverse-parallel diodes, and the DC smoothing reactors have been replaced by DC smoothing capacitors. Such converters derive their name from the discrete, two voltage levels at the AC output of each phase that correspond to the electrical potentials of the positive and negative DC terminals. Pulse-width modulation (PWM) is usually used to improve the harmonic distortion of the converter.

Some HVDC systems have been built with *three level converters*, but today most new VSC HVDC systems are being built with some form of *multi-level converter*, most commonly the *Modular Multi-Level Converter* (MMC), in which each valve consists of a number of independent converter sub-modules, each containing its own storage capacitor. The IGBTs in each sub-module either bypass the capacitor or connect it into the circuit, allowing the valve to synthesize a stepped voltage with very low levels of harmonic distortion.

Converter Transformers

At the AC side of each converter, a bank of transformers, often three physically separated single-phase transformers, isolate the station from the AC supply, to provide a local earth, and to ensure the correct eventual DC voltage. The output of these transformers is then connected to the converter.

Converter transformers for LCC HVDC schemes are quite specialised because of the high levels of harmonic currents which flow through them, and because the secondary winding insulation experiences a permanent DC voltage, which affects the design of the insulating structure (valve side requires more solid insulation) inside the tank. In LCC systems, the transformer(s) also need to provide the 30° phase shift needed for harmonic cancellation.

Converter transformers for VSC HVDC systems are usually simpler and more conventional in design than those for LCC HVDC systems.

Reactive Power

A major drawback of HVDC systems using line-commutated converters is that the converters inherently consume reactive power. The AC current flowing into the converter from the AC system lags behind the AC voltage so that, irrespective of the direction of active power flow, the converter always absorbs reactive power, behaving in the same way as a shunt reactor. The reactive power absorbed is at least 0.5 MVAr/MW under ideal conditions and can be higher than this when the converter is operating at higher than usual firing or extinction angle, or reduced DC voltage.

Although at HVDC converter stations connected directly to power stations some of the reactive power may be provided by the generators themselves, in most cases the reactive power consumed by the converter must be provided by banks of shunt capacitors connected at the AC terminals of the converter. The shunt capacitors are usually connected directly to the grid voltage but in some cases may be connected to a lower voltage via a tertiary winding on the converter transformer.

Since the reactive power consumed depends on the active power being transmitted, the shunt capacitors usually need to be sub-divided into a number of switchable banks (typically 4 per converter) in order to prevent a surplus of reactive power being generated at low transmitted power.

The shunt capacitors are almost always provided with tuning reactors and, where necessary, damping resistors so that they can perform a dual role as harmonic filters.

Voltage-source converters, on the other hand, can either produce or consume reactive power on demand, with the result that usually no separate shunt capacitors are needed (other than those required purely for filtering).

Harmonics and Filtering

All power electronic converters generate some degree of harmonic distortion on the AC and DC systems to which they are connected, and HVDC converters are no exception.

With the recently developed Modular Multi-Level Converter (MMC), levels of harmonic distortion may be practically negligible, but with line-commutated converters and simpler types of voltage-source converters, considerable harmonic distortion may be produced on both the AC and DC sides of the converter. As a result, harmonic filters are nearly always required at the AC terminals of such converters, and in HVDC transmission schemes using overhead lines, may also be required on the DC side.

Filters for Line-Commutated Converters

The basic building-block of a line-commutated HVDC converter is the *six-pulse bridge*. This arrangement produces very high levels of harmonic distortion by acting as a current source injecting harmonic currents of order $6n\pm1$ into the AC system and generating harmonic voltages of order $6n$ superimposed on the DC voltage.

It is very costly to provide harmonic filters capable of suppressing such harmonics, so a variant known as the *twelve-pulse bridge* (consisting of two six-pulse bridges in series with a 30° phase shift between them) is nearly always used. With the twelve-pulse arrangement, harmonics are still produced but only at orders $12n\pm1$ on the AC side and $12n$ on the DC side. The task of suppressing such harmonics is still challenging, but manageable.

Line-commutated converters for HVDC are usually provided with combinations of harmonic filters designed to deal with the 11th and 13th harmonics on the AC side, and 12th harmonic on the DC side. Sometimes, high-pass filters may be provided to deal with 23rd, 25th, 35th, 37th... on the AC side and 24th, 36th... on the DC side. Sometimes, the AC filters may also need to provide damping at lower-order, *non-characteristic* harmonics such as 3rd or 5th harmonics.

The task of designing AC harmonic filters for HVDC converter stations is complex and computationally intensive, since in addition to ensuring that the converter does not produce an unacceptable level of voltage distortion on the AC system, it must be ensured that the harmonic filters do not resonate with some component elsewhere in the AC system. A detailed knowledge of the *harmonic impedance* of the AC system, at a wide range of frequencies, is needed in order to design the AC filters.

DC filters are only required for HVDC transmission systems involving overhead lines. Voltage distortion is not a problem in its own right, since consumers do not connect directly to the DC terminals of the system, so the main design criterion for the DC filters is to ensure that the harmonic currents flowing in the DC lines do not induce interference in nearby open-wire telephone lines. With the rise in digital mobile telecommunication systems, which are much less susceptible to interference, DC filters are becoming less important for HVDC systems.

Filters for Voltage-Sourced Converters

Some types of voltage-sourced converters may produce such low levels of harmonic distortion that no filters are required at all. However, converter types such as the *two-level* converter, used with pulse-width modulation (PWM), still require some filtering, albeit less than on line-commutated converter systems.

With such converters, the harmonic spectrum is generally shifted to higher frequencies than with line-commutated converters. This usually allows the filter equipment to be smaller. The dominant harmonic frequencies are sidebands of the PWM frequency and multiples thereof. In HVDC applications, the PWM frequency is typically around 1–2 kHz.

Configurations

Monopole

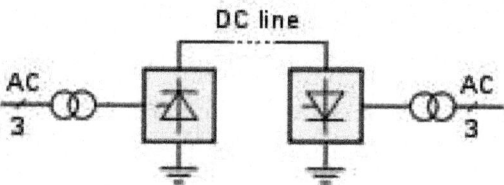

Fig. : Block diagram of a monopole system with earth return.

In a common configuration, called monopole, one of the terminals of the rectifier is connected to earth ground. The other terminal, at a potential high above or below ground, is connected to a transmission line. The earthed terminal may be connected to the corresponding connection at the inverting station by means of a second conductor.

Monopole and Earth Return

If no metallic conductor is installed, current flows in the earth and/or sea between two specially designed earth electrodes. This arrangement is a type of single wire earth return system.

The electrodes are usually located some tens of kilometres from the stations and are connected to the stations via a medium-voltage electrode line. The design of the electrodes themselves depends on whether they are located on land, on the shore or at sea. For the monopolar configuration with earth return, the earth current flow is unidirectional, which means that the design of one of the electrodes (the cathode) can be relatively simple, although the design of anode electrode is quite complex.

For long-distance transmission, earth return can be considerably cheaper than alternatives using a dedicated neutral conductor, but it can lead to problems such as :

- Electrochemical corrosion of long buried metal objects such as pipelines.
- Underwater earth-return electrodes in seawater may produce chlorine or otherwise affect water chemistry.
- An unbalanced current path may result in a net magnetic field, which can affect magnetic navigational compasses for ships passing over an underwater cable.

Monopole and Metallic Return

These effects can be eliminated with installation of a metallic return conductor between the two ends of the monopolar transmission line. Since one terminal of the converters is connected to earth, the return conductor need not be insulated for the full transmission voltage which makes it less costly than the high-voltage conductor. The decision of whether or not to use a metallic return conductor is based upon economic, technical and environmental factors.

Modern monopolar systems for pure overhead lines carry typically 1.5 GW. If underground or underwater cables are used, the typical value is 600 MW.

Most monopolar systems are designed for future bipolar expansion. Transmission line towers may be designed to carry two conductors, even if only one is used initially for the monopole transmission system. The second conductor is either unused, used as electrode line or connected in parallel with the other (as in case of Baltic-Cable).

Symmetrical Monopole

An alternative is to use two high-voltage conductors, operating at ± half of the DC voltage, with only a single converter at each end. In this arrangement, known as the *symmetrical monopole*, the converters are only earthed via a high impedance and there is no earth current. The symmetrical monopole arrangement is uncommon with line-commutated converters (the NorNed interconnection being a rare example) but is very common with Voltage Sourced Converters when cables are used.

Bipolar

In bipolar transmission a pair of conductors is used, each at a high potential with respect to ground, in opposite polarity. Since these conductors must be insulated for the full voltage, transmission line cost is higher than a monopole with a return conductor. However, there are a number of advantages to bipolar transmission which can make it an attractive option.

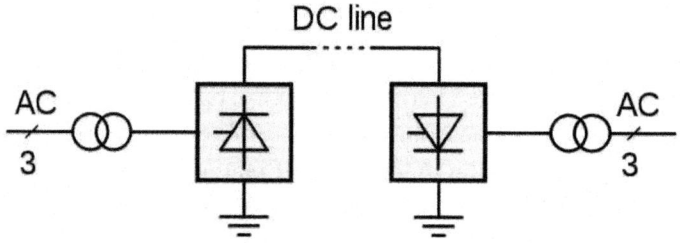

Fig. : Block diagram of a bipolar system that also has an earth return.

- Under normal load, negligible earth-current flows, as in the case of monopolar transmission with a metallic earth-return. This reduces earth return loss and environmental effects.
- When a fault develops in a line, with earth return electrodes installed at each end of the line, approximately half the rated power can continue to flow using the earth as a return path, operating in monopolar mode.
- Since for a given total power rating each conductor of a bipolar line carries only half the current of monopolar lines, the cost of the second conductor is reduced compared to a monopolar line of the same rating.
- In very adverse terrain, the second conductor may be carried on an independent set of transmission towers, so that some power may continue to be transmitted even if one line is damaged.

A bipolar system may also be installed with a metallic earth return conductor.

Bipolar systems may carry as much as 4 GW at voltages of ±660 kV with a single converter per pole, as on the Ningdong–Shandong project in China. With a power rating of 2000 MW per twelve-pulse converter, the converters for that project were (as of 2010) the most powerful HVDC converters ever built. Even higher powers can be achieved by connecting two or more twelve-pulse converters in series in each pole, as is used in the Xiangjiaba–Shanghai project in China, which uses two twelve-pulse converter bridges in each pole, each rated at 400 kV DC and 1600 MW.

Submarine cable installations initially commissioned as a monopole may be upgraded with additional cables and operated as a bipole.

A block diagram of a bipolar HVDC transmission system, between two stations designated A and B. AC – represents an alternating current network CON – represents a converter valve, either rectifier or inverter, TR represents a power transformer, DCTL is the direct-current transmission line conductor, DCL is a direct-current filter inductor, BP represents a bypass switch, and PM represent power factor correction and harmonic filter networks required at both ends of the link. The DC transmission line may be very short in a back-to-back link, or extend

AC and DC Power

hundreds of miles (km) overhead, underground or underwater. One conductor of the DC line may be replaced by connections to earth ground.

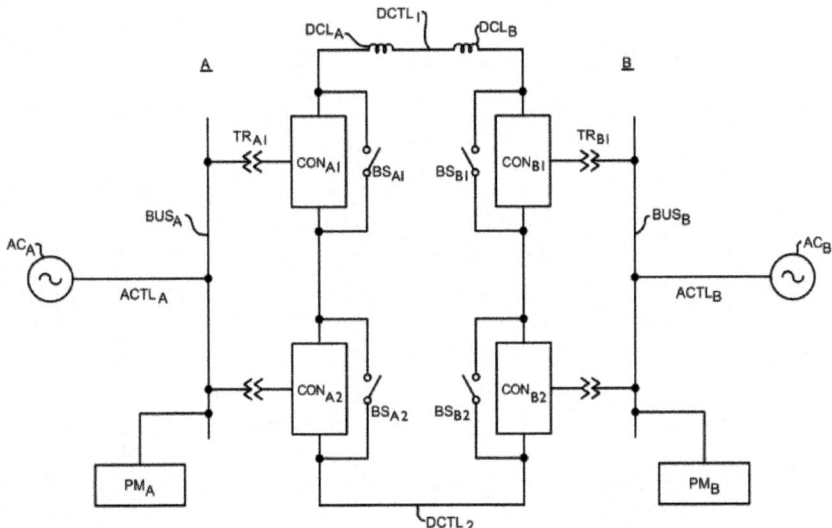

A bipolar scheme can be implemented so that the polarity of one or both poles can be changed. This allows the operation as two parallel monopoles. If one conductor fails, transmission can still continue at reduced capacity. Losses may increase if ground electrodes and lines are not designed for the extra current in this mode. To reduce losses in this case, intermediate switching stations may be installed, at which line segments can be switched off or parallelized. This was done at Inga–Shaba HVDC.

Back to Back

A **back-to-back station** (or B2B for short) is a plant in which both converters are in the same area, usually in the same building. The length of the direct current line is kept as short as possible. HVDC back-to-back stations are used for

- coupling of electricity grids of different frequencies (as in Japan and South America; and the GCC interconnection between UAE [50 Hz] and Saudi Arabia [60 Hz] completed in 2009)
- coupling two networks of the same nominal frequency but no fixed phase relationship (as until 1995/96 in Etzenricht, Dürnrohr, Vienna, and the Vyborg HVDC scheme).
- different frequency and phase number (for example, as a replacement for traction current converter plants).

The DC voltage in the intermediate circuit can be selected freely at HVDC back-to-back stations because of the short conductor length. The DC voltage is usually selected to be as low as possible, in order to build a small valve hall and to reduce the number of thyristors connected in series in each valve. For this reason, at HVDC back-to-back stations, valves with the highest available current rating (in some cases, up to 4,500 A) used.

Multi-Terminal Systems

The most common configuration of an HVDC link consists of two converter stations connected by an overhead power line or undersea cable.

Multi-terminal HVDC links, connecting more than two points, are rare. The configuration of multiple terminals can be series, parallel, or hybrid (a mixture of series and parallel). Parallel configuration tends to be used for large capacity stations, and series for lower capacity stations. An example is the 2,000 MW Quebec - New England Transmission system opened in 1992, which is currently the largest multi-terminal HVDC system in the world.

Multi-terminal systems are difficult to realise using line commutated converters because reversals of power are effected by reversing the polarity of DC voltage, which affects all converters connected to the system. With Voltage Sourced Converters, power reversal is achieved instead by reversing the direction of current, making parallel-connected multi-terminals systems much easier to control. For this reason, multi-terminal systems are expected to become much more common in the near future.

China is expanding its grid to keep up with increased power demand, while addressing environmental targets. China Southern Power Grid started a three terminals VSC HVDC pilot project in 2011. The project has designed ratings of ±160kV/200MW-100MW-50MW and will be used to bring wind power generated on Nanao island into the mainland Guangdong power grid through 32 km of combination of HVDC land cables, sea cables and overheard lines. This project was been put into operation on December 19, 2013.

In India, the multi-terminal North-East Agra project is planned for commissioning in 2016. It will be rated 8000 MW, 800 kV using four bipolar lines, and will tranmmit power from two conveter stations in the east to a converter at Agra, a distance of 1728 km.

Tripole

A scheme patented in 2004 is intended for conversion of existing AC transmission lines to HVDC. Two of the three circuit conductors are operated as a bipole. The third conductor is used as a parallel monopole, equipped with reversing valves (or parallel valves connected in reverse polarity). The parallel monopole periodically relieves current from one pole or the other, switching polarity over a span of several minutes. The bipole conductors would be loaded to either 1.37 or 0.37 of their thermal limit, with the parallel monopole always carrying ±1 times its thermal limit current. The combined RMS heating effect is as if each of the conductors is always carrying 1.0 of its rated current. This allows heavier currents to be carried by the bipole conductors, and full use of the installed third conductor for energy transmission. High currents can be circulated through the line conductors even when load demand is low, for removal of ice.

As of 2012, no tri-pole conversions are in operation, although a transmission line in India has been converted to bipole HVDC (HVDC Sileru-Barsoor).

Other Arrangements

Cross-Skagerrak consists of 3 poles, from which 2 are switched in parallel and the third uses an opposite polarity with a higher transmission voltage. A similar arrangement was the HVDC Inter-Island in New Zealand after a capacity upgrade in 1992, in which the two original converters (using mercury-arc valves) were parallel-switched feeding the same pole and a new third (thyristor) converter installed with opposite polarity and higher operation voltage. This configuration ended in 2012 when the two old converters were replaced with a single, new, thyristor converter.

Corona Discharge

Corona discharge is the creation of ions in a fluid (such as air) by the presence of a strong electric field. Electrons are torn from neutral air, and either the positive ions or the electrons are attracted to the conductor, while the charged particles drift. This effect can cause considerable power loss, create audible and radio-frequency interference, generate toxic compounds such as oxides of nitrogen and ozone, and bring forth arcing.

Both AC and DC transmission lines can generate coronas, in the former case in the form of oscillating particles, in the latter a constant wind. Due to the space charge formed around the conductors, an HVDC system may have about half the loss per unit length of a high voltage AC system carrying the same amount of power. With monopolar transmission the choice of polarity of the energized conductor leads to a degree of control over the corona discharge. In particular, the polarity of the ions emitted can be controlled, which may have an environmental impact on ozone creation. Negative coronas generate considerably more ozone than positive coronas, and generate it further *downwind* of the power line, creating the potential for health effects. The use of a *positive* voltage will reduce the ozone impacts of monopole HVDC power lines.

Applications

Overview

The controllability of current-flow through HVDC rectifiers and inverters, their application in connecting unsynchronized networks, and their applications in efficient submarine cables mean that HVDC interconnections are often used at national or regional boundaries for the exchange of power (in North America, HVDC connections divide much of Canada and the United States into several electrical regions that cross-national borders, although the purpose of these connections is still to connect unsynchronized AC grids to each other). Offshore windfarms also require undersea cables, and their turbines are unsynchronized. In very long-distance connections between two locations, such as power transmission from a large hydroelectric power plant at a remote site to an urban area, HVDC transmission systems may appropriately be used; several schemes of these kind have been built. For interconnections to Siberia, Canada, and the Scandinavian North, the decreased line-costs of HVDC also make it applicable.

AC Network Interconnections

AC transmission lines can interconnect only synchronized AC networks with the same frequency with limits on the allowable phase difference between the two ends of the line. Many areas that wish to share power have unsynchronized networks. The power grids of the UK, Northern Europe and continental Europe are not united into a single synchronized network. Japan has 50 Hz and 60 Hz networks. Continental North America, while operating at 60 Hz throughout, is divided into regions which are unsynchronised : East, West, Texas, Quebec, and Alaska. Brazil and Paraguay, which share the enormous Itaipu Dam hydroelectric plant, operate on 60 Hz and 50 Hz respectively. However, HVDC systems make it possible to interconnect unsynchronized AC networks, and also add the possibility of controlling AC voltage and reactive power flow.

A generator connected to a long AC transmission line may become unstable and fall out of synchronization with a distant AC power system. An HVDC transmission link may make it economically feasible to use remote generation sites. Wind farms located off-shore may use HVDC systems to collect power from multiple unsynchronized generators for transmission to the shore by an underwater cable.

In general, however, an HVDC power line will interconnect two AC regions of the power-distribution grid. Machinery to convert between AC and DC power adds a considerable cost in power transmission. The conversion from AC to DC is known as rectification, and from DC to AC as inversion. Above a certain break-even distance (about 50 km for submarine cables, and perhaps 600–800 km for overhead cables), the lower cost of the HVDC electrical conductors outweighs the cost of the electronics.

The conversion electronics also present an opportunity to effectively manage the power grid by means of controlling the magnitude and direction of power flow. An additional advantage of the existence of HVDC links, therefore, is potential increased stability in the transmission grid.

Renewable Electricity Superhighways

A number of studies have highlighted the potential benefits of very wide area super grids based on HVDC since they can mitigate the effects of intermittency by averaging and smoothing the outputs of large numbers of geographically dispersed wind farms or solar farms. Czisch's study concludes that a grid covering the fringes of Europe could bring 100% renewable power (70% wind, 30% biomass) at close to today's prices. There has been debate over the technical feasibility of this proposal and the political risks involved in energy transmission across a large number of international borders.

The construction of such green power superhighways is advocated in a white paper that was released by the American Wind Energy Association and the Solar Energy Industries Association in 2009. Clean Line Energy Partners is developing four HVDC lines in the U.S. for long distance electric power transmission.

In January 2009, the European Commission proposed €300 million to subsidize the development of HVDC links between Ireland, Britain, the Netherlands, Germany, Denmark, and Sweden, as part of a wider €1.2 billion package supporting links to offshore wind farms and cross-border interconnectors throughout Europe. Meanwhile the recently founded Union of the Mediterranean has embraced a Mediterranean Solar Plan to import large amounts of concentrating solar power into Europe from North Africa and the Middle East.

ALTERNATING CURRENT

In **alternating current** (**AC**, also **ac**), the flow of electric charge periodically reverses direction. In direct current (**DC**, also **dc**), the flow of electric charge is only in one direction. The abbreviations *AC* and *DC* are often used to mean simply *alternating* and *direct*, as when they modify *current* or *voltage*.

AC is the form in which electric power is delivered to businesses and residences. The usual waveform of an AC power circuit is a sine wave. In certain applications, different waveforms are used, such as triangular or square waves. Audio and radio signals carried on electrical wires are also examples of alternating current. In these applications, an important goal is often the recovery of information encoded (or modulated) onto the AC signal.

Transmission, Distribution, and Domestic Power Supply

AC voltage may be increased or decreased with a transformer. Use of a higher voltage leads to significantly more efficient transmission of power. The power losses in a conductor are a product of the square of the current and the resistance of the conductor, described by the formula

$$P_L = I^2 R.$$

This means that when transmitting a fixed power on a given wire, if the current is doubled, the power loss will be four times greater.

The power transmitted is equal to the product of the current and the voltage (assuming no phase difference); that is,

$$P_T = IV.$$

Thus, the same amount of power can be transmitted with a lower current by increasing the voltage. It is therefore advantageous when transmitting large amounts of power to distribute the power with high voltages (often hundreds of kilovolts).

However, high voltages also have disadvantages, the main one being the increased insulation required, and generally increased difficulty in their safe handling. In a power plant, power is generated at a convenient voltage for the design of a generator, and then stepped up to a high voltage for transmission. Near the loads, the transmission voltage is stepped down to the voltages used by equipment. Consumer voltages vary depending on the country and size of load, but generally motors and lighting are built to use up to a few hundred volts between phases.

The utilization voltage delivered to equipment such as lighting and motor loads is standardized, with an allowable range of voltage over which equipment is expected to operate. Standard power utilization voltages and percentage tolerance vary in the different mains power systems found in the world.

Modern high-voltage direct-current (HVDC) electric power transmission systems contrast with the more common alternating-current systems as a means for the efficient bulk transmission of electrical power over long distances. HVDC systems, however, tend to be more expensive and less efficient over shorter distances than transformers. Transmission with high voltage direct current was not feasible when Edison, Westinghouse and Tesla were designing their power systems, since there was then no way to economically convert AC power to DC and back again at the necessary voltages.

Three-phase electrical generation is very common. The simplest case is three separate coils in the generator stator that are physically offset by an angle of 120° to each other. Three current waveforms are produced that are equal in magnitude and 120° out of phase to each other. If coils are added opposite to these (60° spacing), they generate the same phases with reverse polarity and so can be simply wired together.

In practice, higher "pole orders" are commonly used. For example, a 12-pole machine would have 36 coils (10° spacing). The advantage is that lower speeds can be used. For example, a 2-pole machine running at 3600 rpm and a 12-pole machine running at 600 rpm produce the same frequency. This is much more practical for larger machines.

If the load on a three-phase system is balanced equally among the phases, no current flows through the neutral point. Even in the worst-case unbalanced (linear) load, the neutral current will not exceed the highest of the phase currents. Non-linear loads (*e.g.*, computers) may require an oversized neutral bus and neutral conductor in the upstream distribution panel to handle harmonics. Harmonics can cause neutral conductor current levels to exceed that of one or all phase conductors.

For three-phase at utilization voltages a four-wire system is often used. When stepping down three-phase, a transformer with a Delta (3-wire) primary and a Star

(4-wire, center-earthed) secondary is often used so there is no need for a neutral on the supply side.

For smaller customers (just how small varies by country and age of the installation) only a single phase and the neutral or two phases and the neutral are taken to the property. For larger installations all three phases and the neutral are taken to the main distribution panel. From the three-phase main panel, both single and three-phase circuits may lead off.

Three-wire single-phase systems, with a single center-tapped transformer giving two live conductors, is a common distribution scheme for residential and small commercial buildings in North America. This arrangement is sometimes incorrectly referred to as "two phase". A similar method is used for a different reason on construction sites in the UK. Small power tools and lighting are supposed to be supplied by a local center-tapped transformer with a voltage of 55 V between each power conductor and earth. This significantly reduces the risk of electric shock in the event that one of the live conductors becomes exposed through an equipment fault whilst still allowing a reasonable voltage of 110 V between the two conductors for running the tools.

A third wire, called the bond (or earth) wire, is often connected between non-current-carrying metal enclosures and earth ground. This conductor provides protection from electric shock due to accidental contact of circuit conductors with the metal chassis of portable appliances and tools. Bonding all non-current-carrying metal parts into one complete system ensures there is always a low electrical impedance path to ground sufficient to carry any fault current for as long as it takes for the system to clear the fault. This low impedance path allows the maximum amount of fault current, causing the overcurrent protection device (breakers, fuses) to trip or burn out as quickly as possible, bringing the electrical system to a safe state. All bond wires are bonded to ground at the main service panel, as is the Neutral/Identified conductor if present.

AC Power Supply Frequencies

The frequency of the electrical system varies by country; most electric power is generated at either 50 or 60 hertz. Some countries have a mixture of 50 Hz and 60 Hz supplies, notably electricity power transmission in Japan.

A low frequency eases the design of electric motors, particularly for hoisting, crushing and rolling applications, and commutator-type traction motors for applications such as railways. However, low frequency also causes noticeable flicker in arc lamps and incandescent light bulbs. The use of lower frequencies also provided the advantage of lower impedance losses, which are proportional to frequency. The original Niagara Falls generators were built to produce 25 Hz power, as a compromise between low frequency for traction and heavy induction motors, while still allowing incandescent lighting to operate (although with noticeable flicker). Most of the 25 Hz residential and commercial customers for Niagara Falls power were converted to 60 Hz by the late 1950s, although some

25 Hz industrial customers still existed as of the start of the 21st century. 16.7 Hz power (formerly 16 2/3 Hz) is still used in some European rail systems, such as in Austria, Germany, Norway, Sweden and Switzerland.

Off-shore, military, textile industry, marine, computer mainframe, aircraft, and spacecraft applications sometimes use 400 Hz, for benefits of reduced weight of apparatus or higher motor speeds.

Effects at High Frequencies

A direct current flows uniformly throughout the cross-section of a uniform wire. An alternating current of any frequency is forced away from the wire's center, toward its outer surface. This is because the acceleration of an electric charge in an alternating current produces waves of electromagnetic radiation that cancel the propagation of electricity toward the center of materials with high conductivity. This phenomenon is called skin effect.

At very high frequencies the current no longer flows *in* the wire, but effectively flows *on* the surface of the wire, within a thickness of a few skin depths. The skin depth is the thickness at which the current density is reduced by 63%. Even at relatively low frequencies used for power transmission (50–60 Hz), non-uniform distribution of current still occurs in sufficiently thick conductors. For example, the skin depth of a copper conductor is approximately 8.57 mm at 60 Hz, so high current conductors are usually hollow to reduce their mass and cost.

Since the current tends to flow in the periphery of conductors, the effective cross-section of the conductor is reduced. This increases the effective AC resistance of the conductor, since resistance is inversely proportional to the cross-sectional area. The AC resistance often is many times higher than the DC resistance, causing a much higher energy loss due to ohmic heating (also called I^2R loss).

Techniques for Reducing AC Resistance

For low to medium frequencies, conductors can be divided into stranded wires, each insulated from one other, and the relative positions of individual strands specially arranged within the conductor bundle. Wire constructed using this technique is called Litz wire. This measure helps to partially mitigate skin effect by forcing more equal current throughout the total cross-section of the stranded conductors. Litz wire is used for making high-Q inductors, reducing losses in flexible conductors carrying very high currents at lower frequencies, and in the windings of devices carrying higher radio frequency current (up to hundreds of kilohertz), such as switch-mode power supplies and radio frequency transformers.

Techniques for Reducing Radiation Loss

As written above, an alternating current is made of electric charge under periodic acceleration, which causes radiation of electromagnetic waves. Energy that is radiated is lost. Depending on the frequency, different techniques are used to minimize the loss due to radiation.

Twisted Pairs

At frequencies up to about 1 GHz, pairs of wires are twisted together in a cable, forming a twisted pair. This reduces losses from electromagnetic radiation and inductive coupling. A twisted pair must be used with a balanced signalling system, so that the two wires carry equal but opposite currents. Each wire in a twisted pair radiates a signal, but it is effectively cancelled by radiation from the other wire, resulting in almost no radiation loss.

Coaxial Cables

Coaxial cables are commonly used at audio frequencies and above for convenience. A coaxial cable has a conductive wire inside a conductive tube, separated by a dielectric layer. The current flowing on the inner conductor is equal and opposite to the current flowing on the inner surface of the tube. The electromagnetic field is thus completely contained within the tube, and (ideally) no energy is lost to radiation or coupling outside the tube. Coaxial cables have acceptably small losses for frequencies up to about 5 GHz. For microwave frequencies greater than 5 GHz, the losses (due mainly to the electrical resistance of the central conductor) become too large, making waveguides a more efficient medium for transmitting energy. Coaxial cables with an air rather than solid dielectric are preferred as they transmit power with lower loss.

Waveguides

Waveguides are similar to coax cables, as both consist of tubes, with the biggest difference being that the waveguide has no inner conductor. Waveguides can have any arbitrary cross-section, but rectangular cross-sections are the most common. Because waveguides do not have an inner conductor to carry a return current, waveguides cannot deliver energy by means of an electric current, but rather by means of a *guided* electromagnetic field. Although surface currents do flow on the inner walls of the waveguides, those surface currents do not carry power. Power is carried by the guided electromagnetic fields. The surface currents are set up by the guided electromagnetic fields and have the effect of keeping the fields inside the waveguide and preventing leakage of the fields to the space outside the waveguide.

Waveguides have dimensions comparable to the wavelength of the alternating current to be transmitted, so they are only feasible at microwave frequencies. In addition to this mechanical feasibility, electrical resistance of the non-ideal metals forming the walls of the waveguide cause dissipation of power (surface currents flowing on lossy conductors dissipate power). At higher frequencies, the power lost to this dissipation becomes unacceptably large.

Fiber Optics

At frequencies greater than 200 GHz, waveguide dimensions become impractically small, and the ohmic losses in the waveguide walls become large. Instead, fiber optics, which are a form of dielectric waveguides, can be used. For such frequencies, the concepts of voltages and currents are no longer used.

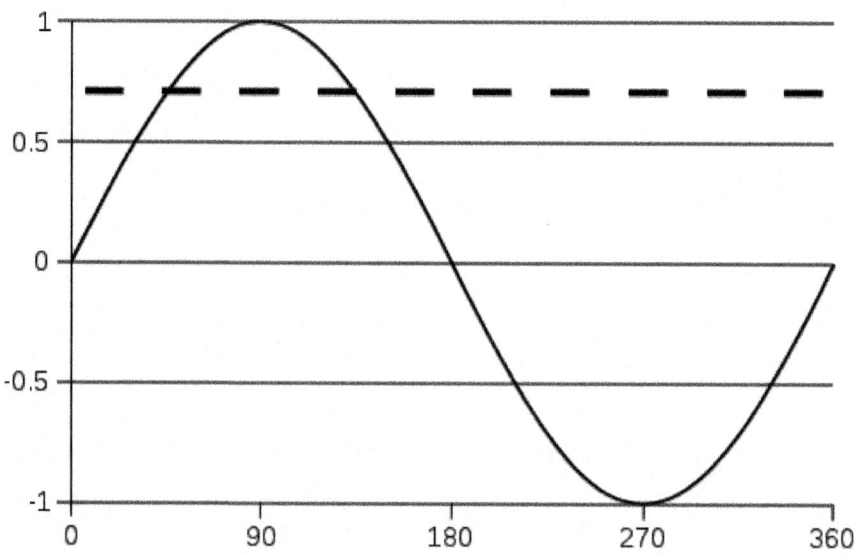

Fig. : A sine wave, over one cycle (360°). The dashed line represents the root mean square (RMS) value at about 0.707.

Mathematics of AC Voltages

Alternating currents are accompanied (or caused) by alternating voltages. An AC voltage v can be described mathematically as a function of time by the following equation :

$$v(t) = V_{peak} \cdot \sin(\omega t),$$

where :
- V_{peak} is the peak voltage (unit : volt),
- ω is the angular frequency (unit : radians per second)
   - The angular frequency is related to the physical frequency, f (unit = hertz), which represents the number of cycles per second, by the equation $\omega = 2\pi f$.
- t is the time (unit : second).

The peak-to-peak value of an AC voltage is defined as the difference between its positive peak and its negative peak. Since the maximum value of sin (x) is +1 and the minimum value is −1, an AC voltage swings between $+V_{peak}$ and $-V_{peak}$.

AC and DC Power

The peak-to-peak voltage, usually written as V_{pp} or V_{p-p}, is therefore $V_{peak} - (-V_{peak}) = 2V_{peak}$.

Power and Root Mean Square

The relationship between voltage and the power delivered is

$$p(t) = \frac{v^2(t)}{R}$$ where R represents a load resistance.

Rather than using instantaneous power, $p(t)$, it is more practical to use a time averaged power (where the averaging is performed over any integer number of cycles). Therefore, AC voltage is often expressed as a root mean square (RMS) value, written as V_{rms}, because

$$P_{time}\text{ averaged} = \frac{V_{rms}^2}{R}.$$

For a sinusoidal voltage:

$$V_{rms} = \sqrt{\frac{1}{T}\int_0^T [V_{pk}\sin(\omega t + \phi)]^2\, dt}$$

$$= V_{pk}\sqrt{\frac{1}{2T}\int_0^T [1 - \cos(2\omega t + 2\phi)]\, dt}$$

$$= V_{pk}\sqrt{\frac{1}{2T}\int_0^T dt}$$

$$= \frac{V_{pk}}{\sqrt{2}}$$

The factor $\sqrt{2}$ is called the crest factor, which varies for different waveforms.

- For a triangle waveform centered about zero
$$V_{rms} = \frac{V_{peak}}{\sqrt{3}}.$$
- For a square waveform centered about zero
$$V_{rms} = V_{peak}.$$
- For an arbitrary periodic waveform $v(t)$ of period T:
$$V_{rms} = \sqrt{\frac{1}{T}\int_0^T v^2(t)\, dt}.$$

Example

To illustrate these concepts, consider a 230 V AC mains supply used in many countries around the world. It is so called because its root mean square value is 230 V. This means that the time-averaged power delivered is equivalent to the power delivered by a DC voltage of 230 V. To determine the peak voltage (amplitude), we can rearrange the above equation to:

$$V_{peak} = \sqrt{2}\, V_{rms}.$$

For 230 V AC, the peak voltage V_{peak} is therefore 230 × $\sqrt{2}$, which is about 325 V. The peak-to-peak value V_{p-p} of the 230 V AC is double that, at about 650 V.

History

The first alternator to produce alternating current was a dynamo electric generator based on Michael Faraday's principles constructed by the French instrument maker Hippolyte Pixii in 1832. Pixii later added a commutator to his device to produce the (then) more commonly used direct current. The earliest recorded practical application of alternating current is by Guillaume Duchenne, inventor and developer of electrotherapy. In 1855, he announced that AC was superior to direct current for electrotherapeutic triggering of muscle contractions.

Alternating current technology had first developed in Europe due to the work of Guillaume Duchenne (1850s), The Hungarian Ganz Works (1870s), Sebastian Ziani de Ferranti (1880s), Lucien Gaulard, and Galileo Ferraris.

In 1876, Russian engineer Pavel Yablochkov invented a lighting system based on a set of induction coils where the primary windings were connected to a source of AC. The secondary windings could be connected to several 'electric candles' (arc lamps) of his own design. The coils Yablochkov employed functioned essentially as transformers. In 1878, the Ganz factory, Budapest, Hungary, began manufacturing equipment for electric lighting and, by 1883, had installed over fifty systems in Austria-Hungary. Their AC systems used arc and incandescent lamps, generators, and other equipment.

A power transformer developed by Lucien Gaulard and John Dixon Gibbs was demonstrated in London in 1881, and attracted the interest of Westinghouse. They also exhibited the invention in Turin in 1884.

The Advantage of Dc Systems Over the Early AC Systems

During the initial years of electricity distribution, Edison's direct current was the standard for the United States, and Edison did not want to lose all his patent royalties. Direct current worked well with incandescent lamps, which were the principal load of the day, and with motors. Direct-current systems could be directly used with storage batteries, providing valuable load-levelling and backup power during interruptions of generator operation. Direct-current generators could be easily paralleled, allowing economical operation by using smaller machines during periods of light load and improving reliability. At the introduction of Edison's system, no practical AC motor was available. Edison had invented a meter to allow customers to be billed for energy proportional to consumption, but this meter worked only with direct current. A bipolar open-core power transformer developed by Lucien Gaulard and John Dixon Gibbs was demonstrated in London in 1881, and attracted the interest of Westinghouse. They also exhibited the invention in Turin in 1884.

However these early induction coils with open magnetic circuits are inefficient at transferring power to loads. Until about 1880, the paradigm for AC power transmission from a high voltage supply to a low voltage load was a series circuit. Open-core transformers with a ratio near 1 : 1 were connected with their primaries in series to allow use of a high voltage for transmission while presenting a low voltage to the lamps. The inherent flaw in this method was that turning off a single lamp (or other electric device) affected the voltage supplied to all others on the same circuit. Many adjustable transformer designs were introduced to compensate for this problematic characteristic of the series circuit, including those employing methods of adjusting the core or bypassing the magnetic flux around part of a coil. The direct current systems did not have these drawbacks, giving it significant advantages over early AC systems.

The Ganz AC System Counter

In the autumn of 1884, Károly Zipernowsky, Ottó Bláthy and Miksa Déri (ZBD), three engineers associated with the Ganz factory, had determined that open-core devices were impracticable, as they were incapable of reliably regulating voltage. In their joint 1885 patent applications for novel transformers (later called ZBD transformers), they described two designs with closed magnetic circuits where copper windings were either a) wound around iron wire ring core or b) surrounded by iron wire core. The two designs were the first application of the two basic transformer constructions in common use to this day, which can as a class all be termed as either core form or shell form (or alternatively, core type or shell type), as in a) or b), respectively. The Ganz factory had also in the autumn of 1884 made delivery of the world's first five high-efficiency AC transformers, the first of these units having been shipped on September 16, 1884. This first unit had been manufactured to the following specifications : 1,400 W, 40 Hz, 120 : 72 V, 11.6 : 19.4 A, ratio 1.67 : 1, one-phase, shell form. In both designs, the magnetic flux linking the primary and secondary windings traveled almost entirely within the confines of the iron core, with no intentional path through air. The new transformers were 3.4 times more efficient than the open-core bipolar devices of Gaulard and Gibbs.

The ZBD patents included two other major interrelated innovations : one concerning the use of parallel connected, instead of series connected, utilization loads, the other concerning the ability to have high turns ratio transformers such that the supply network voltage could be much higher (initially 1,400 to 2,000 V) than the voltage of utilization loads (100 V initially preferred). When employed in parallel connected electric distribution systems, closed-core transformers finally made it technically and economically feasible to provide electric power for lighting in homes, businesses and public spaces. Bláthy had suggested the use of closed cores, Zipernowsky had suggested the use of parallel shunt connections, and Déri had performed the experiments; The other essential milestone was the introduction of 'voltage source, voltage intensive' (VSVI) systems' by the invention of constant voltage generators in 1885. Ottó Bláthy also invented the AC electricity meter to complete the competition of AC and DC technology. Transformers today are designed on the principles discovered by the three engineers. They also popularized the word 'transformer' to describe a device for altering the emf of an electric current, although the term had already been in use by 1882. In 1886, the ZBD engineers designed, and the Ganz factory supplied electrical equipment for, the world's first power station that used AC generators to power a parallel connected common electrical network, the steam-powered Rome-Cerchi power plant. The reliability of the AC technology received impetus after the Ganz Works electrified a large European metropolis : Rome in 1886.

Sebastian Ziani de Ferranti went into this business in 1882 when he set up a shop in London designing various electrical devices. Ferranti believed in the success of alternating current power distribution early on, and was one of the few experts in this system in the UK. In 1887 the London Electric Supply Corporation (LESCo) hired Ferranti for the design of their power station at Deptford. He designed the building, the generating plant and the distribution system. On its completion in 1891 it was the first truly modern power station, supplying high-voltage AC power that was then "stepped down" for consumer use on each street. This basic system remains in use today around the world. Many homes all over the world still have electric meters with the Ferranti AC patent stamped on them.

William Stanley, Jr. designed one of the first practical devices to transfer AC power efficiently between isolated circuits. Using pairs of coils wound on a common iron core, his design, called an induction coil, was an early transformer. The AC power system used today developed rapidly after 1886, and included contributions by Nikola Tesla (licensed to George Westinghouse) and Carl Wilhelm Siemens. AC systems overcame the limitations of the direct current system used by Thomas Edison to distribute electricity efficiently over long distances even though Edison attempted to discredit alternating current as too dangerous during the War of Currents.

The first commercial power plant in the United States using three-phase alternating current was at the Mill Creek No. 1 Hydroelectric Plant near Redlands, California, in 1893 designed by Almirian Decker. Decker's design incorporated 10,000-volt three-phase transmission and established the standards for the complete system of generation, transmission and motors used today.

The Ames Hydroelectric Generating Plant (spring of 1891) and the original Niagara Falls Adams Power Plant (August 25, 1895) were among the first hydroelectric AC-power plants.

The Jaruga Hydroelectric Power Plant in Croatia was set in operation on 28 August 1895. The two generators (42 Hz, 550 kW each) and the transformers were produced and installed by the Hungarian company Ganz. The transmission line from the power plant to the City of Šibenik was 11.5 kilometers (7.1 mi) long on wooden towers, and the municipal distribution grid 3000 V/110 V included six transforming stations.

Alternating current circuit theory developed rapidly in the latter part of the 19th and early 20th century. Notable contributors to the theoretical basis of alternating current calculations include Charles Steinmetz, Oliver Heaviside, and many others. Calculations in unbalanced three-phase systems were simplified by the symmetrical components methods discussed by Charles Legeyt Fortescue in 1918.

# Chapter 3

POWER MEASUREMENT FUNDAMENTALS

Electrical measurements are the methods, devices and calculations used to measure electrical quantities. Measurement of electrical quantities may be done to measure electrical parameters of a system. Using transducers, physical properties such as temperature, pressure, flow, force, and many others can be converted into electrical signals, which can then be conveniently measured and recorded. High-precision laboratory measurements of electrical quantities are used in experiments to determine fundamental physical properties such as the charge of the electron or the speed of light, and in the definition of the units for electrical measurements, with precision in some cases on the order of a few parts per million. Less precise measurements are required every day in industrial practice. Electrical measurements are a branch of the science of metrology.

VOLTAGE

Voltage, electrical potential difference, electric tension or **electric pressure** (denoted ΔV) and measured in units of electric potential : volts, or joules per coulomb is the electric potential difference between two points, or the difference in electric potential energy of a unit charge transported between two points. Voltage is equal to the work done per unit charge against a static electric field to move the charge between two points. A voltage may represent either a source of energy (electromotive force), or lost, used, or stored energy (potential drop). A voltmeter can be used to measure the voltage (or potential difference) between two points in a system; usually a common reference potential such as the ground of the system is used as one of the points. Voltage can be caused by static electric fields, by electric current through a magnetic field, by time-varying magnetic fields, or some combination of these three.

Definition

Given two points in the space, called A and B, voltage is the difference of electric potentials between those two points. From the definition of electric potential it follows that :

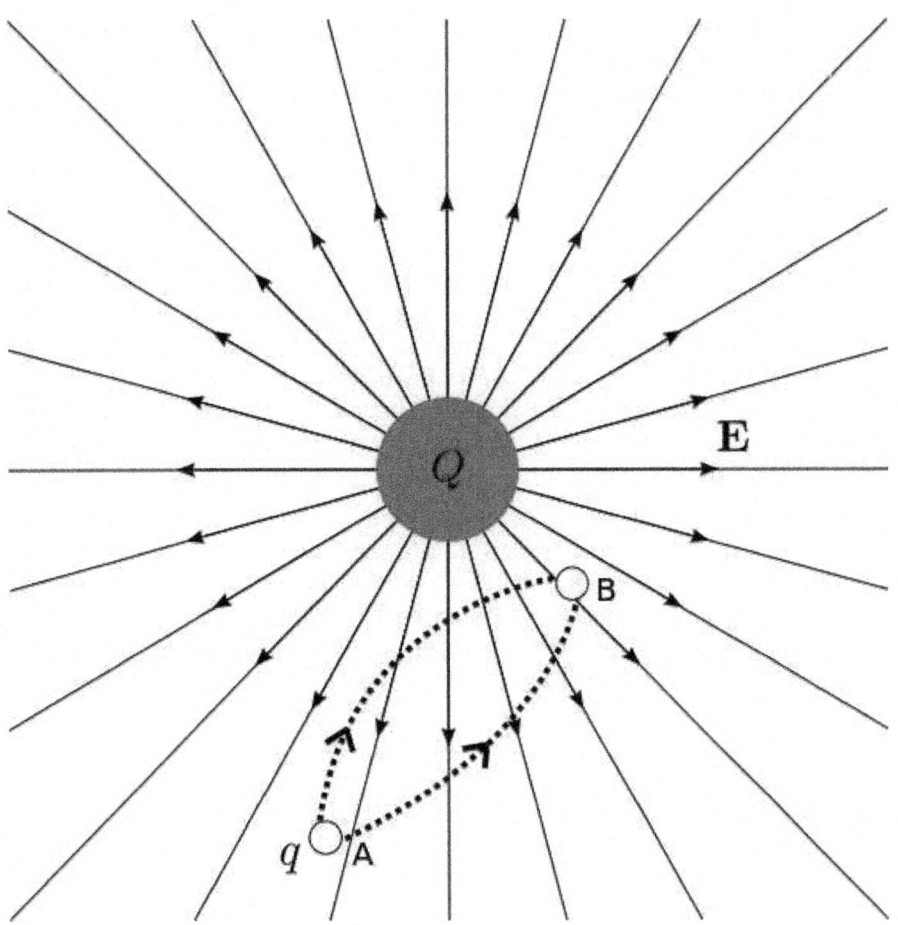

Fig. : In a static field, the work is independent of the path.

$$\Delta V_{BA} = V_B - V_A = -\int_{r_0}^{B} \vec{E} \cdot \vec{dl} - \left(-\int_{r_0}^{A} \vec{E} \cdot \vec{dl}\right)$$

$$= \int_{B}^{r_0} \vec{E} \cdot \vec{dl} + \int_{r_0}^{A} \vec{E} \cdot \vec{dl} = \int_{B}^{A} \vec{E} \cdot \vec{dl}$$

Voltage is electric potential energy per unit charge, measured in joules per coulomb (= volts). It is often referred to as "electric potential", which then must be distinguished from electric potential energy by noting that the "potential" is a "per-unit-charge" quantity. Like mechanical potential energy, the zero of potential can be chosen at any point, so the difference in voltage is the quantity which is physically meaningful. The difference in voltage measured when moving from point A to point B is equal to the work which would have to be done, per unit charge, against the electric field to move the charge from A to B. The voltage between the two ends of a path is the total energy required to move a small electric charge along that path, divided by the magnitude of the charge. Mathematically this is expressed as the line integral of the electric field and the time rate of change of magnetic field along that path. In the general case, both a static (unchanging) electric field and a dynamic (time-varying) electromagnetic field must be included in determining the voltage between two points.

Historically this quantity has also been called "tension" and "pressure". Pressure is now obsolete but tension is still used, for example within the phrase "high tension" (HT) which is commonly used in thermionic valve (vacuum tube) based electronics.

Voltage is defined so that negatively charged objects are pulled towards higher voltages, while positively charged objects are pulled towards lower voltages. Therefore, the conventional current in a wire or resistor always flows from higher voltage to lower voltage. Current can flow from lower voltage to higher voltage, but only when a source of energy is present to "push" it against the opposing electric field. For example, inside a battery, chemical reactions provide the energy needed for current to flow from the negative to the positive terminal.

Technically, in a material the electric field is not the only factor determining charge flow, and different materials naturally develop electric potential differences at equilibrium (Galvani potentials). The electric potential of a material is not even a well defined quantity, since it varies on the sub-atomic scale. A more convenient definition of 'voltage' can be found instead in the concept of Fermi level. In this case the voltage between two bodies is the thermodynamic work required to move a unit of charge between them. This definition is practical since a real voltmeter actually measures this work, not differences in electric potential.

Hydraulic Analogy

A simple analogy for an electric circuit is water flowing in a closed circuit of pipework, driven by a mechanical pump. This can be called a "water circuit". Potential difference between two points corresponds to the pressure difference between two points. If the pump creates a pressure difference between two points, then water flowing from one point to the other will be able to do work, such as driving a turbine. Similarly, work can be done by an electric current driven by the potential difference provided by a battery. For example, the voltage provided by a sufficiently-charged automobile battery can "push" a large current through the windings of an automobile's starter motor. If the pump isn't working, it produces no pressure difference, and the turbine will not rotate. Likewise, if the automobile's battery is very weak or "dead" (or "flat"), then it will not turn the starter motor.

The hydraulic analogy is a useful way of understanding many electrical concepts. In such a system, the work done to move water is equal to the pressure multiplied by the volume of water moved. Similarly, in an electrical circuit, the work done to move electrons or other charge-carriers is equal to "electrical pressure" multiplied by the quantity of electrical charges moved. In relation to "flow", the larger the "pressure difference" between two points (potential difference or water pressure difference), the greater the flow between them (electric current or water flow).

Applications

Specifying a voltage measurement requires explicit or implicit specification of the points across which the voltage is measured. When using a voltmeter to measure potential difference, one electrical lead of the voltmeter must be connected to the first point, one to the second point.

A common use of the term "voltage" is in describing the voltage dropped across an electrical device (such as a resistor). The voltage drop across the device can be understood as the difference between measurements at each terminal of the device with respect to a common reference point (or ground). The voltage drop is the difference between the two readings. Two points in an electric circuit that are connected by an ideal conductor without resistance and not within a changing magnetic field have a voltage of zero. Any two points with the same potential may be connected by a conductor and no current will flow between them.

Addition of Voltages

The voltage between A and C is the sum of the voltage between A and B and the voltage between B and C. The various voltages in a circuit can be computed using Kirchhoff's circuit laws.

When talking about alternating current (AC) there is a difference between instantaneous voltage and average voltage. Instantaneous voltages can be added for direct current (DC) and AC, but average voltages can be meaningfully added only when they apply to signals that all have the same frequency and phase.

Measuring Instruments

Instruments for measuring voltages include the voltmeter, the potentiometer, and the oscilloscope. The voltmeter works by measuring the current through a fixed resistor, which, according to Ohm's Law, is proportional to the voltage across the resistor. The potentiometer works by balancing the unknown voltage against a known voltage in a bridge circuit. The cathode-ray oscilloscope works by amplifying the voltage and using it to deflect an electron beam from a straight path, so that the deflection of the beam is proportional to the voltage.

Typical Voltages

A common voltage for flashlight batteries is 1.5 volts (DC). A common voltage for automobile batteries is 12 volts (DC).

Common voltages supplied by power companies to consumers are 110 to 120 volts (AC) and 220 to 240 volts (AC). The voltage in electric power transmission lines used to distribute electricity from power stations can be several hundred times greater than consumer voltages, typically 110 to 1200 kV (AC).

The voltage used in overhead lines to power railway locomotives is between 12 kV and 50 kV (AC).

Galvani Potential *versus* Electrochemical Potential

Inside a conductive material, the energy of an electron is affected not only by the average electric potential, but also by the specific thermal and atomic environment that it is in. When a voltmeter is connected between two different types of metal, it measures not the electrostatic potential difference, but instead something else that is affected by thermodynamics. The quantity measured by a voltmeter is the negative of difference of electrochemical potential of electrons (Fermi level) divided by electron charge, while the pure unadjusted electrostatic potential (not measurable with voltmeter) is sometimes called Galvani potential. The terms "voltage" and "electric potential" are a bit ambiguous in that, in practice, they can refer to *either* of these in different contexts.

ELECTRIC CURRENT

An **electric current** is a flow of electric charge. In electric circuits this charge is often carried by moving electrons in a wire. It can also be carried by ions in an electrolyte, or by both ions and electrons such as in a plasma.

The SI unit for measuring an electric current is the ampere, which is the flow of electric charges through a surface at the rate of one coulomb per second. Electric current can be measured using an ammeter.

Electric currents cause many effects, notably heating, but also induce magnetic fields, which are widely used for motors, inductors and generators.

Symbol

The conventional symbol for current is I, which originates from the French phrase *intensité de courant*, or in English *current intensity*. This phrase is frequently used when discussing the value of an electric current, but modern practice often shortens this to simply *current*. The I symbol was used by André-Marie Ampère, after whom the unit of electric current is named, in formulating the eponymous Ampère's force law which he discovered in 1820. The notation travelled from France to Britain, where it became standard, although at least one journal did not change from using C to I until 1896.

Conventions

A flow of positive charges gives the same electric current, and has the same effect in a circuit, as an equal flow of negative charges in the opposite direction. Since current can be the flow of either positive or negative charges, or both, a convention for the direction of current which is independent of the type of charge carriers is needed. The direction of *conventional current* is arbitrarily defined to be the same as the direction of the flow of positive charges.

In metals, which make up the wires and other conductors in most electrical circuits, the positive charges are immobile, and the charge carriers are electrons. Because the electrons carry negative charge, their motion in a metal conductor is in the direction opposite to that of conventional current.

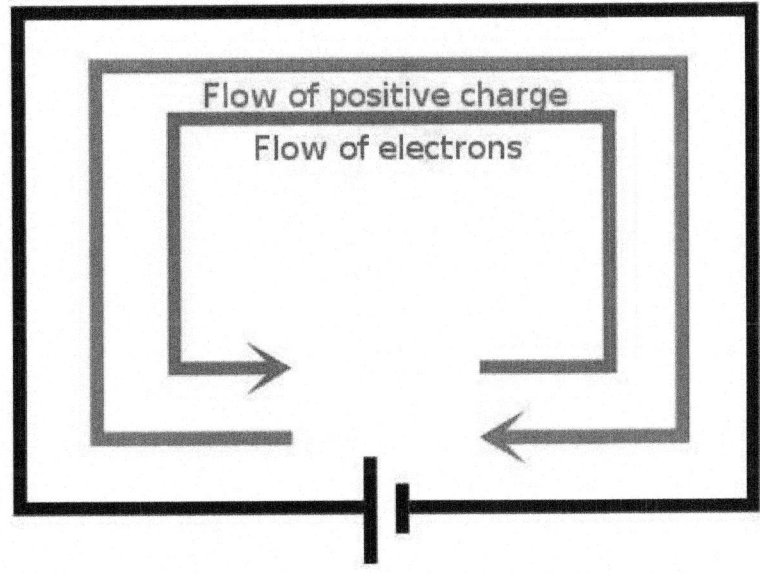

Fig. : The electrons, the charge carriers in an electrical circuit, flow in the opposite direction of the *conventional* electric current.

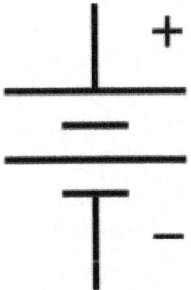

Fig. : The symbol for a battery in a circuit diagram.

Reference Direction

When analyzing electrical circuits, the actual direction of current through a specific circuit element is usually unknown. Consequently, each circuit element is assigned a current variable with an arbitrarily chosen *reference direction*. This is usually indicated on the circuit diagram with an arrow next to the current variable. When the circuit is solved, the circuit element currents may have positive or negative values. A negative value means that the actual direction of current through that circuit element is opposite that of the chosen reference direction. In electronic circuits, the reference current directions are often chosen so that all currents are toward ground. This often corresponds to conventional current direction, because in many circuits the power supply voltage is positive with respect to ground.

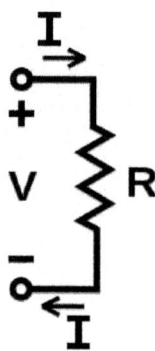

Fig. : V, I, and R, the parameters of Ohm's law.

Ohm's Law

Ohm's law states that the current through a conductor between two points is directly proportional to the potential difference across the two points. Introducing the constant of proportionality, the resistance, one arrives at the usual mathematical equation that describes this relationship :

$$I = \frac{V}{R},$$

where I is the current through the conductor in units of amperes, V is the potential difference measured *across* the conductor in units of volts, and R is the resistance of the conductor in units of ohms. More specifically, Ohm's law states that the R in this relation is constant, independent of the current.

The law was named after the German physicist Georg Ohm, who, in a treatise published in 1827, described measurements of applied voltage and current through simple electrical circuits containing various lengths of wire. He presented a slightly more complex equation than the one above to explain his experimental results. The above equation is the modern form of Ohm's law.

In physics, the term *Ohm's law* is also used to refer to various generalizations of the law originally formulated by Ohm. The simplest example of this is :

$$J = \sigma E,$$

where J is the current density at a given location in a resistive material, E is the electric field at that location, and σ is a material dependent parameter called the conductivity. This reformulation of Ohm's law is due to Gustav Kirchhoff.

History

In January 1781, before Georg Ohm's work, Henry Cavendish experimented with Leyden jars and glass tubes of varying diameter and length filled with salt solution. He measured the current by noting how strong a shock he felt as he completed the circuit with his body. Cavendish wrote that the "velocity" (current) varied directly as the "degree of electrification" (voltage). He did not communicate his results to other scientists at the time, and his results were unknown until Maxwell published them in 1879.

Ohm did his work on resistance in the years 1825 and 1826, and published his results in 1827 as the book *Die galvanische Kette, mathematisch bearbeitet* ("The galvanic circuit investigated mathematically"). He drew considerable inspiration from Fourier's work on heat conduction in the theoretical explanation of his work. For experiments, he initially used voltaic piles, but later used a thermocouple as this provided a more stable voltage source in terms of internal resistance and constant potential difference. He used a galvanometer to measure current, and knew that the voltage between the thermocouple terminals was proportional to the junction temperature. He then added test wires of varying length, diameter, and material to complete the circuit. He found that his data could be modelled through the equation

$$x = \frac{a}{b+l},$$

where x was the reading from the galvanometer, l was the length of the test conductor, a depended only on the thermocouple junction temperature, and b was a constant of the entire setup. From this, Ohm determined his law of proportionality and published his results.

Ohm's law was probably the most important of the early quantitative descriptions of the physics of electricity. We consider it almost obvious today. When Ohm first published his work, this was not the case; critics reacted to his treatment of the subject with hostility. They called his work a "web of naked fancies" and the German Minister of Education proclaimed that "a professor who preached such heresies was unworthy to teach science." The prevailing scientific philosophy in Germany at the time asserted that experiments need not be performed to develop an understanding of nature because nature is so well ordered, and that scientific truths may be deduced through reasoning alone. Also, Ohm's brother Martin, a mathematician, was battling the German educational system. These factors hindered the acceptance of Ohm's work, and his work did not become widely accepted until the 1840s. Fortunately, Ohm received recognition for his contributions to science well before he died.

In the 1850s, Ohm's law was known as such and was widely considered proved, and alternatives, such as "Barlow's law", were discredited, in terms of real applications to telegraph system design, as discussed by Samuel F. B. Morse in 1855.

While the old term for electrical conductance, the mho (the inverse of the resistance unit ohm), is still used, a new name, the siemens, was adopted in 1971, honouring Ernst Werner von Siemens. The siemens is preferred in formal papers.

In the 1920s, it was discovered that the current through a practical resistor actually has statistical fluctuations, which depend on temperature, even when voltage and resistance are exactly constant; this fluctuation, now known as Johnson–Nyquist noise, is due to the discrete nature of charge. This thermal effect implies that measurements of current and voltage that are taken over sufficiently short periods of time will yield ratios of V/I that fluctuate from the value of R implied by the time average or ensemble average of the measured current; Ohm's law remains correct for the average current, in the case of ordinary resistive materials.

Ohm's work long preceded Maxwell's equations and any understanding of frequency-dependent effects in AC circuits. Modern developments in electromagnetic theory and circuit theory do not contradict Ohm's law when they are evaluated within the appropriate limits.

Scope

Ohm's law is an empirical law, a generalization from many experiments that have shown that current is approximately proportional to electric field for most materials. It is less fundamental than Maxwell's equations and is not always obeyed. Any given material will break down under a strong-enough electric field, and some materials of interest in electrical engineering are "non-ohmic" under weak fields.

Ohm's law has been observed on a wide range of length scales. In the early 20th century, it was thought that Ohm's law would fail at the atomic scale, but experiments have not borne out this expectation. As of 2012, researchers have demonstrated that Ohm's law works for silicon wires as small as four atoms wide and one atom high.

Microscopic Origins

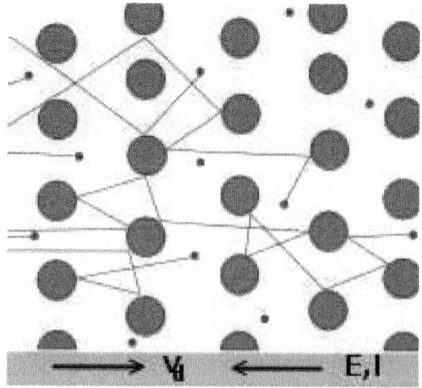

Fig. : Drude Model electrons (shown here in blue) constantly bounce among heavier, stationary crystal ions (shown in red).

The dependence of the current density on the applied electric field is essentially quantum mechanical in nature; A qualitative description leading to Ohm's law can be based upon classical mechanics using the Drude model developed by Paul Drude in 1900.

The Drude model treats electrons (or other charge carriers) like pinballs bouncing among the ions that make up the structure of the material. Electrons will be accelerated in the opposite direction to the electric field by the average electric field at their location. With each collision, though, the electron is deflected in a random direction with a velocity that is much larger than the velocity gained by the electric field. The net result is that electrons take a zigzag path due to the collisions, but generally drift in a direction opposing the electric field.

The drift velocity then determines the electric current density and its relationship to E and is independent of the collisions. Drude calculated the average drift velocity from $p = -eE\tau$ where p is the average momentum, $-e$ is the charge of the electron and τ is the average time between the collisions. Since both the momentum and the current density are proportional to the drift velocity, the current density becomes proportional to the applied electric field; this leads to Ohm's law.

Hydraulic Analogy

A hydraulic analogy is sometimes used to describe Ohm's law. Water pressure, measured by pascals (or PSI), is the analog of voltage because establishing a water pressure difference between two points along a (horizontal) pipe causes water to flow. Water flow rate, as in liters per second, is the analog of current, as in coulombs per second. Finally, flow restrictors — such as apertures placed in pipes between points where the water pressure is measured — are the analog of resistors. We say that the rate of water flow through an aperture restrictor is proportional to the difference in water pressure across the restrictor. Similarly, the rate of flow of electrical charge, that is, the electric current, through an electrical resistor is proportional to the difference in voltage measured across the resistor.

Flow and pressure variables can be calculated in fluid flow network with the use of the hydraulic ohm analogy. The method can be applied to both steady and transient flow situations. In the linear laminar flow region, Poiseuille's law describes the hydraulic resistance of a pipe, but in the turbulent flow region the pressure–flow relations become non-linear.

The hydraulic analogy to Ohm's law has been used, for example, to approximate blood flow through the circulatory system.

Circuit Analysis

In circuit analysis, three equivalent expressions of Ohm's law are used interchangeably :

$$I = \frac{V}{R} \text{ or } V = IR \text{ or } R = \frac{V}{I}.$$

Each equation is quoted by some sources as the defining relationship of Ohm's law, or all three are quoted, or derived from a proportional form, or even just the two that do not correspond to Ohm's original statement may sometimes be given.

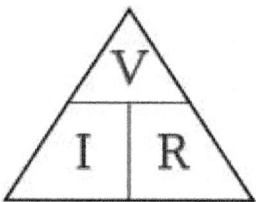

Fig. : Ohm's law triangle.

The interchangeability of the equation may be represented by a triangle, where V (voltage) is placed on the top section, the I (current) is placed to the left section, and the R (resistance) is placed to the right. The line that divides the left and right sections indicate multiplication, and the divider between the top and bottom sections indicates division (hence the division bar).

Resistive Circuits

Resistors are circuit elements that impede the passage of electric charge in agreement with Ohm's law, and are designed to have a specific resistance value R. In a schematic diagram the resistor is shown as a zig-zag symbol. An element (resistor or conductor) that behaves according to Ohm's law over some operating range is referred to as an *ohmic device* (or an *ohmic resistor*) because Ohm's law and a single value for the resistance suffice to describe the behaviour of the device over that range.

Ohm's law holds for circuits containing only resistive elements (no capacitances or inductances) for all forms of driving voltage or current, regardless of whether the driving voltage or current is constant (DC) or time-varying such as AC. At any instant of time Ohm's law is valid for such circuits.

Resistors which are in *series* or in *parallel* may be grouped together into a single "equivalent resistance" in order to apply Ohm's law in analyzing the circuit. This application of Ohm's law is illustrated with examples in "How to Analyze Resistive Circuits Using Ohm's Law" on wikiHow.

Reactive Circuits with Time-Varying Signals

When reactive elements such as capacitors, inductors, or transmission lines are involved in a circuit to which AC or time-varying voltage or current is applied, the relationship between voltage and current becomes the solution to a differential equation, so Ohm's law (as defined above) does not directly apply since that form contains only resistances having value R, not complex impedances which may contain capacitance ("C") or inductance ("L").

Power Measurement Fundamentals

Equations for time-invariant AC circuits take the same form as Ohm's law, however, the variables are generalized to complex numbers and the current and voltage waveforms are complex exponentials.

In this approach, a voltage or current waveform takes the form Ae^{st}, where t is time, s is a complex parameter, and A is a complex scalar. In any linear time-invariant system, all of the currents and voltages can be expressed with the same s parameter as the input to the system, allowing the time-varying complex exponential term to be canceled out and the system described algebraically in terms of the complex scalars in the current and voltage waveforms.

The complex generalization of resistance is impedance, usually denoted Z; it can be shown that for an inductor,

$$Z = sL$$

and for a capacitor,

$$Z = \frac{1}{sC}.$$

We can now write,

$$V = I \cdot Z$$

where V and I are the complex scalars in the voltage and current respectively and Z is the complex impedance.

This form of Ohm's law, with Z taking the place of R, generalizes the simpler form. When Z is complex, only the real part is responsible for dissipating heat.

In the general AC circuit, Z varies strongly with the frequency parameter s, and so also will the relationship between voltage and current.

For the common case of a steady sinusoid, the s parameter is taken to be $j\omega$, corresponding to a complex sinusoid $Ae^{j\omega t}$. The real parts of such complex current and voltage waveforms describe the actual sinusoidal currents and voltages in a circuit, which can be in different phases due to the different complex scalars.

Linear Approximations

Ohm's law is one of the basic equations used in the analysis of electrical circuits. It applies to both metal conductors and circuit components (resistors) specifically made for this behaviour. Both are ubiquitous in electrical engineering. Materials and components that obey Ohm's law are described as "ohmic" which means they produce the same value for resistance (R = V/I) regardless of the value of V or I which is applied and whether the applied voltage or current is DC (direct current) of either positive or negative polarity or AC (alternating current).

In a true ohmic device, the same value of resistance will be calculated from R = V/I regardless of the value of the applied voltage V. That is, the ratio of V/I is constant, and when current is plotted as a function of voltage the curve is *linear* (a straight line). If voltage is forced to some value V, then that voltage V divided by measured current I will equal R. Or if the current is forced to some value I, then the measured voltage V divided by that current I is also R. Since the plot of I versus V is a straight line, then it is also true that for any set of two different voltages V_1 and V_2 applied across a given device of resistance R, producing currents $I_1 = V_1/R$ and $I_2 = V_2/R$, that the ratio $(V_1-V_2)/(I_1-I_2)$ is also a constant equal to R. The operator "delta" (Δ) is used to represent a difference in a quantity, so we can write $\Delta V = V_1-V_2$ and $\Delta I = I_1-I_2$. Summarizing, for any truly ohmic device having resistance R, $V/I = \Delta V/\Delta I = R$ for any applied voltage or current or for the difference between any set of applied voltages or currents.

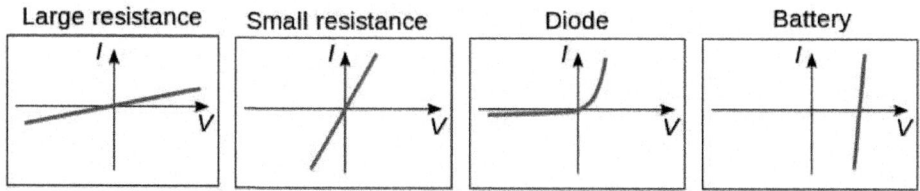

Fig.: The I–V curves of four devices: Two resistors, a diode, and a battery. The two resistors follow Ohm's law: The plot is a straight line through the origin. The other two devices do *not* follow Ohm's law.

There are, however, components of electrical circuits which do not obey Ohm's law; that is, their relationship between current and voltage (their I–V curve) is *non-linear* (or non-ohmic). An example is the p-n junction diode (curve at right). The current does not increase linearly with applied voltage for a diode. One can determine a value of current (I) for a given value of applied voltage (V) from the curve, but not from Ohm's law, since the value of "resistance" is not constant as a function of applied voltage. Further, the current only increases significantly if the applied voltage is positive, not negative. The ratio *V/I* for some point along the non-linear curve is sometimes called the *static*, or *chordal*, or DC, resistance, but The value of total *V* over total *I* varies depending on the particular point along the non-linear curve which is chosen. This means the "DC resistance" V/I at some point on the curve is not the same as what would be determined by applying an AC signal having peak amplitude ΔV volts or ΔI amps centered at that same point along the curve and measuring $\Delta V/\Delta I$. However, in some diode applications, the AC signal applied to the device is small and it is possible to analyze the circuit in terms of the *dynamic, small-signal,* or *incremental* resistance, defined as the one over the slope of the V–I curve at the average value (DC operating point) of the voltage (that is, one over the derivative of current with respect to voltage). For sufficiently small signals, the dynamic resistance allows the Ohm's law small signal resistance to be calculated as approximately one over the slope of a line drawn tangentially to the V-I curve at the DC operating point.

Temperature Effects

Ohm's law has sometimes been stated as, "for a conductor in a given state, the electromotive force is proportional to the current produced." That is, that the resistance, the ratio of the applied electromotive force (or voltage) to the current, "does not vary with the current strength." The qualifier "in a given state" is usually interpreted as meaning "at a constant temperature," since the resistivity of materials is usually temperature dependent. Because the conduction of current is related to Joule heating of the conducting body, according to Joule's first law, the temperature of a conducting body may change when it carries a current. The dependence of resistance on temperature therefore makes resistance depend upon the current in a typical experimental setup, making the law in this form difficult to directly verify. Maxwell and others worked out several methods to test the law experimentally in 1876, controlling for heating effects.

Relation to Heat Conductions

Ohm's principle predicts the flow of electrical charge (*i.e.* current) in electrical conductors when subjected to the influence of voltage differences; Jean-Baptiste-Joseph Fourier's principle predicts the flow of heat in heat conductors when subjected to the influence of temperature differences.

The same equation describes both phenomena, the equation's variables taking on different meanings in the two cases. Specifically, solving a heat conduction (Fourier) problem with *temperature* (the driving "force") and *flux of heat* (the rate of flow of the driven "quantity", *i.e.* heat energy) variables also solves an analogous electrical conduction (Ohm) problem having *electric potential* (the driving "force") and *electric current* (the rate of flow of the driven "quantity", *i.e.* charge) variables.

The basis of Fourier's work was his clear conception and definition of thermal conductivity. He assumed that, all else being the same, the flux of heat is strictly proportional to the gradient of temperature. Although undoubtedly true for small temperature gradients, strictly proportional behaviour will be lost when real materials (*e.g.* ones having a thermal conductivity that is a function of temperature) are subjected to large temperature gradients.

A similar assumption is made in the statement of Ohm's law : other things being alike, the strength of the current at each point is proportional to the gradient of electric potential. The accuracy of the assumption that flow is proportional to the gradient is more readily tested, using modern measurement methods, for the electrical case than for the heat case.

Other Versions

Ohm's law, in the form above, is an extremely useful equation in the field of electrical/electronic engineering because it describes how voltage, current and resistance are interrelated on a "macroscopic" level, that is, commonly, as circuit elements in an electrical circuit. Physicists who study the electrical properties of matter at the microscopic level use a closely related and more general vector equation, sometimes also referred to as Ohm's law, having variables that are closely related to the V, I, and R scalar variables of Ohm's law, but which are each functions of position within the conductor. Physicists often use this continuum form of Ohm's Law :

$$E = \rho J$$

where "E" is the electric field vector with units of volts per meter (analogous to "V" of Ohm's law which has units of volts), "J" is the current density vector with units of amperes per unit area (analogous to "I" of Ohm's law which has units of amperes), and "ρ" (Greek "rho") is the resistivity with units of ohm meters (analogous to "R" of Ohm's law which has units of ohms). The above equation is sometimes written as $J = \sigma E$ where "σ" (Greek "sigma") is the conductivity which is the reciprocal of ρ.

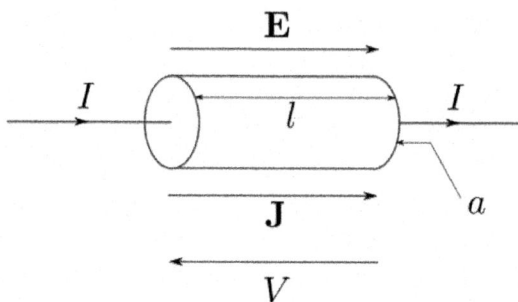

Fig.: Current flowing through a uniform cylindrical conductor (such as a round wire) with a uniform field applied.

The potential difference between two points is defined as :

$$DV = - \int E \cdot dl$$

with *dl* the element of path along the integration of electric field vector **E**. If the applied E field is uniform and oriented along the length of the conductor, then defining the voltage V in the usual convention of being opposite in direction to the field, and with the understanding that the voltage V is measured differentially across the length of the conductor allowing us to drop the Δ symbol, the above vector equation reduces to the scalar equation :

$$V = El \text{ or } E = \frac{V}{l}.$$

Since the E field is uniform in the direction of wire length, for a conductor having uniformly consistent resistivity ρ, the current density J will also be uniform in any cross-sectional area and oriented in the direction of wire length, so we may write :

$$J = \frac{I}{a}.$$

Substituting the above 2 results (for E and J respectively) into the continuum form shown at the beginning of this section :

Power Measurement Fundamentals

$$\frac{V}{l} = \frac{I}{u}\rho \text{ or } V = I\,\rho\frac{l}{u}.$$

The electrical resistance of a uniform conductor is given in terms of resistivity by :

$$R = \rho\frac{l}{a}$$

where l is the length of the conductor in SI units of meters, a is the cross-sectional area (for a round wire $a = \pi r^2$ if r is radius) in units of meters squared, and ρ is the resistivity in units of ohm meters.

After substitution of R from the above equation into the equation preceding it, the continuum form of Ohm's law for a uniform field (and uniform current density) oriented along the length of the conductor reduces to the more familiar form :

$$\underline{V} = I\,R.$$

A perfect crystal lattice, with low enough thermal motion and no deviations from periodic structure, would have no resistivity, but a real metal has crystallographic defects, impurities, multiple isotopes, and thermal motion of the atoms. Electrons scatter from all of these, resulting in resistance to their flow.

The more complex generalized forms of Ohm's law are important to condensed matter physics, which studies the properties of matter and, in particular, its electronic structure. In broad terms, they fall under the topic of constitutive equations and the theory of transport coefficients.

Magnetic Effects

If an external **B**-field is present and the conductor is not at rest but moving at velocity **v**, then an extra term must be added to account for the current induced by the Lorentz force on the charge carriers.

$$\mathbf{J} = \sigma(\mathbf{E} + \mathbf{V} \times \mathbf{B})$$

In the rest frame of the moving conductor this term drops out because $v = 0$. There is no contradiction because the electric field in the rest frame differs from the E-field in the lab frame : $\mathbf{E}' = \mathbf{E} + \mathbf{v} \times \mathbf{B}$.

If the current **J** is alternating because the applied voltage or E-field varies in time, then reactance must be added to resistance to account for self-inductance. The reactance may be strong if the frequency is high or the conductor is coiled.

Occurrences

Natural observable examples of electrical current include lightning, static electricity, and the solar wind, the source of the polar auroras.

Man-made occurrences of electric current include the flow of conduction electrons in metal wires such as the overhead power lines that deliver electrical energy across long distances and the smaller wires within electrical and electronic equipment. Eddy currents are electric currents that occur in conductors exposed to changing magnetic fields. Similarly, electric currents occur, particularly in the surface, of conductors exposed to electromagnetic waves. When oscillating electric currents flow at the correct voltages within radio antennas, radio waves are generated.

In electronics, other forms of electric current include the flow of electrons through resistors or through the vacuum in a vacuum tube, the flow of ions inside a battery or a neuron, and the flow of holes within a semi-conductor.

Current Measurement

Current can be measured using an ammeter.

At the circuit level, there are various techniques that can be used to measure current:

- Shunt resistors
- Hall effect current sensor transducers
- Transformers (however DC cannot be measured)
- Magnetoresistive field sensors

Joule Heating

Joule heating, also known as **ohmic heating** and **resistive heating**, is the process by which the passage of an electric current through a conductor releases heat. The amount of heat released is proportional to the square of the current such that

$$Q \propto I^2 . R$$

This relationship is known as **Joule's first law**. The SI unit of energy was subsequently named the joule and given the symbol J. The commonly known unit of power, the watt, is equivalent to one joule per second. Joule heating is independent of the direction of current, unlike heating due to the Peltier effect.

Background

History

Resistive heating was first studied by James Prescott Joule in 1841. Joule immersed a length of wire in a fixed mass of water and measured the temperature rise due to a known current flowing through the wire for a 30 minute period. By varying the current and the length of the wire he deduced that the heat produced was proportional to the square of the current multiplied by the electrical resistance of the wire.

Microscopic Description

Joule heating is caused by interactions between the moving particles that form the current (usually, but not always, electrons) and the atomic ions that make up

Power Measurement Fundamentals

the body of the conductor. Charged particles in an electric circuit are accelerated by an electric field but give up some of their kinetic energy each time they collide with an ion. The increase in the kinetic or vibrational energy of the ions manifests itself as heat and a rise in the temperature of the conductor. Hence energy is transferred from the electrical power supply to the conductor and any materials with which it is in thermal contact.

Power Loss and Noise

Joule heating is referred to as *ohmic heating* or *resistive heating* because of its relationship to Ohm's Law. It forms the basis for the large number of practical applications involving electric heating. However, in applications where heating is an unwanted by-product of current use (*e.g.*, load losses in electrical transformers) the diversion of energy is often referred to as *resistive loss*. The use of high voltages in electric power transmission systems is specifically designed to reduce such losses in cabling by operating with commensurately lower currents. The ring circuits, or ring mains, used in UK homes are another example, where power is delivered to outlets at lower currents, thus reducing Joule heating in the wires. Joule heating does not occur in superconducting materials, as these materials have zero electrical resistance in the superconducting state.

Resistors create electrical noise, called Johnson–Nyquist noise. There is an intimate relationship between Johnson–Nyquist noise and Joule heating, explained by the fluctuation-dissipation theorem.

Formulas

Direct Current

The most general and fundamental formula for Joule heating is :

$$P = VI$$

where :

- P is the power (energy per unit time) converted from electrical energy to thermal energy,
- I is the current travelling through the resistor or other element,
- V is the voltage drop across the element.

The explanation of this formula ($P=VI$) is :

(*Energy dissipated per unit time*) = (*Energy dissipated per charge passing through resistor*) × (*Charge passing through resistor per unit time*)

When Ohm's law is also applicable, the formula can be written in other equivalent forms :

$$P = IV = I^2 R = V^2/R$$

where R is the resistance.

Alternating Current

When current varies, as it does in AC circuits,

$$P(t) = I(t) V(t)$$

where t is time and P is the instantaneous power being converted from electrical energy to heat. Far more often, the *average* power is of more interest than the instantaneous power:

$$P_{avg} = I_{rms} V_{rms} = I^2_{rms} R = V^2_{rms}/R$$

where "avg" denotes average (mean) over one or more cycles, and "rms" denotes root mean square.

These formulas are valid for an ideal resistor, with zero reactance. If the reactance is non-zero, the formulas are modified:

$$P_{avg} = I_{rms} V_{rms} \cos \phi = I^2_{rms} Re(Z) = V^2_{rms} Re(Y^*)$$

where ϕ is the phase difference between current and voltage, Re means real part, Z is the complex impedance, and Y^* is the complex conjugate of the admittance (equal to $1/Z^*$).

Differential Form

In plasma physics, the Joule heating often needs to be calculated at a particular location in space. The differential form of the Joule heating equation gives the power per unit volume.

$$P = J \cdot E$$

Here, J is the current density, and E is the electric field.

Reason for High-Voltage Transmission of Electricity

In electric power transmission, high voltage is used to reduce Joule heating of the overhead power lines. The valuable electric energy is intended to be used by consumers, not for heating the power lines. Therefore this Joule heating is referred to as a type of *transmission loss*.

A given quantity of electric power can be transmitted through a transmission line either at low voltage and high current, or with a higher voltage and lower current. Transformers can convert a high transmission voltage to a lower voltage for use by customer loads. Since the power lost in the wires is proportional to the conductor resistance and the square of the current, using low current at high voltage reduces the loss in the conductors due to Joule heating (or alternatively allows smaller conductors to be used for the same relative loss).

Applications

There are many practical uses of Joule heating. Some of the commonest are as follows :

- An incandescent light bulb glows when the filament is heated by Joule heating, so hot that it glows white with thermal radiation (also called blackbody radiation).
- Electric stoves and other electric heaters usually work by Joule heating.
- Soldering irons and cartridge heaters are very often heated by Joule heating.
- Electric fuses rely on the fact that if enough current flows, enough heat will be generated to melt the fuse wire.
- Electronic cigarettes usually work by Joule heating, vapourizing propylene glycol and vegetable glycerine.
- Thermistors and resistance thermometers are resistors whose resistance changes when the temperature changes. These are sometimes used in conjunction with Joule heating (also called self-heating in this context) : If a large current is running through the non-linear resistor, the resistor's temperature rises and therefore its resistance changes. Therefore, these components can be used in a circuit-protection role similar to fuses, or for feedback in circuits, or for many other purposes. In general, self-heating can turn a resistor into a non-linear and hysteretic circuit element.

Less-Common Applications

- Food processing equipment may make use of Joule heating in food production. In this case, the food material serves as an electrical resistor, and heat is released internally.

Heating Efficiency

As a heating technology, Joule heating has a coefficient of performance of 1.0, meaning that every 1 watt of electrical power is converted to 1 joule of heat. By comparison, a heat pump can have a coefficient of more than 1.0 since it also absorbs additional heating energy from the environment, moving this thermal energy to where it is needed.

The definition of the efficiency of a heating process requires defining the boundaries of the system to be considered. When heating a building, the overall efficiency is different when considering heating effect per unit of electric energy delivered on the customer's side of the meter, compared to the overall efficiency when also considering the losses in the power plant and transmission of power.

Hydraulic Equivalent

In the energy balance of groundwater flow a hydraulic equivalent of Joule's law is used :

$$\frac{dE}{dx} = \frac{v_x^2}{K}$$

where :

dE/dx = loss of hydraulic energy (E) due to friction of flow in x-direction per unit of time (m/day) – comparable to Q/t

v_x = flow velocity in x-direction (m/day) – comparable to I

K = hydraulic conductivity of the soil (m/day) – the hydraulic conductivity is inversely proportional to the hydraulic resistance which compares to R.

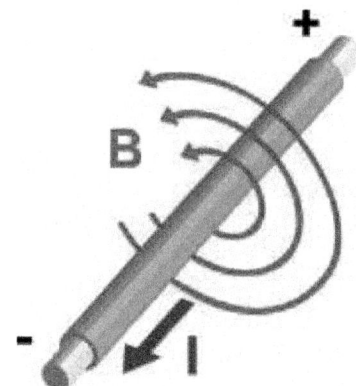

Fig. : According to Ampère's law, an electric current produces a magnetic field.

Electromagnetism

Electromagnet

Electric current produces a magnetic field. The magnetic field can be visualized as a pattern of circular field lines surrounding the wire that persists as long as there is current.

Magnetism can also produce electric currents. When a changing magnetic field is applied to a conductor, an Electromotive force (EMF) is produced, and when there is a suitable path, this causes current.

Electric current can be directly measured with a galvanometer, but this method involves breaking the electrical circuit, which is sometimes inconvenient. Current can also be measured without breaking the circuit by detecting the magnetic field associated with the current. Devices used for this include Hall effect sensors, current clamps, current transformers, and Rogowski coils.

Radio Waves

When an electric current flows in a suitably shaped conductor at radio frequencies radio waves can be generated. These travel at the speed of light and can cause electric currents in distant conductors.

Conduction Mechanisms in Various Media

In metallic solids, electric charge flows by means of electrons, from lower to higher electrical potential. In other media, any stream of charged objects (ions, for example) may constitute an electric current. To provide a definition of current that is independent of the type of charge carriers flowing, *conventional current* is defined to be in the same direction as positive charges. So in metals where the charge carriers (electrons) are negative, conventional current is in the opposite direction as the electrons. In conductors where the charge carriers are positive, conventional current is in the same direction as the charge carriers.

In a vacuum, a beam of ions or electrons may be formed. In other conductive materials, the electric current is due to the flow of both positively and negatively charged particles at the same time. In still others, the current is entirely due to positive charge flow. For example, the electric currents in electrolytes are flows of positively and negatively charged ions. In a common lead-acid electrochemical cell, electric currents are composed of positive hydrogen ions (protons) flowing in one direction, and negative sulfate ions flowing in the other. Electric currents in sparks or plasma are flows of electrons as well as positive and negative ions. In ice and in certain solid electrolytes, the electric current is entirely composed of flowing ions.

Metals

A solid conductive metal contains mobile, or free electrons, which function as conduction electrons. These electrons are bound to the metal lattice but no longer to an individual atom. Metals are particularly conductive because there are a large number of these free electrons, typically one per atom in the lattice. Even with no external electric field applied, these electrons move about randomly due to thermal energy but, on average, there is zero net current within the metal. At room temperature, the average speed of these random motions is 10^6 metres per second. Given a surface through which a metal wire passes, electrons move in both directions across the surface at an equal rate. As George Gamow wrote in his popular science book, *One, Two, Three...Infinity* (1947), "The metallic substances differ from all other materials by the fact that the outer shells of their atoms are bound rather loosely, and often let one of their electrons go free. Thus the interior of a metal is filled up with a large number of unattached electrons that travel aimlessly around like a crowd of displaced persons. When a metal wire is subjected to electric force applied on its opposite ends, these free electrons rush in the direction of the force, thus forming what we call an electric current."

When a metal wire is connected across the two terminals of a DC voltage source such as a battery, the source places an electric field across the conductor. The moment contact is made, the free electrons of the conductor are forced to drift toward the positive terminal under the influence of this field. The free electrons are therefore the charge carrier in a typical solid conductor.

For a steady flow of charge through a surface, the current I (in amperes) can be calculated with the following equation :

$$I = \frac{Q}{t},$$

where Q is the electric charge transferred through the surface over a time t. If Q and t are measured in coulombs and seconds respectively, I is in amperes.

More generally, electric current can be represented as the rate at which charge flows through a given surface as :

$$I = \frac{dQ}{dt}.$$

Electrolytes

Electric currents in electrolytes are flows of electrically charged particles (ions). For example, if an electric field is placed across a solution of Na^+ and Cl^- (and conditions are right) the sodium ions move towards the negative electrode (cathode), while the chloride ions move towards the positive electrode (anode). Reactions take place at both electrode surfaces, absorbing each ion.

Water-ice and certain solid electrolytes called proton conductors contain positive hydrogen ions or "protons" which are mobile. In these materials, electric currents are composed of moving protons, as opposed to the moving electrons found in metals.

In certain electrolyte mixtures, brightly coloured ions are the moving electric charges. The slow progress of the colour makes the current visible.

Gases and Plasmas

In air and other ordinary gases below the breakdown field, the dominant source of electrical conduction is via relatively few mobile ions produced by radioactive gases, ultra-violet light, or cosmic rays. Since the electrical conductivity is low, gases are dielectrics or insulators. However, once the applied electric field approaches the breakdown value, free electrons become sufficiently accelerated by the electric field to create additional free electrons by colliding, and ionizing, neutral gas atoms or molecules in a process called avalanche breakdown. The breakdown process forms a plasma that contains enough mobile electrons and positive ions to make it an electrical conductor. In the process, it forms a light emitting conductive path, such as a spark, arc or lightning.

Plasma is the state of matter where some of the electrons in a gas are stripped or "ionized" from their molecules or atoms. A plasma can be formed by high temperature, or by application of a high electric or alternating magnetic field as noted above. Due to their lower mass, the electrons in a plasma accelerate more quickly in response to an electric field than the heavier positive ions, and hence carry the bulk of the current. The free ions recombine to create new chemical compounds (for example, breaking atmospheric oxygen into single oxygen [$O_2 \rightarrow 2O$], which then recombine creating ozone [O_3]).

Vacuum

Since a "perfect vacuum" contains no charged particles, it normally behaves as a perfect insulator. However, metal electrode surfaces can cause a region of the vacuum to become conductive by injecting free electrons or ions through either field electron emission or thermionic emission. Thermionic emission occurs when the thermal energy exceeds the metal's work function, while field electron emission occurs when the electric field at the surface of the metal is high enough to cause tunneling, which results in the ejection of free electrons from the metal into the vacuum. Externally heated electrodes are often used to generate an electron cloud as in the filament or indirectly heated cathode of vacuum tubes. Cold electrodes can also spontaneously produce electron clouds via thermionic emission when small incandescent regions (called **cathode spots** or **anode spots**) are formed. These are incandescent regions of the electrode surface that are created by a localized high current. These regions may be initiated by field electron emission, but are then sustained by localized thermionic emission once a vacuum arc forms. These small electron-emitting regions can form quite rapidly, even explosively, on a metal surface subjected to a high electrical field. Vacuum tubes and sprytrons are some of the electronic switching and amplifying devices based on vacuum conductivity.

Super-conductivity

Super-conductivity is a phenomenon of exactly zero electrical resistance and expulsion of magnetic fields occurring in certain materials when cooled below a characteristic critical temperature. It was discovered by Heike Kamerlingh Onnes on April 8, 1911 in Leiden. Like ferromagnetism and atomic spectral lines, Super-conductivity is a quantum mechanical phenomenon. It is characterized by the Meissner effect, the complete ejection of magnetic field lines from the interior of the super-conductor as it transitions into the superconducting state. The occurrence of the Meissner effect indicates that Super-conductivity cannot be understood simply as the idealization of *perfect conductivity* in classical physics.

Semi-conductor

In a semi-conductor it is sometimes useful to think of the current as due to the flow of positive "holes" (the mobile positive charge carriers that are places where the semi-conductor crystal is missing a valence electron). This is the case in a p-type semi-conductor. A semi-conductor has electrical conductivity intermediate in magnitude between that of a conductor and an insulator. This means a conductivity roughly in the range of 10^{-2} to 10^4 siemens per centimeter (S·cm^{-1}).

In the classic crystalline semi-conductors, electrons can have energies only within certain bands (*i.e.* ranges of levels of energy). Energetically, these bands are located between the energy of the ground state, the state in which electrons are tightly bound to the atomic nuclei of the material, and the free electron energy, the latter describing the energy required for an electron to escape entirely from the material. The energy bands each correspond to a large number of discrete quantum states of the electrons, and most of the states with low energy (closer to the nucleus) are occupied, up to a particular band called the *valence band*. Semi-conductors and insulators are distinguished from metals because the valence band in any given metal is nearly filled with electrons under usual operating conditions, while very few (semi-conductor) or virtually none (insulator) of them are available in the *conduction band*, the band immediately above the valence band.

The ease with which electrons in the semi-conductor can be excited from the valence band to the conduction band depends on the band gap between the bands. The size of this energy bandgap serves as an arbitrary dividing line (roughly 4 eV) between semi-conductors and insulators.

With covalent bonds, an electron moves by hopping to a neighbouring bond. The Pauli exclusion principle requires the electron to be lifted into the higher anti-bonding state of that bond. For delocalized states, for example, in one dimension – that is in a nanowire, for every energy there is a state with electrons flowing in one direction and another state with the electrons flowing in the other. For a net current to flow, more states for one direction than for the other direction must be occupied. For this to occur, energy is required, as in the semi-conductor the next higher states lie above the band gap. Often this is stated as : full bands do not contribute to the electrical conductivity. However, as the temperature of a semi-conductor rises above absolute zero, there is more energy in the semi-conductor to spend on lattice vibration and on exciting electrons into the conduction band. The current-carrying electrons in the conduction band are known as "free electrons", although they are often simply called "electrons" if context allows this usage to be clear.

Current Density and Ohm's Law

Current density is a measure of the density of an electric current. It is defined as a vector whose magnitude is the electric current per cross-sectional area. In SI units, the current density is measured in amperes per square metre.

$$I = \int \vec{J} \cdot d\vec{A}$$

where I is current in the conductor, $\vec{J}$ is the current density, and $d\vec{A}$ is the differential cross-sectional area vector.

The current density (current per unit area) $\vec{J}$ in materials with finite resistance is directly proportional to the electric field $\vec{E}$ in the medium. The proportionality constant is call the conductivity σ of the material, whose value depends on the material concerned and, in general, is dependent on the temperature of the material :

$$\vec{J} = \sigma \vec{E}$$

The reciprocal of the conductivity σ of the material is called the resistivity ρ of the material and the above equation, when written in terms of resistivity becomes:

$$\vec{J} = \frac{\vec{E}}{\rho} \text{ or}$$

$$\vec{E} = \rho \vec{J}$$

Power Measurement Fundamentals

Conduction in semi-conductor devices may occur by a combination of drift and diffusion, which is proportional to diffusion constant D and charge density α_q. The current density is then:

$$J = \sigma E + Dq\Delta n.$$

with q being the elementary charge and n the electron density. The carriers move in the direction of decreasing concentration, so for electrons a positive current results for a positive density gradient. If the carriers are holes, replace electron density n by the negative of the hole density p.

In linear anisotropic materials, σ, ρ and D are tensors.

In linear materials such as metals, and under low frequencies, the current density across the conductor surface is uniform. In such conditions, Ohm's law states that the current is directly proportional to the potential difference between two ends (across) of that metal (ideal) resistor (or other ohmic device):

$$I = \frac{V}{R},$$

where I is the current, measured in amperes; V is the potential difference, measured in volts; and R is the resistance, measured in ohms. For alternating currents, especially at higher frequencies, skin effect causes the current to spread unevenly across the conductor cross-section, with higher density near the surface, thus increasing the apparent resistance.

Drift Speed

The mobile charged particles within a conductor move constantly in random directions, like the particles of a gas. In order for there to be a net flow of charge, the particles must also move together with an average drift rate. Electrons are the charge carriers in metals and they follow an erratic path, bouncing from atom-to-atom, but generally drifting in the opposite direction of the electric field. The speed at which they drift can be calculated from the equation:

$$I = nAvQ,$$

where:

  I is the electric current

  n is number of charged particles per unit volume (or charge carrier density)

  A is the cross-sectional area of the conductor

  v is the drift velocity, and

  Q is the charge on each particle.

Typically, electric charges in solids flow slowly. For example, in a copper wire of cross-section 0.5 mm², carrying a current of 5 A, the drift velocity of the electrons is on the order of a millimetre per second. To take a different example, in the near-vacuum inside a cathode ray tube, the electrons travel in near-straight lines at about a tenth of the speed of light.

Any accelerating electric charge, and therefore any changing electric current, gives rise to an electromagnetic wave that propagates at very high speed outside the surface of the conductor. This speed is usually a significant fraction of the speed of light, as can be deduced from Maxwell's Equations, and is therefore many times faster than the drift velocity of the electrons. For example, in AC power lines, the waves of electromagnetic energy propagate through the space between the wires, moving from a source to a distant load, even though the electrons in the wires only move back and forth over a tiny distance.

The ratio of the speed of the electromagnetic wave to the speed of light in free space is called the velocity factor, and depends on the electromagnetic properties of the conductor and the insulating materials surrounding it, and on their shape and size.

The magnitudes (but, not the natures) of these three velocities can be illustrated by an analogy with the three similar velocities associated with gases.

- The low drift velocity of charge carriers is analogous to air motion; in other words, winds.
- The high speed of electromagnetic waves is roughly analogous to the speed of sound in a gas (these waves move through the medium much faster than any individual particles do)
- The random motion of charges is analogous to heat – the thermal velocity of randomly vibrating gas particles.

ELECTRICAL RESISTANCE AND CONDUCTANCE

The **electrical resistance** of an electrical conductor is the opposition to the passage of an electric current through that conductor. The inverse quantity is **electrical conductance**, the ease with which an electric current passes. Electrical resistance shares some conceptual parallels with the mechanical notion of friction. The SI unit of electrical resistance is the ohm (Ω), while electrical conductance is measured in siemens (S).

An object of uniform cross-section has a resistance proportional to its resistivity and length and inversely proportional to its cross-sectional area. All materials show some resistance, except for superconductors, which have a resistance of zero.

The resistance (R) of an object is defined as the ratio of voltage across it (V) to current through it (I), while the conductance (G) is the inverse :

$$R = \frac{dE}{dx} \quad G = \frac{v_x^2}{K} \quad G = \frac{Q}{t}$$

For a wide variety of materials and conditions, V and I are directly proportional to each other, and therefore R and G are constant (although they can depend on other factors like temperature or strain). This proportionality is called Ohm's law, and materials that satisfy it are called "Ohmic" materials.

In other cases, such as a diode or battery, V and I are *not* directly proportional, or in other words the I–V curve is not a straight line through the origin, and Ohm's law does not hold. In this case, resistance and conductance are less useful concepts, and more difficult to define. The ratio V/I is sometimes still useful, and is referred to as a "chordal resistance" or "static resistance", as it corresponds to the inverse slope of a chord between the origin and an I–V curve. In other situations, the derivative $\frac{dV}{dI}$ may be most useful; this is called the "differential resistance".

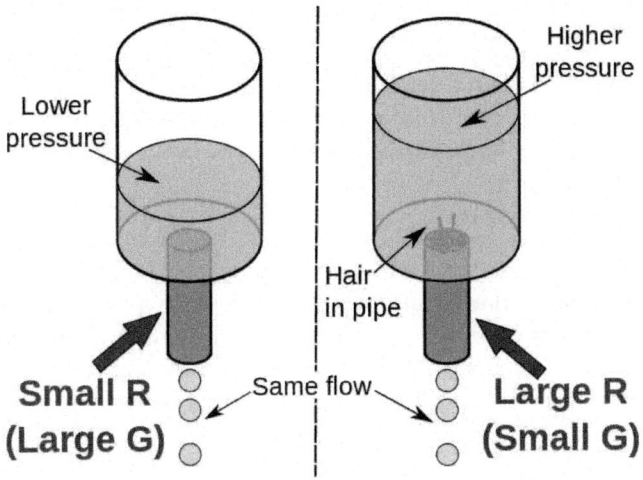

Fig. : The hydraulic analogy compares electric current flowing through circuits to water flowing through pipes. When a pipe (left) is filled with hair (right), it takes a larger pressure to achieve the same flow of water. Pushing electric current through a large resistance is like pushing water through a pipe clogged with hair : It requires a larger push (electromotive force) to drive the same flow (electric current).

Introduction

In the hydraulic analogy, current flowing through a wire (or resistor) is like water flowing through a pipe, and the voltage drop across the wire is like the pressure drop that pushes water through the pipe. Conductance is proportional to how much flow occurs for a given pressure, and resistance is proportional to how much pressure is required to achieve a given flow. (Conductance and resistance are reciprocals.)

The voltage *drop* (*i.e.*, difference in voltage between one side of the resistor and the other), not the voltage itself, provides the driving force pushing current through a resistor. In hydraulics, it is similar : The pressure *difference* between two sides of a pipe, not the pressure itself, determines the flow through it. For example, there may be a large water pressure above the pipe, which tries to push water down through the pipe. But there may be an equally large water pressure below the pipe, which tries to push water back up through the pipe. If these pressures are equal, no water flows.

Two properties—geometry (shape) and material—mostly determine the resistance and conductance of a wire, resistor, or other element.

Geometry is important because it is more difficult to push water through a long, narrow pipe than a wide, short pipe. In the same way, a long, thin copper wire has higher resistance (lower conductance) than a short, thick copper wire.

Materials are important as well. A pipe filled with hair restricts the flow of water more than a clean pipe of the same shape and size. In a similar way, electrons can flow freely and easily through a copper wire, but cannot as easily flow through a steel wire of the same shape and size, and they essentially cannot flow at all through an insulator like rubber, regardless of its shape. The difference between, copper, steel, and rubber is related to their microscopic structure and electron configuration, and is quantified by a property called resistivity.

Conductors and Resistors

Substances electricity can flow through are called conductors. A piece of conducting material of a particular resistance meant for use in a circuit is called a resistor. Conductors are made of high-conductivity materials such as metals, in particular copper and aluminium. Resistors, on the other hand, are made of a wide variety of materials depending on factors such as the desired resistance, amount of energy that it needs to dissipate, precision, and costs.

Ohm's law

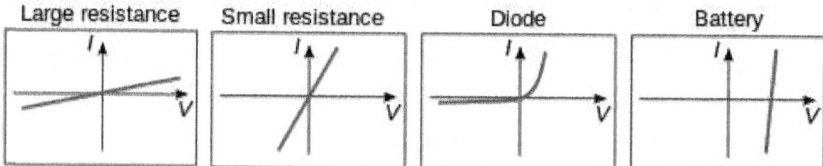

Fig.: The current-voltage characteristics of four devices: Two resistors, a diode, and a battery. The horizontal axis is voltage drop, the vertical axis is current. Ohm's law is satisfied when the graph is a straight line through the origin. Therefore, the two resistors are "ohmic", but the diode and battery are not.

Ohm's law is an empirical law relating the voltage V across an element to the current I through it :

$$V \propto I$$

(V is directly proportional to I). This law is not always true : For example, it is false for diodes, batteries, etc. However, it *is* true to a very good approximation for wires and resistors (assuming that other conditions, including temperature, are held fixed). Materials or objects where Ohm's law is true are called *ohmic*, whereas objects that do not obey Ohm's law are *non-ohmic*.

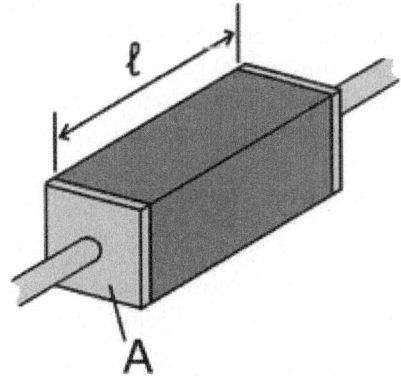

Fig. : A piece of resistive material with electrical contacts on both ends.

Relation to Resistivity and Conductivity

The resistance of a given object depends primarily on two factors : What material it is made of, and its shape. For a given material, the resistance is inversely proportional to the cross-sectional area; for example, a thick copper wire has lower resistance than an otherwise-identical thin copper wire. Also, for a given material, the resistance is proportional to the length; for example, a long copper wire has higher resistance than an otherwise-identical short copper wire. The resistance R and conductance G of a conductor of uniform cross-section, therefore, can be computed as

$$R = \rho \frac{\ell}{A},$$
$$G = \sigma \frac{A}{\ell}.$$

where ℓ is the length of the conductor, measured in metres [m], A is the cross-section area of the conductor measured in square metres [m²], σ (sigma) is the electrical conductivity measured in siemens per meter (S·m⁻¹), and ρ (rho) is the electrical resistivity (also called *specific electrical resistance*) of the material, measured in ohm-metres (Ω·m). The resistivity and conductivity are proportionality constants, and therefore depend only on the material the wire is made of, not the geometry of the wire. Resistivity and conductivity are reciprocals : $\rho = 1/\sigma$. Resistivity is a measure of the material's ability to oppose electric current.

This formula is not exact : It assumes the current density is totally uniform in the conductor, which is not always true in practical situations. However, this formula still provides a good approximation for long thin conductors such as wires.

Another situation for which this formula is not exact is with alternating current (AC), because the skin effect inhibits current flow near the center of the conductor. Then, the *geometrical* cross-section is different from the *effective* cross-section in which current actually flows, so resistance is higher than expected. Similarly, if two conductors near each other carry AC current, their resistances increase due to the proximity effect. At commercial power frequency, these effects are significant for large conductors carrying large currents, such as busbars in an electrical substation, or large power cables carrying more than a few hundred amperes.

What Determines Resistivity?

The resistivity of different materials varies by an enormous amount: For example, the conductivity of teflon is about 10^{30} times lower than the conductivity of copper. Why is there such a difference? Loosely speaking, a metal has large numbers of "delocalized" electrons that are not stuck in any one place, but free to move across large distances, whereas in an insulator (like teflon), each electron is tightly bound to a single molecule, and a great force is required to pull it away. Semi-conductors lie between these two extremes.

Resistivity varies with temperature. In semi-conductors, resistivity also changes when light is shining on it.

Measuring Resistance

An instrument for measuring resistance is called an ohmmeter. Simple ohmmeters cannot measure low resistances accurately because the resistance of their measuring leads causes a voltage drop that interferes with the measurement, so more accurate devices use four-terminal sensing.

Typical Resistances

Component	Resistance (Ω)
1 meter of copper wire with 1mm diameter	0.02
1 km overhead power line (typical)	0.03
AA battery (typical internal resistance)	0.1
Incandescent light bulb filament (typical)	200-1000
Human body	1000 to 100,000

Static and Differential Resistance

The IV curve of a non-ohmic device (purple). The **static resistance** at point A is the inverse slope of line B through the origin. The **differential resistance** at A is the inverse slope of tangent line C.

Many electrical elements, such as diodes and batteries do *not* satisfy Ohm's law. These are called *non-ohmic* or *non-linear*, and are characterized by an *I–V* curve, which is *not* a straight line through the origin.

Electrical Power Systems Technology

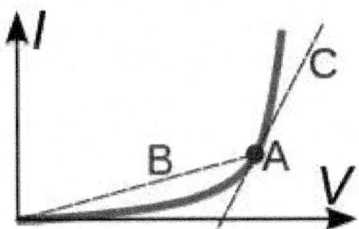

Resistance and conductance can still be defined for non-ohmic elements. However, unlike ohmic resistance, non-linear resistance is not constant but varies with the voltage or current through the device; its operating point. There are two types :

- **Static resistance** (also called *chordal* or *DC resistance*) - This corresponds to the usual definition of resistance; the voltage divided by the current

$$R_{static} = \frac{V}{I}.$$

It is the slope of the line (chord) from the origin through the point on the curve. Static resistance determines the power dissipation in an electrical component. Points on the *IV* curve located in the 2nd or 4th quadrants, for which the slope of the chordal line is negative, have *negative static resistance*. Passive devices, which have no source of energy, cannot have negative static resistance. However active devices such as transistors or op-amps can synthesize negative static resistance with feedback, and it is used in some circuits such as gyrators.

- **Differential resistance** (also called *dynamic, incremental* or *small signal resistance*) - Differential resistance is the derivative of the voltage with respect to the current; the slope of the *IV* curve at a point

$$R_{diff} = \frac{dV}{dI}.$$

If the *IV* curve is non-monotonic (with peaks and troughs), the curve has a negative slope in some regions — so in these regions the device has *negative differential resistance*. Devices with negative differential resistance can amplify a signal applied to them, and are used to make amplifiers and oscillators. These include tunnel diodes, Gunn diodes, IMPATT diodes, magnetron tubes, and unijunction transistors.

AC Circuits

Impedance and Admittance

The voltage (red) and current (blue) versus time (horizontal axis) for a capacitor (top) and inductor (bottom). Since the amplitude of the current and voltage sinusoids are the same, the absolute value of impedance is 1 for both the capacitor and the inductor (in whatever units the graph is using). On the other hand, the phase difference between current and voltage is -90° for the capacitor; therefore, the complex phase of the impedance of the capacitor is -90°. Similarly, the phase difference between current and voltage is +90° for the inductor; therefore, the complex phase of the impedance of the inductor is +90°.

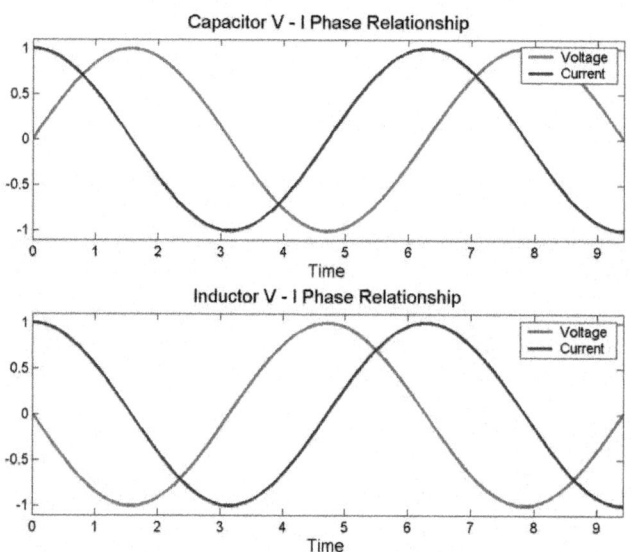

When an alternating current flows through a circuit, the relation between current and voltage across a circuit element is characterized not only by the ratio of their magnitudes, but also the difference in their phases. For example, in an ideal resistor, the moment when the voltage reaches its maximum, the current also reaches its maximum (current and voltage are oscillating in phase). But for a capacitor or inductor, the maximum current flow occurs as the voltage passes through zero and *vice-versa*. Complex numbers are used to keep track of both the phase and magnitude of current and voltage :

$$V(t) = \text{Re}(V_0 e^{j\omega t}), \ I(t) = \text{Re}(I_0 e^{j\omega t}), \ Z = \frac{V_0}{I_0}, \ Y = \frac{I_0}{V_0}$$

where :
- t is time,
- $V(t)$ and $I(t)$ are, respectively, voltage and current as a function of time,
- V_0, I_0, Z, and Y are complex numbers,
- Z is called impedance,
- Y is called admittance,
- Re indicates real part,
- ω is the angular frequency of the AC current,
- $j = \sqrt{-1}$ is the imaginary unit.

The impedance and admittance may be expressed as complex numbers that can be broken into real and imaginary parts :

$$Z = R + jX, Y = G + jB$$

where R and G are resistance and conductance respectively, X is reactance, and B is susceptance. For ideal resistors, Z and Y reduce to R and G respectively, but for AC networks containing capacitors and inductors, X and B are non-zero.

$Z = 1/Y$ for AC circuits, just as $R = 1/G$ for DC circuits.

Frequency Dependence of Resistance

Another complication of AC circuits is that the resistance and conductance can be frequency-dependent. One reason, mentioned above is the skin effect (and the related proximity effect). Another reason is that the resistivity itself may depend on frequency.

Energy Dissipation and Joule Heating

Resistors (and other elements with resistance) oppose the flow of electric current; therefore, electrical energy is required to push current through the resistance. This electrical energy is dissipated, heating the resistor in the process. This is called *Joule heating* (after James Prescott Joule), also called *ohmic heating* or *resistive heating*.

The dissipation of electrical energy is often undesired, particularly in the case of transmission losses in power lines. High voltage transmission helps reduce the losses by reducing the current for a given power.

On the other hand, Joule heating is sometimes useful, for example, in electric stoves and other electric heaters (also called *resistive heaters*). As another example, incandescent lamps rely on Joule heating : the filament is heated to such a high temperature that it glows "white hot" with thermal radiation (also called incandescence).

The formula for Joule heating is :

$$P = I^2R$$

where P is the power (energy per unit time) converted from electrical energy to thermal energy, R is the resistance, and I is the current through the resistor.

Dependence of Resistance on Other Conditions

Temperature Dependence

Near room temperature, the resistivity of metals typically increases as temperature is increased, while the resistivity of semi-conductors typically decreases as temperature is increased. The resistivity of insulators and electrolytes may increase or decrease depending on the system. For the detailed behaviour and explanation.

As a consequence, the resistance of wires, resistors, and other components often change with temperature. This effect may be undesired, causing an electronic circuit to malfunction at extreme temperatures. In some cases, however, the effect is put to good use. When temperature-dependent resistance of a component is used purposefully, the component is called a resistance thermometer or thermistor. (A resistance thermometer is made of metal, usually platinum, while a thermistor is made of ceramic or polymer.)

Resistance thermometers and thermistors are generally used in two ways. First, they can be used as thermometers : By measuring the resistance, the temperature of the environment can be inferred. Second, they can be used in conjunction with Joule heating (also called self-heating) : If a large current is running through the resistor, the resistor's temperature rises and therefore its resistance changes. Therefore, these components can be used in a circuit-protection role similar to fuses, or for feedback in circuits, or for many other purposes. In general, self-heating can turn a resistor into a non-linear and hysteretic circuit element.

If the temperature T does not vary too much, a linear approximation is typically used :

$$R(T) = R_0[1 + \alpha(T - T_0)]$$

where α is called the *temperature coefficient of resistance*, T_0 is a fixed reference temperature (usually room temperature), and R_0 is the resistance at temperature T_0. The parameter α is an empirical parameter fitted from measurement data. Because the linear approximation is only an approximation, α is different for different reference temperatures. For this reason it is usual to specify the temperature that α was measured at with a suffix, such as α_{15}, and the relationship only holds in a range of temperatures around the reference.

The temperature coefficient α is typically $+3\times10^{-3}$ K^{-1} to $+6\times10^{-3}$ K^{-1} for metals near room temperature. It is usually negative for semi-conductors and insulators, with highly variable magnitude.

Strain Dependence

Just as the resistance of a conductor depends upon temperature, the resistance of a conductor depends upon strain. By placing a conductor under tension (a form of stress that leads to strain in the form of stretching of the conductor), the length of the section of conductor under tension increases and its cross-sectional area decreases. Both these effects contribute to increasing the resistance of the strained section of conductor. Under compression (strain in the opposite direction), the resistance of the strained section of conductor decreases.

Light Illumination Dependence

Some resistors, particularly those made from semi-conductors, exhibit *photoconductivity*, meaning that their resistance changes when light is shining on them. Therefore, they are called *photoresistors* (or *light dependent resistors*). These are a common type of light detector.

Superconductivity

Superconductors are materials that have exactly zero resistance and infinite conductance, because they can have V=0 and I≠0. This also means there is no joule heating, or in other words no dissipation of electrical energy. Therefore, if superconductive wire is made into a closed loop, current flows around the loop forever. Superconductors require cooling to temperatures near 4 K with liquid helium for most metallic superconductors like NbSn alloys, or cooling to temperatures near 77K with liquid nitrogen for the expensive, brittle and delicate ceramic high temperature superconductors. Nevertheless, there are many technological applications of superconductivity, including superconducting magnets.

ELECTRICAL REACTANCE

In electrical and electronic systems, **reactance** is the opposition of a circuit element to a change of electric current or voltage, due to that element's inductance or capacitance. A built-up electric field resists the change of voltage on the element, while a magnetic field resists the change of current. The notion of reactance is similar to electrical resistance, but they differ in several respects.

An ideal resistor has zero reactance, while ideal inductors and capacitors consist entirely of reactance. The magnitude of the reactance of an inductor is proportional to frequency, while the magnitude of the reactance of a capacitor is inversely proportional to frequency.

Analysis

In phasor analysis, reactance is used to compute amplitude and phase changes of sinusoidal alternating current going through the circuit element. It is denoted by the symbol X.

Both reactance X and resistance R are components of impedance Z.

$$Z = R + jX$$

where :
- Z is the impedance, measured in ohms.
- R is the resistance, measured in ohms.
- X is the reactance, measured in ohms.
- $j = \sqrt{-1}$.

Both capacitive reactance X_C and inductive reactance X_L contribute to the total reactance X.

$$X = X_L - X_C = \omega L - \frac{1}{\omega C}$$

where
- X_C is the capacitive reactance, measured in ohms
- X_L is the inductive reactance, measured in ohms.

Although X_L and X_C are both positive by convention, the capacitive reactance X_C makes a negative contribution to total reactance.

Hence,

- If $X > 0$, the reactance is said to be inductive.
- If $X = 0$, then the impedance is purely resistive.
- If $X < 0$, the reactance is said to be capacitive.

Capacitive Reactance

Capacitive reactance is an opposition to the change of voltage across an element. Capacitive reactance X_C is inversely proportional to the signal frequency f (or angular frequency ω) and the capacitance C.

$$X_C = \frac{1}{\omega C} = \frac{1}{2\pi f C}$$

A capacitor consists of two conductors separated by an insulator, also known as a dielectric.

At low frequencies a capacitor is open circuit, as no current flows in the dielectric. A DC voltage applied across a capacitor causes positive charge to accumulate on one side and negative charge to accumulate on the other side; the electric field due to the accumulated charge is the source of the opposition to the current. When the potential associated with the charge exactly balances the applied voltage, the current goes to zero.

Driven by an AC supply, a capacitor will only accumulate a limited amount of charge before the potential difference changes polarity and the charge dissipates. The higher the frequency, the less charge will accumulate and the smaller the opposition to the current.

Inductive reactance

Inductive reactance is an opposition to the change of current through an element. Inductive reactance X_L is proportional to the sinusoidal signal frequency f and the inductance L.

$$X_L = \omega L = 2\pi f L$$

The average current flowing in an inductance L in series with a sinusoidal AC voltage source of RMS amplitude A and frequency f is equal to :

$$I_L = \frac{A}{\omega L} = \frac{A}{2\pi f L}$$

The average current flowing in an inductance L in series with a square wave AC voltage source of RMS amplitude A and frequency f is equal to :

$$I_L = \frac{A\pi^2}{8\omega L} = \frac{A\pi}{16 f L}$$

Electrical Power Systems Technology

making it appear as if the inductive reactance to a square wave was

$$X_L = \frac{16}{\pi} fL$$

Any conductor of finite dimensions has inductance; the inductance is made larger by the multiple turns in an electromagnetic coil. Faraday's law of electromagnetic induction gives the counter-emf ε (voltage opposing current) due to a rate-of-change of magnetic flux density B through a current loop.

$$\varepsilon = \frac{d\Phi_B}{dt}$$

For an inductor consisting of a coil with N loops this gives.

$$\varepsilon = -N\frac{d\Phi_B}{dt}$$

The counter-emf is the source of the opposition to current flow. A constant direct current has a zero rate-of-change, and sees an inductor as a short-circuit (it is typically made from a material with a low resistivity). An alternating current has a time-averaged rate-of-change that is proportional to frequency, this causes the increase in inductive reactance with frequency.

Phase Relationship

The phase of the voltage across a purely reactive device (a capacitor with an infinite resistance or an inductor with a resistance of zero) *lags* the current by $\pi/2$ radians for a capacitive reactance and *leads* the current by $\pi/2$ radians for an inductive reactance. Note that without knowledge of both the resistance and reactance the relationship between voltage and current cannot be determined.

The origin of the different signs for capacitive and inductive reactance is the phase factor in the impedance.

$$\tilde{Z}_C = \frac{1}{\omega C}e^{j\left(-\frac{\pi}{2}\right)} = -j\left(\frac{1}{\omega C}\right) = --jX_C$$

$$\tilde{Z}_L = \omega L e^{j\frac{\pi}{2}} = j\omega L = j Z_L$$

For a reactive component the sinusoidal voltage across the component is in quadrature (a $\pi/2$ phase difference) with the sinusoidal current through the component. The component alternately absorbs energy from the circuit and then returns energy to the circuit, thus a pure reactance does not dissipate power.

MAGNETIC FLUX

In physics, specifically electromagnetism, the magnetic flux (often denoted Φ or Φ_B) through a surface is the surface integral of the normal component of the magnetic field B passing through that surface. The SI unit of magnetic flux is the weber (Wb) (in derived units : volt-seconds), and the CGS unit is the maxwell.

Magnetic flux is usually measured with a fluxmeter, which contains measuring coils and electronics, that evaluates the change of voltage in the measuring coils to calculate the magnetic flux.

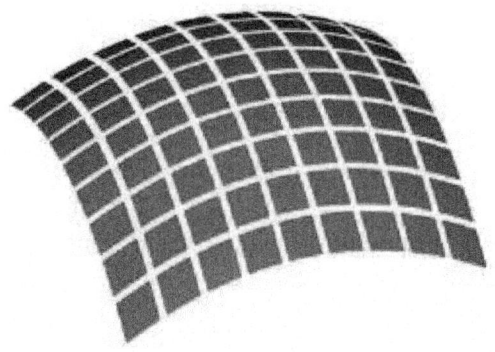

Description

The magnetic flux through a surface when the magnetic field is variable relies on splitting the surface into small surface elements, over which the magnetic field can be considered to be locally constant. The total flux is then a formal summation of these surface elements.

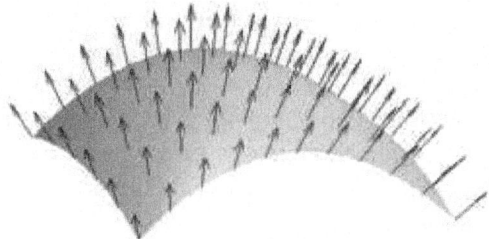

Each point on a surface is associated with a direction, called the surface normal; the magnetic flux through a point is then the component of the magnetic field along this direction.

The magnetic interaction is described in terms of a vector field, where each point in space (and time) is associated with a vector that determines what force a moving charge would experience at that point. Since a vector field is quite difficult to visualize at first, in elementary physics one may instead visualize this field with field lines. The magnetic flux through some surface, in this simplified picture, is proportional to the number of field lines passing through that surface (in some contexts, the flux may be defined to be precisely the number of field lines passing through that surface; although technically misleading, this distinction is not important). Note that the magnetic flux is the *net* number of field lines passing through that surface; that is, the number passing through in one direction minus the number passing through in the other direction. In more advanced physics, the field line analogy is dropped and the magnetic flux is properly defined as the surface integral of the normal component of the magnetic field passing through a surface. If the magnetic field is constant, the magnetic flux passing through a surface of vector area **S** is

$$\Phi_B = \mathbf{B} \cdot \mathbf{S} = BS \cos \theta,$$

where **B** is the magnitude of the magnetic field (the magnetic flux density) having the unit of Wb/m² (tesla), **S** is the area of the surface, and θ is the angle between the magnetic field lines and the normal (perpendicular) to **S**. For a varying magnetic field, we first consider the magnetic flux through an infinitesimal area element d**S**, where we may consider the field to be constant :

$$d\Phi_B = \mathbf{B} \cdot d\mathbf{S}.$$

A generic surface, S, can then be broken into infinitesimal elements and the total magnetic flux through the surface is then the surface integral

$$\Phi_B = \iint_S B \cdot dS.$$

From the definition of the magnetic vector potential **A** and the fundamental theorem of the curl the magnetic flux may also be defined as :

$$\Phi_B = \oint_{\partial S} A \cdot d\ell$$

where the line integral is taken over the boundary of the surface S, which is denoted ∂S.

Magnetic Flux Through a Closed Surface

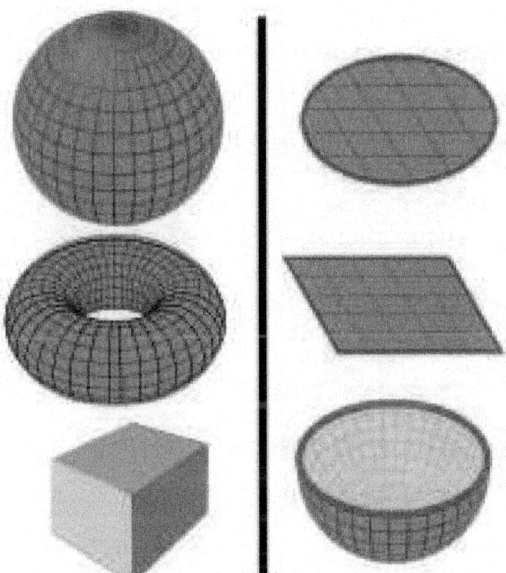

Fig. : Some examples of closed surfaces (left) and open surfaces (right). Left : Surface of a sphere, surface of a torus, surface of a cube. Right : Disk surface, square surface, surface of a hemisphere. (The surface is blue, the boundary is red.)

Gauss's law for magnetism, which is one of the four Maxwell's equations, states that the total magnetic flux through a closed surface is equal to zero. (A "closed surface" is a surface that completely encloses a volume(s) with no holes.) This law is a consequence of the empirical observation that magnetic monopoles have never been found.

In other words, Gauss's law for magnetism is the statement:

$$\Phi_B = \oiint_S B \cdot dS = 0$$

for any closed surface S.

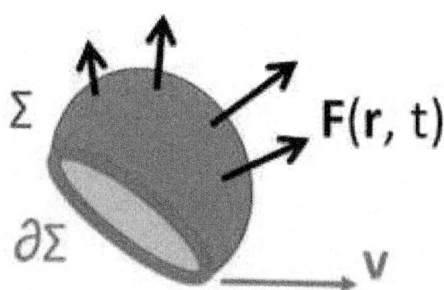

Fig.: For an open surface Σ, the electromotive force along the surface boundary, $\partial\Sigma$, is a combination of the boundary's motion, with velocity **v**, through a magnetic field **B** (illustrated by the generic F field in the diagram) and the induced electric field caused by the changing magnetic field.

Magnetic Flux Through an Open Surface

While the magnetic flux through a closed surface is always zero, the magnetic flux through an open surface need not be zero and is an important quantity in electromagnetism. For example, a change in the magnetic flux passing through a loop of conductive wire will cause an electromotive force, and therefore an electric current, in the loop. The relationship is given by Faraday's law:

$$\varepsilon = \oint_{\partial\Sigma} (E + v \times B) \cdot d\ell = -\frac{d\Phi_B}{dt},$$

where:

ε is the electromotive force (EMF),

Φ_B is the magnetic flux through the open surface Σ,

$\partial\Sigma$ is the boundary of the open surface Σ; note that the surface, in general, may be in motion and deforming, and so is generally a function of time. The electromotive force is induced along this boundary.

$d\ell$ is an infinitesimal vector element of the contour $\partial\Sigma$,

v is the velocity of the boundary $\partial\Sigma$,

E is the electric field,

B is the magnetic field.

The two equations for the EMF are, firstly, the work per unit charge done against the Lorentz force in moving a test charge around the (possibly moving) surface boundary $\partial\Sigma$ and, secondly, as the change of magnetic flux through the open surface Σ. This equation is the principle behind an electrical generator.

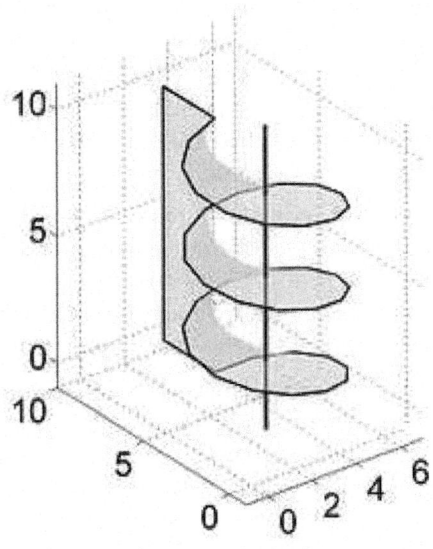

Fig. : Area defined by an electric coil with three turns.

Comparison with Electric Flux

By way of contrast, Gauss's law for electric fields, another of Maxwell's equations, is

$$\Phi_E = \oiint_S E \cdot dS = \frac{Q}{\epsilon_0}$$

where :

E is the electric field,

S is any closed surface,

Q is the total electric charge inside the surface S,

ϵ_0 is the electric constant (a universal constant, also called the "permittivity of free space").

Note that the flux of **E** through a closed surface is *not* always zero; this indicates the presence of "electric monopoles", that is, free positive or negative charges.

ELECTRIC CHARGE

Electric charge is the physical property of matter that causes it to experience a force when close to other electrically charged matter. There are two types of electric charges – positive and negative. Positively charged substances are repelled from other positively charged substances, but attracted to negatively charged substances; negatively charged substances are repelled from negative and attracted to positive. An object will be negatively charged if it has an excess of electrons, and will otherwise be positively charged or uncharged. The SI derived unit of electric charge is the coulomb (C), although in electrical engineering it is also common to use the ampere-hour (Ah), and in chemistry it is common to use the elementary charge (e) as a unit. The symbol Q is often used to denote a charge. The study of how charged substances interact is classical electrodynamics, which is accurate insofar as quantum effects can be ignored.

The *electric charge* is a fundamental conserved property of some subatomic particles, which determines their electromagnetic interaction. Electrically charged matter is influenced by, and produces, electromagnetic fields. The interaction between a moving charge and an electromagnetic field is the source of the electromagnetic force, which is one of the four fundamental forces.

Twentieth-century experiments demonstrated that electric charge is *quantized*; that is, it comes in integer multiples of individual small units called the elementary charge, e, approximately equal to 1.602×10^{-19} coulombs (except for particles called quarks, which have charges that are integer multiples of $e/3$). The proton has a charge of e, and the electron has a charge of $-e$. The study of charged particles, and how their interactions are mediated by photons, is quantum electrodynamics.

Overview

Charge is the fundamental property of forms of matter that exhibit electrostatic attraction or repulsion in the presence of other matter. Electric charge is a characteristic property of many subatomic particles. The charges of free-standing particles are integer multiples of the elementary charge e; we say that electric charge is *quantized*. Michael Faraday, in his electrolysis experiments, was the first to note the discrete nature of electric charge. Robert Millikan's oil-drop experiment demonstrated this fact directly, and measured the elementary charge.

By convention, the charge of an electron is -1, while that of a proton is $+1$. Charged particles whose charges have the same sign repel one another, and particles whose charges have different signs attract. Coulomb's law quantifies the electrostatic force between two particles by asserting that the force is proportional to the product of their charges, and inversely proportional to the square of the distance between them.

The charge of an anti-particle equals that of the corresponding particle, but with opposite sign. Quarks have fractional charges of either $-1/3$ or $+2/3$, but free-standing quarks have never been observed (the theoretical reason for this fact is asymptotic freedom).

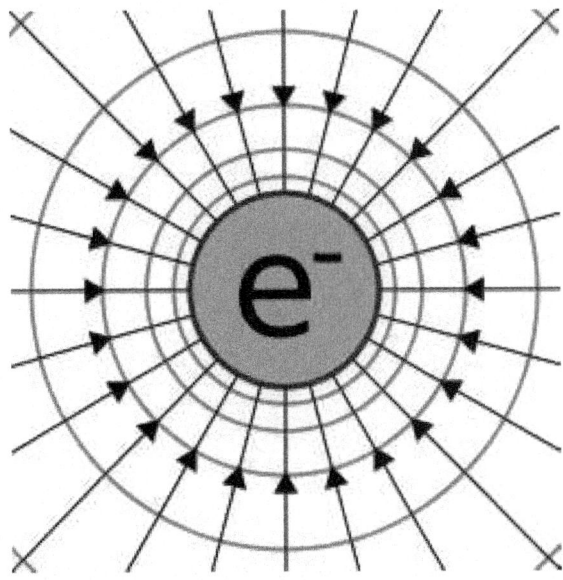

Fig.: Diagram showing field lines and equipotentials around an electron, a negatively charged particle. In an electrically neutral atom, the number of electrons is equal to the number of protons (which are positively charged), resulting in a net zero overall charge.

The electric charge of a macroscopic object is the sum of the electric charges of the particles that make it up. This charge is often small, because matter is made of atoms, and atoms typically have equal numbers of protons and electrons, in which case their charges cancel out, yielding a net charge of zero, thus making the atom neutral.

An *ion* is an atom (or group of atoms) that has lost one or more electrons, giving it a net positive charge (cation), or that has gained one or more electrons, giving it a net negative charge (anion). *Monatomic ions* are formed from single atoms, while *polyatomic ions* are formed from two or more atoms that have been bonded together, in each case yielding an ion with a positive or negative net charge.

During the formation of macroscopic objects, usually the constituent atoms and ions will combine in such a manner that they form structures composed of neutral *ionic compounds* electrically bound to neutral atoms. Thus macroscopic objects tend toward being neutral overall, but macroscopic objects are rarely perfectly net neutral.

There are times when macroscopic objects contain ions distributed throughout the material, rigidly bound in place, giving an overall net positive or negative charge to the object. Also, macroscopic objects made of conductive elements, can more or less easily (depending on the element) take on or give off electrons, and then maintain a net negative or positive charge indefinitely. When the net electric charge of an object is non-zero and motionless, the phenomenon is known as static electricity. This can easily be produced by rubbing two dissimilar materials together, such as rubbing amber with fur or glass with silk. In this way non-conductive materials can be charged to a significant degree, either positively or negatively. Charge taken from one material is moved to the other material, leaving an opposite charge of the same magnitude behind. The law of *conservation of charge* always applies, giving the object from which a negative charge has been taken a positive charge of the same magnitude, and *vice-versa*.

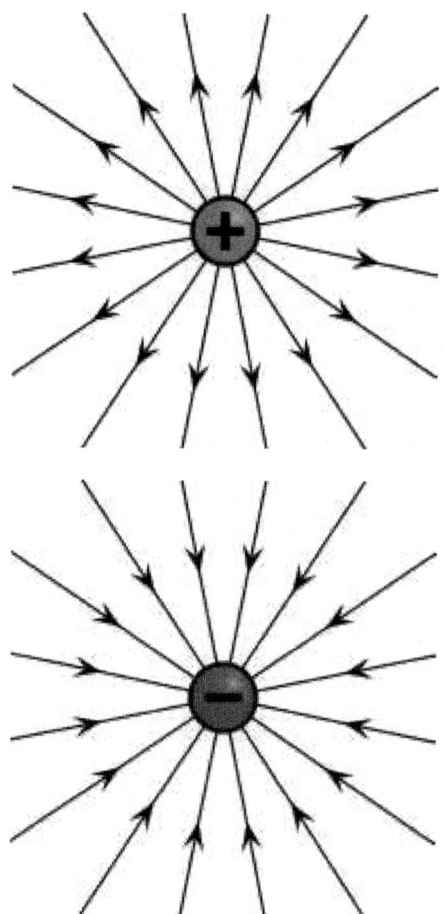

Fig. : Electric field induced by a positive electric charge (left) and a field induced by a negative electric charge (right).

Even when an object's net charge is zero, charge can be distributed non-uniformly in the object (*e.g.*, due to an external electromagnetic field, or bound polar molecules). In such cases the object is said to be polarized. The charge due to polarization is known as bound charge, while charge on an object produced by electrons gained or lost from outside the object is called *free charge*. The motion of electrons in conductive metals in a specific direction is known as electric current.

Units

The SI unit of quantity of electric charge is the coulomb, which is equivalent to about 6.242×10^{18} e (e is the charge of a proton). Hence, the charge of an electron is approximately -1.602×10^{-19} C. The coulomb is defined as the quantity of charge that has passed through the cross-section of an electrical conductor carrying one ampere within one second. The symbol Q is often used to denote a quantity of electricity or charge. The quantity of electric charge can be directly measured with an electrometer, or indirectly measured with a ballistic galvanometer.

After finding the quantized character of charge, in 1891 George Stoney proposed the unit 'electron' for this fundamental unit of electrical charge. This was before the discovery of the particle by J.J. Thomson in 1897. The unit is today treated as nameless, referred to as "elementary charge", "fundamental unit of charge", or simply as "e". A measure of charge should be a multiple of the elementary charge e, even if at large scales, charge seems to behave as a real quantity. In some contexts it is meaningful to speak of fractions of a charge; for example, in the charging of a capacitor, or in the fractional quantum Hall effect.

History

As reported by the ancient Greek philosopher Thales of Miletus around 600 B.C., charge (or *electricity*) could be accumulated by rubbing fur on various substances, such as amber. The Greeks noted that the charged amber buttons could attract light objects such as hair. They also noted that if they rubbed the amber for long enough, they could even get an electric spark to jump. This property derives from the triboelectric effect.

In 1600, the English scientist William Gilbert returned to the subject in *De Magnete*, and coined the New Latin word *electricus* from ηλεκτρον (*elektron*), the Greek word for "amber", which soon gave rise to the English words "electric" and "electricity." He was followed in 1660 by Otto von Guericke, who invented what was probably the first electrostatic generator. Other European pioneers were Robert Boyle, who in 1675 stated that electric attraction and repulsion can act across a vacuum; Stephen Gray, who in 1729 classified materials as conductors and insulators; and C. F. du Fay, who proposed in 1733 that electricity comes in two varieties that cancel each other, and expressed this in terms of a two-fluid theory. When glass was rubbed with silk, du Fay said that the glass was charged with *vitreous electricity*, and, when amber was rubbed with fur, the amber was said to be charged with *resinous electricity*. In 1839, Michael Faraday showed that the apparent division between static electricity, current electricity, and bioelectricity was incorrect, and all were a consequence of the behaviour of a single kind of electricity appearing in opposite polarities. It is arbitrary which polarity is called positive and which is called negative. Positive charge can be defined as the charge left on a glass rod after being rubbed with silk.

One of the foremost experts on electricity in the 18th century was Benjamin Franklin, who argued in favour of a one-fluid theory of electricity. Franklin imagined electricity as being a type of invisible fluid present in all matter; for example, he believed that it was the glass in a Leyden jar that held the accumulated charge. He posited that rubbing insulating surfaces together caused this fluid to change location, and that a flow of this fluid constitutes an electric current. He also posited that when matter contained too little of the fluid it was "negatively" charged, and when it had an excess it was "positively" charged. For a reason that was not recorded, he identified the term "positive" with vitreous electricity and "negative" with resinous electricity. William Watson arrived at the same explanation at about the same time.

Static Electricity and Electric Current

Static electricity and electric current are two separate phenomena, both involving electric charge, and may occur simultaneously in the same object. Static electricity is a reference to the electric charge of an object and the related electrostatic discharge when two objects are brought together that are not at equilibrium. An electrostatic discharge creates a change in the charge of each of the two objects. In contrast, electric current is the flow of electric charge through an object, which produces no net loss or gain of electric charge.

Electrification by Friction

Let a piece of glass and a piece of resin, neither of which exhibiting any electrical properties, be rubbed together and left with the rubbed surfaces in contact. They will still exhibit no electrical properties. Let them be separated. They will now attract each other.

If a second piece of glass be rubbed with a second piece of resin, and if the piece be then separated and suspended in the neighbourhood of the former pieces of glass and resin, it may be observed :

1. that the two pieces of glass repel each other.
2. that each piece of glass attracts each piece of resin.
3. that the two pieces of resin repel each other.

These phenomena of attraction and repulsion are called electrical phenomena, and the bodies that exhibit them are said to be 'electrified', or to be 'charged with electricity'.

Bodies may be electrified in many other ways, as well as by friction.

The electrical properties of the two pieces of glass are similar to each other but opposite to those of the two pieces of resin : The glass attracts what the resin repels and repels what the resin attracts.

If a body electrified in any manner whatsoever behaves as the glass does, that is, if it repels the glass and attracts the resin, the body is said to be 'vitreously' electrified, and if it attracts the glass and repels the resin it is said to be 'resinously' electrified. All electrified bodies are found to be either vitreously or resinously electrified.

It is the established convention of the scientific community to define the vitreous electrification as positive, and the resinous electrification as negative. The exactly opposite properties of the two kinds of electrification justify our indicating them by opposite signs, but the application of the positive sign to one rather than to the other kind must be considered as a matter of arbitrary convention, just as it is a matter of convention in mathematical diagram to reckon positive distances towards the right hand.

No force, either of attraction or of repulsion, can be observed between an electrified body and a body not electrified.

Actually, all bodies are electrified, but may appear not to be so by the relative similar charge of neighbouring objects in the environment. An object further electrified + or – creates an equivalent or opposite charge by default in neighbouring objects, until those charges can equalize. The effects of attraction can be observed in high-voltage experiments, while lower voltage effects are merely weaker and therefore less obvious. The attraction and repulsion forces are codified by Coulomb's Law (attraction falls off at the square of the distance, which has a corollary for acceleration in a gravitational field, suggesting that gravitation may be merely electrostatic phenomenon between relatively weak charges in terms of scale).

It is now known that the Franklin/Watson model was fundamentally correct. There is only one kind of electrical charge, and only one variable is required to keep track of the amount of charge. On the other hand, just knowing the charge is not a complete description of the situation. Matter is composed of several kinds of electrically charged particles, and these particles have many properties, not just charge.

The most common charge carriers are the positively charged proton and the negatively charged electron. The movement of any of these charged particles constitutes an electric current. In many situations, it suffices to speak of the *conventional current* without regard to whether it is carried by positive charges moving in the direction of the conventional current and/or by negative charges moving in the opposite direction. This macroscopic view-point is an approximation that simplifies electromagnetic concepts and calculations.

At the opposite extreme, if one looks at the microscopic situation, one sees there are many ways of carrying an electric current, including : a flow of electrons; a flow of electron "holes" that act like positive particles; and both negative and positive particles (ions or other charged particles) flowing in opposite directions in an electrolytic solution or a plasma.

Beware that, in the common and important case of metallic wires, the direction of the conventional current is opposite to the drift velocity of the actual charge carriers, *i.e.*, the electrons. This is a source of confusion for beginners.

Properties

Aside from the properties described in articles about electromagnetism, charge is a relativistic invariant. This means that any particle that has charge Q, no matter how fast it goes, always has charge Q. This property has been experimentally verified by showing that the charge of *one* helium nucleus (two protons and two neutrons bound together in a nucleus and moving around at high speeds) is the same as *two* deuterium nuclei (one proton and one neutron bound together, but moving much more slowly than they would if they were in a helium nucleus).

Charge Conservation

In physics, **charge conservation** is the principle that electric charge can neither be created nor destroyed. The net quantity of electric charge, the amount of positive charge minus the amount of negative charge in the universe, is always *conserved*. The first written statement of the principle was by American scientist and statesman Benjamin Franklin in 1747.

> it is now discovered and demonstrated, both here and in Europe, that the Electrical Fire is a real Element, or Species of Matter, not *created* by the Friction, but *collected* only.
>
> — Benjamin Franklin, *Letter to Cadwallader Colden, 5 June 1747*

Charge conservation is a physical law that states that the change in the amount of electric charge in any volume of space is exactly equal to the amount of charge flowing into the volume minus the amount of charge flowing out of the volume. In essence, charge conservation is an accounting relationship between the amount of charge in a region and the flow of charge into and out of that region.

Mathematically, we can state the law as a continuity equation :

$$Q(t_2) = Q(t^1) + Q_{IN} - Q_{OUT}.$$

$Q(t)$ is the quantity of electric charge in a specific volume at time t, Q_{IN} is the amount of charge flowing into the volume between time t_1 and t_2, and Q_{OUT} is the amount of charge flowing out of the volume during the same time period.

This does not mean that individual positive and negative charges cannot be created or destroyed. Electric charge is carried by subatomic particles such as electrons and protons, which can be created and destroyed. In particle physics, charge conservation means that in elementary particle reactions that create charged particles, equal numbers of positive and negative particles are always created, keeping the net amount of charge unchanged. Similarly, when particles are destroyed, equal numbers of positive and negative charges are destroyed.

Although conservation of charge requires that the total quantity of charge in the universe is constant, it leaves open the question of what that quantity is. Most evidence indicates that the net charge in the universe is zero; that is, there are equal quantities of positive and negative charge.

Formal Statement of the Law

Vector calculus can be used to express the law in terms of charge density ρ (in coulombs per cubic meter) and electric current density **J** (in amperes per square meter) :

$$\frac{\partial \rho}{\partial t} + \Delta \cdot \mathbf{J} = 0.$$

The term on the left is the rate of change of the charge density ρ at a point. The term on the right is the divergence of the current density **J**. *The equation equates these two factors, which says that the only way for the charge density at a point to change is for a current of charge to flow into or out of the point. This statement is equivalent to a conservation of four-current.*

Mathematical Derivation

The net current **into** a volume is

$$I = -\iint_S \mathbf{J} \cdot d\mathbf{S}$$

where $S = \partial V$ is the boundary of V oriented by outward-pointing normals, and $d\mathbf{S}$ is shorthand for $\mathbf{N}dS$, the outward pointing normal of the boundary ∂V. Here *J is the current density (charge per unit area per unit time) at the surface of the volume. The vector points in the direction of the current.*

From the Divergence theorem this can be written

$$I = -\iiint_V (\nabla \cdot \mathbf{J}) \, dvV.$$

Charge conservation requires that the net current into a volume must necessarily equal the net change in charge within the volume.

$$\frac{dq}{dt} = \iiint_V (\nabla \cdot \mathbf{J}) \, dvV.$$

Charge is related to charge density by the relation

$$q = \iiint_V \rho \, dV.$$

This yields

$$0 = \iiint_V \left(\frac{\partial \rho}{\partial t} + \Delta \cdot \mathbf{J} \right) dV.$$

Since this is true for every volume, we have in general

$$\frac{\partial \rho}{\partial t} + \Delta \cdot \mathbf{J} = 0.$$

Connection to Gauge Invariance

Charge conservation can also be understood as a consequence of symmetry through Noether's theorem, a central result in theoretical physics that asserts that each conservation law is associated with a symmetry of the underlying physics. The symmetry that is associated with charge conservation is the global gauge invariance of the electromagnetic field. This is related to the fact that the electric and magnetic fields are not changed by different choices of the value representing the zero point of electrostatic potential ϕ. However the full symmetry is more complicated, and also involves the vector potential **A**. The full statement of gauge invariance is that the physics of an electromagnetic field are unchanged when the scalar and vector potential are shifted by the gradient of an arbitrary scalar field X:

$$\phi' = \phi - \frac{\partial_X}{\partial t} \qquad \mathbf{A}' = \mathbf{A} + \Delta_X.$$

In quantum mechanics the scalar field is equivalent to a phase shift in the wavefunction of the charged particle :

$$\psi' = e^{ix}\psi$$

so gauge invariance is equivalent to the well known fact that changes in the phase of a wavefunction are unobservable, and only changes in the magnitude of the wavefunction result in changes to the probability function $|\psi|^2$. This is the ultimate theoretical origin of charge conservation.

Gauge invariance is a very important, well established property of the electromagnetic field and has many testable consequences. The theoretical justification for charge conservation is greatly strengthened by being linked to this symmetry. For example, local gauge invariance also requires that the photon be massless, so the good experimental evidence that the photon has zero mass is also strong evidence that charge is conserved.

Even if gauge symmetry is exact, however, there might be apparent electric charge non-conservation if charge could leak from our normal 3-dimensional space into hidden extra dimensions.

Experimental Evidence

The best experimental tests of electric charge conservation are searches for particle decays that would be allowed if electric charge is not always conserved. No such decays have ever been seen. The best experimental test comes from searches for the energetic photon from an electron decaying into a neutrino and a single photon :

e → ν + γ   mean lifetime is greater than 4.6×10^{26} years (90% Confidence Level),

but there are theoretical arguments that such single-photon decays will never occur even if charge is not conserved. Charge disappearance tests are sensitive to decays without energetic photons, other unusual charge violating processes such as an electron spontaneously changing into a positron, and to electric charge moving into other dimensions. The best experimental bounds on charge disappearance are :

e → *anything*   mean lifetime is greater than 6.4×1024 years (68% CL)
n → p + ν + ν   charge non-conserving decays are less than 8×10^{-27}
(68% CL) of all neutron decays

ELECTRIC FIELD

An electric field is generated by electrically charged particles and time-varying magnetic fields. The electric field describes the electric force experienced by a motionless positively charged test particle at any point in space relative to the source(s) of the field. The concept of an electric field was introduced by Michael Faraday.

Qualitative Description

The electric field is a vector field. The field vector at a given point is defined as the force vector per unit charge that would be exerted on a stationary test charge at that point. An electric field is generated by electric charge, as well as by a time-varying magnetic field. Electric fields contain electrical energy with energy density proportional to the square of the field amplitude. The electric field is to charge as gravitational acceleration is to mass. The SI units of the field are newtons per coulomb (N·C^{-1}) or, equivalently, volts per metre (V·m^{-1}), which in terms of SI base units are kg·m·s^{-3}·A^{-1}.

An electric field that changes with time, such as due to the motion of charged particles producing the field, influences the local magnetic field. That is : the electric and magnetic fields are not separate phenomena; what one observer perceives as an electric field, another observer in a different frame of reference perceives as a mixture of electric and magnetic fields. For this reason, one speaks of "electromagnetism" or "electromagnetic fields". In quantum electrodynamics, disturbances in the electromagnetic fields are called photons.

Definition

Electric Field

Consider a point charge q with position (x,y,z). Now suppose the charge is subject to a force $\mathbf{F}_{on\,q}$ due to other charges. Since this force varies with the position of the charge and by Coulomb's Law it is defined at all points in space, $\mathbf{F}_{on\,q}$ is a continuous function of the charge's position. This suggests that there is some property of the space that causes the force which is exerted on the charge q. This property is called the electric field and it is defined by

$$\mathbf{E}(x,y,x) \equiv \frac{\mathbf{F}_{on\,q}(x,y,z)}{q}$$

Notice that the magnitude of the electric field has dimensions of Force/Charge. Mathematically, the E field can be thought of as a function that associates a vector with every point in space. Each such vector's magnitude is proportional to how much force a charge at that point would "feel" if it were present and this force would have the same direction as the electric field vector at that point. It is also important to note that the electric field defined above is caused by a configuration of *other* electric charges. This means that the charge q in the equation above is not the charge that is *creating* the electric field, but rather, being acted upon *by* it. This definition does not give a means of computing the electric field caused by a group of charges.

Superposition

Array of Discrete Point Charges

Electric fields satisfy the superposition principle. If more than one charge is present, the total electric field at any point is equal to the vector sum of the separate electric fields that each point charge would create in the absence of the others. That is,

$$E = \sum_i E_i = E_1 + E_2 + E_3 + \dots$$

where E_i is the electric field created by the *i*-th point charge.

At any point of interest, the total E-field due to N point charges is simply the superposition of the E-fields due to each point charge, given by

$$E = \sum_{i=1}^{N} E_i = \frac{1}{4\pi\varepsilon_0} \sum_{i=1}^{N} \frac{Q_i}{r_i^2} \hat{r}i.$$

where Q_i is the electric charge of the *i*-th point charge, $\hat{r}i$ the corresponding unit vector of r_i, which is the position of charge Q_i with respect to the point of interest.

Continuum of Charges

It holds for an infinite number of infinitesimally small elements of charges – *i.e.* a continuous distribution of charge. By taking the limit as N approaches infinity in the previous equation, the electric field for a continuum of charges can be given by the integral:

$$E = \int_V dE = \frac{1}{4\pi\varepsilon_0} \int_V \frac{\rho}{r^2} \hat{r} dV = \frac{1}{4\pi\varepsilon_0} \int_V \frac{\rho}{r^3} r\, dV$$

where ρ is the charge density (the amount of charge per unit volume), ε_0 the permittivity of free space, and dV is the differential volume element. This integral is a volume integral over the region of the charge distribution.

The equations above express the electric field of point charges as derived from Coulomb's law, which is a special case of Gauss's Law. While Coulomb's law is only true for stationary point charges, Gauss's law is true for all charges either in static form or in motion. Gauss's Law establishes a more fundamental relationship between the distribution of electric charge in space and the resulting electric field. It is one of Maxwell's equations governing electromagnetism.

Gauss's law allows the E-field to be calculated in terms of a continuous distribution of charge density. In differential form, it can be stated as

$$\Delta \cdot E = \frac{\rho}{\varepsilon_0}$$

where ∇· is the divergence operator, ρ is the total charge density, including free and bound charge, in other words all the charge present in the system (per unit volume).

Electrostatic Fields

Electrostatic fields are E-fields which do not change with time, which happens when the charges are stationary.

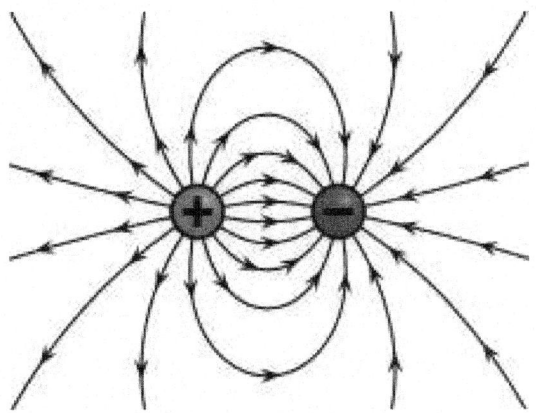

Fig.: Illustration of the electric field surrounding a positive (red) and a negative (blue) charge.

The electric field **E** at a point **r**, that is, **E(r)**, is equal to the negative gradient of the electric potential Φ(**r**), a scalar field at the same point :

$$E = -\nabla\Phi$$

where ∇ is the gradient operator. This is equivalent to the force definition above, since electric potential Φ is defined by the electric potential energy U per unit (test) positive charge :

$$\Phi = \frac{U}{q}$$

and force is the negative of potential energy gradient :

$$F = -\nabla U$$

If several spatially distributed charges generate such an electric potential, *e.g.* in a solid, an electric field gradient may also be defined.

Uniform Fields

A uniform field is one in which the electric field is constant at every point. It can be approximated by placing two conducting plates parallel to each other and maintaining a voltage (potential difference) between them; it is only an approximation because of edge effects. Ignoring such effects, the equation for the magnitude of the electric field E is :

$$E = -\frac{\Delta\phi}{d}$$

where $\Delta\phi$ is the potential difference between the plates and d is the distance separating the plates. The negative sign arises as positive charges repel, so a positive charge will experience a force away from the positively charged plate, in the opposite direction to that in which the voltage increases. In micro- and nanoapplications, for instance in relation to semi-conductors, a typical magnitude of an electric field is in the order of 1 volt/μm achieved by applying a voltage of the order of 1 volt between conductors spaced 1 μm apart.

Parallels between Electrostatic and Gravitational Fields

Coulomb's law, which describes the interaction of electric charges :

$$\mathbf{F} = q\left(\frac{Q}{4\pi\varepsilon_0}\frac{\hat{\mathbf{r}}}{|\mathbf{r}|^2}\right) = q\mathbf{E}$$

is similar to Newton's law of universal gravitation :

$$\mathbf{F} = m\left(-GM\frac{\hat{\mathbf{r}}}{|\mathbf{r}|^2}\right) = m\mathbf{g} \ .$$

This suggests similarities between the electric field **E** and the gravitational field **g**, so sometimes mass is called "gravitational charge".

Similarities between electrostatic and gravitational forces :
1. Both act in a vacuum.
2. Both are central and conservative.
3. Both obey an inverse-square law (both are inversely proportional to square of r).

Differences between electrostatic and gravitational forces :
1. Electrostatic forces are much greater than gravitational forces for natural values of charge and mass. For instance, the ratio of the electrostatic force to the gravitational force between two electrons is about 10^{42}.
2. Gravitational forces are attractive for like charges, whereas electrostatic forces are repulsive for like charges.
3. There are not negative gravitational charges (no negative mass) while there are both positive and negative electric charges. This difference, combined with the previous two, implies that gravitational forces are always attractive, while electrostatic forces may be either attractive or repulsive.

Electrodynamic Fields

Electrodynamic fields are E-fields which do change with time, when charges are in motion.

ELECTRICAL POWER SYSTEMS TECHNOLOGY

An electric field can be produced not only by a static charge, but also by a changing magnetic field (in which case it is a non-conservative field). The electric field is then given by :

$$E = -\nabla\varphi - \frac{\partial \mathbf{A}}{\partial t}$$

in which **B** satisfies

$$\mathbf{B} = \nabla \times \mathbf{A}$$

and $\nabla\times$ denotes the curl. The vector field **B** is the magnetic flux density and the vector **A** is the magnetic vector potential. Taking the curl of the electric field equation we obtain,

$$\nabla \times \mathbf{E} = -\frac{\partial \mathbf{B}}{\partial t}$$

which is Faraday's law of induction, another one of Maxwell's equations.

Energy in the Electric Field

The electrostatic field stores energy. The energy density u (energy per unit volume) is given by

$$u = \frac{1}{2}\varepsilon |\mathbf{E}|^2,$$

where ε is the permittivity of the medium in which the field exists, and **E** is the electric field vector (in newtons per coulomb).

The total energy U stored in the electric field in a given volume V is therefore

$$U = \frac{1}{2}\varepsilon \int_V |\mathbf{E}|^2 dV,$$

Further Extensions

Definitive Equation of Vector Fields

In the presence of matter, it is helpful in electromagnetism to extend the notion of the electric field into three vector fields, rather than just one :

$$\mathbf{D} = \varepsilon_0 \mathbf{E} + \mathbf{P}$$

where **P** is the electric polarization – the volume density of electric dipole moments, and **D** is the electric displacement field. Since E and P are defined separately, this equation can be used to define **D**. The physical interpretation of **D** is not as clear as E (effectively the field applied to the material) or P (induced field due to the dipoles in the material), but still serves as a convenient mathematical simplification, since Maxwell's equations can be simplified in terms of free charges and currents.

Constitutive Relation

The E and D fields are related by the permittivity of the material, ε.

For linear, homogeneous, isotropic materials **E** and **D** are proportional and constant throughout the region, there is no position dependence : For inhomogeneous materials, there is a position dependence throughout the material :

$$\mathbf{D}(\mathbf{r}) = \varepsilon \mathbf{E}(\mathbf{r})$$

For anisotropic materials the **E** and **D** fields are not parallel, and so **E** and **D** are related by the permittivity tensor (a 2nd order tensor field), in component form :

$$D_i = \varepsilon_{ij} E_j$$

For non-linear media, **E** and **D** are not proportional. Materials can have varying extents of linearity, homogeneity and isotropy.

ELECTRICITY METER

An **electricity meter** or **energy meter** is a device that measures the amount of electric energy consumed by a residence, business, or an electrically powered device.

Electricity meters are typically calibrated in billing units, the most common one being the kilowatt hour [*kWh*]. Periodic readings of electricity meters establishes billing cycles and energy used during a cycle.

In settings when energy savings during certain periods are desired, meters may measure demand, the maximum use of power in some interval. "Time of day" metering allows electric rates to be changed during a day, to record usage during peak high-cost periods and off-peak, lower-cost, periods. Also, in some areas meters have relays for demand response load shedding during peak load periods.

History

Direct Current (DC)

As commercial use of electric energy spread in the 1880s, it became increasingly important that an electric energy meter, similar to the then existing gas meters, was required to properly bill customers for the cost of energy, instead of billing for a fixed number of lamps per month. Many experimental types of meter were developed. Edison at first worked on a DC electromechanical meter with a direct reading register, but instead developed an electrochemical metering system, which used an electrolytic cell to totalise current consumption. At periodic intervals the plates were removed, weighed, and the customer billed. The electrochemical meter was labour-intensive to read and not well received by customers.

An early type of electrochemical meter used in the United Kingdom was the 'Reason' meter. This consisted of a vertically mounted glass structure with a mercury reservoir at the top of the meter. As current was drawn from the supply, electrochemical action transferred the mercury to the bottom of the column. Like all other DC meters, it recorded ampere-hours. Once the mercury pool was exhausted, the meter became an open circuit. It was therefore necessary for the consumer to pay for a further supply of electricity, whereupon, the supplier's agent would unlock the meter from its mounting and invert it restoring the mercury to the reservoir and the supply.

In 1885 Ferranti offered a mercury motor meter with a register similar to gas meters; this had the advantage that the consumer could easily read the meter and verify consumption. The first accurate, recording electricity consumption meter was a DC meter by Dr. Hermann Aron, who patented it in 1883. Hugo Hirst of the British General Electric Company introduced it commercially into Great Britain from 1888. Unlike their AC counterparts, DC meters did not measure energy. Instead they measured charge in ampere-hours. Since the voltage of the supply should remain substantially constant, the reading of the meter was proportional to actual energy consumed. For example, if a meter recorded that 100 ampere-hours had been consumed on a 200 volt supply, then 20 kilowatt-hours of energy had been supplied. Aron's meter recorded the total charge used over time, and showed it on a series of clock dials.

Alternating Current (AC)

The first specimen of the AC kilowatt-hour meter produced on the basis of Hungarian Ottó Bláthy's patent and named after him was presented by the Ganz Works at the Frankfurt Fair in the autumn of 1889, and the first induction kilowatt-hour meter was already marketed by the factory at the end of the same year. These were the first alternating-current watt-hour meters, known by the name of Bláthy-meters. The AC kilowatt hour meters used at present operate on the same principle as Bláthy's original invention. Also around 1889, Elihu Thomson of the American General Electric company developed a recording watt meter (watt-hour meter) based on an ironless commutator motor. This meter overcame the disadvantages of the electrochemical type and could operate on either alternating or direct current.

In 1894 Oliver Shallenberger of the Westinghouse Electric Corporation applied the induction principle previously used only in AC ampere-hour meters to produce a watt-hour meter of the modern electromechanical form, using an induction disk whose rotational speed was made proportional to the power in the circuit. The Bláthy meter was similar to Shallenberger and Thomson meter in that they are two-phase motor meter. Although the induction meter would only work on alternating current, it eliminated the delicate and troublesome commutator of the Thomson design. Shallenberger fell ill and was unable to refine his initial large and heavy design, although he did also develop a polyphase version.

Unit of Measurement

The most common unit of measurement on the electricity meter is the kilowatt hour [kWh], which is equal to the amount of energy used by a load of one kilowatt over a period of one hour, or 3,600,000 joules. Some electricity companies use the SI megajoule instead.

Demand is normally measured in watts, but averaged over a period, most often a quarter or half hour.

Reactive power is measured in "thousands of volt-ampere reactive-hours", (kvarh). By convention, a "lagging" or inductive load, such as a motor, will have positive reactive power. A "leading", or capacitive load, will have negative reactive power.

Volt-amperes measures all power passed through a distribution network, including reactive and actual. This is equal to the product of root-mean-square volts and amperes.

Distortion of the electric current by loads is measured in several ways. Power factor is the ratio of resistive (or real power) to volt-amperes. A capacitive load has a leading power factor, and an inductive load has a lagging power factor. A purely resistive load (such as a filament lamp, heater or kettle) exhibits a power factor of 1. Current harmonics are a measure of distortion of the wave form. For example, electronic loads such as computer power supplies draw their current at the voltage peak to fill their internal storage elements. This can lead to a significant voltage drop near the supply voltage peak which shows as a flattening of the voltage waveform. This flattening causes odd harmonics which are not permissible if they exceed specific limits, as they are not only wasteful, but may interfere with the operation of other equipment. Harmonic emissions are mandated by law in EU and other countries to fall within specified limits.

Other Units of Measurement

In addition to metering based on the amount of energy used, other types of metering are available.

Meters which measured the amount of charge (coulombs) used, known as ampere-hour meters, were used in the early days of electrification. These were dependent upon the supply voltage remaining constant for accurate measurement of energy usage, which was not a likely circumstance with most supplies.

Some meters measured only the length of time for which charge flowed, with no measurement of the magnitude of voltage or current being made. These were only suited for constant-load applications.

Neither type is likely to be used today.

Types of Meters

1. Voltage coil - many turns of fine wire en cased in plastic, connected in parallel with load.
2. Current coil - three turns of thick wire, connected in series with load.
3. Stator - concentrates and confines magnetic field.
4. Aluminum rotor disc.
5. Rotor brake magnets.

6. Spindle with worm gear.
7. Display dials - note that the 1/10, 10 and 1000 dials rotate clockwise while the 1, 100 and 10000 dials rotate counter-clockwise.

Fig. : Mechanism of electromechanical induction meter.

Electricity meters operate by continuously measuring the instantaneous voltage (volts) and current (amperes) to give energy used (in joules, kilowatt-hours etc.). Meters for smaller services (such as small residential customers) can be connected directly in-line between source and customer. For larger loads, more than about 200 ampere of load, current transformers are used, so that the meter can be located other than in line with the service conductors. The meters fall into two basic categories, electromechanical and electronic.

Electromechanical Meters

The most common type of electricity meter is the electromechanical induction watt-hour meter.

The electromechanical induction meter operates by counting the revolutions of a non-magnetic, but electrically conductive, metal disc which is made to rotate at a speed proportional to the power passing through the meter. The number of revolutions is thus proportional to the energy usage. The voltage coil consumes a small and relatively constant amount of power, typically around 2 watts which is not registered on the meter. The current coil similarly consumes a small amount of power in proportion to the square of the current flowing through it, typically up to a couple of watts at full load, which is registered on the meter.

The disc is acted upon by two sets of coils, which form, in effect, a two phase induction motor. One coil is connected in such a way that it produces a magnetic flux in proportion to the voltage and the other produces a magnetic flux in proportion to the current. The field of the voltage coil is delayed by 90 degrees, due to the coil's inductive nature, and calibrated using a lag coil. This produces eddy currents in the disc and the effect is such that a force is exerted on the disc in proportion to the product of the instantaneous current, voltage and phase angle (power factor) between them. A permanent magnet exerts an opposing force proportional to the speed of rotation of the disc. The equilibrium between these two opposing forces results in the disc rotating at a speed proportional to the power or rate of energy usage. The disc drives a register mechanism which counts revolutions, much like the odometer in a car, in order to render a measurement of the total energy used.

The type of meter described above is used on a single-phase AC supply. Different phase configurations use additional voltage and current coils.

The disc is supported by a spindle which has a worm gear which drives the register. The register is a series of dials which record the amount of energy used. The dials may be of the *cyclometer* type, an odometer-like display that is easy to read where for each dial a single digit is shown through a window in the face of the meter, or of the pointer type where a pointer indicates each digit. With the dial pointer type, adjacent pointers generally rotate in opposite directions due to the gearing mechanism.

The amount of energy represented by one revolution of the disc is denoted by the symbol Kh which is given in units of watt-hours per revolution. The value 7.2 is commonly seen. Using the value of Kh one can determine their power consumption at any given time by timing the disc with a stopwatch.

$$P = \frac{3600 \cdot Kh}{t}.$$

Where :

t = time in seconds taken by the disc to complete one revolution,

P = power in watts.

For example, if Kh = 7.2 as above, and one revolution took place in 14.4 seconds, the power is 1800 watts. This method can be used to determine the power consumption of household devices by switching them on one by one.

Most domestic electricity meters must be read manually, whether by a representative of the power company or by the customer. Where the customer reads the meter, the reading may be supplied to the power company by telephone, post or over the internet. The electricity company will normally require a visit by a company representative at least annually in order to verify customer-supplied readings and to make a basic safety check of the meter.

In an induction type meter, creep is a phenomenon that can adversely affect accuracy, that occurs when the meter disc rotates continuously with potential applied and the load terminals open circuited. A test for error due to creep is called a creep test.

Two standards govern meter accuracy, ANSI C12.20 for North America and IEC 62053.

Electronic Meters

Electronic meters display the energy used on an LCD or LED display, and some can also transmit readings to remote places. In addition to measuring energy used, electronic meters can also record other parameters of the load and supply such as instantaneous and maximum rate of usage demands, voltages, power factor and reactive power used etc. They can also support time-of-day billing, for example, recording the amount of energy used during on-peak and off-peak hours.

Solid-state Design

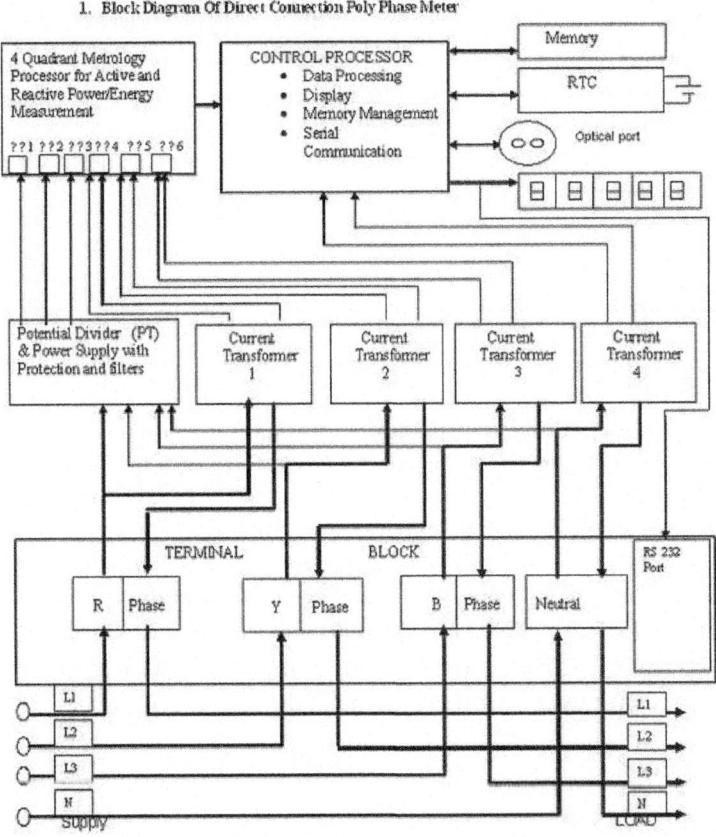

Fig. : Basic block diagram of an electronic energy meter.

As in the block diagram, the meter has a power supply, a metering engine, a processing and communication engine (*i.e.* a micro-controller), and other add-on modules such as RTC, LCD display, communication ports/modules and so on.

The metering engine is given the voltage and current inputs and has a voltage reference, samplers and quantisers followed by an ADC section to yield the digitised equivalents of all the inputs. These inputs are then processed using a digital signal processor to calculate the various metering parameters such as powers, energies etc.

The largest source of long-term errors in the meter is drift in the preamp, followed by the precision of the voltage reference. Both of these vary with temperature as well, and vary wildly because most meters are outdoors. Characterising and compensating for these is a major part of meter design.

The processing and communication section has the responsibility of calculating the various derived quantities from the digital values generated by the metering engine. This also has the responsibility of communication using various protocols and interface with other addon modules connected as slaves to it.

RTC and other add-on modules are attached as slaves to the processing and communication section for various input/output functions. On a modern meter most if not all of this will be implemented inside the microprocessor, such as the real time clock (RTC), LCD controller, temperature sensor, memory and analog to digital converters.

Applications

Multiple Tariff (Variable Rate) Meters

Electricity retailers may wish to charge customers different tariffs at different times of the day to better reflect the costs of generation and transmission. Since it is typically not cost effective to store significant amounts of electricity during a period of low demand for use during a period of high demand, costs will vary significantly depending on the time of day. Low cost generation capacity (baseload) such as nuclear can take many hours to start, meaning a surplus in times of low demand, whereas high cost but flexible generating capacity (such as gas turbines) must be kept available to respond at a moment's notice (spinning reserve) to peak demand, perhaps being used for a few minutes per day, which is very expensive.

Some multiple tariff meters use different tariffs for different amounts of demand. These are usually industrial meters.

Domestic Usage

Domestic variable-rate meters generally permit two to three tariffs ("peak", "off-peak" and "shoulder") and in such installations a simple electromechanical time switch may be used. Historically, these have often been used in conjunction with electrical storage heaters or hot water storage systems.

Multiple tariffs are made easier by time of use (TOU) meters which incorporate or are connected to a time switch and which have multiple registers.

Switching between the tariffs may happen via a radio-activated switch rather than a time switch to prevent tampering with a sealed time switch to obtain cheaper electricity.

United Kingdom

Radio-activated switching is common in the UK, with a nightly data signal sent within the longwave carrier of BBC Radio 4, 198 kHz. The time of off-peak charging is usually seven hours between midnight and 7.00am GMT, and this is designed to power storage heaters and immersion heaters. In the UK, such tariffs are branded *Economy 7* or *White Meter*. The popularity of such tariffs has declined in recent years, at least in the domestic market, because of the (perceived or real) deficiencies of storage heaters and the comparatively low cost of natural gas (although there remain many without the option of gas, whether they are outside the gas supply network or cannot afford the capital cost of a radiator system). An Economy 10 meter is also available, which gives 10 hours of cheap off-peak heating spread out over three timeslots throughout a 24 hour period. This allows multiple top-up boosts to storage heaters, or a good spread of times to run a wet electric heating system on a cheaper electricity rate.

Most meters using *Economy 7* switch the entire electricity supply to the cheaper rate during the 7 hour night time period, not just the storage heater circuit. The downside of this is that the daytime rate will be significantly higher, and standing charges may be a little higher too. For instance, normal rate electricity may be 9p per kWh, whereas *Economy 7*'s daytime rate might be 14 to 17 p per kWh, but only 5.43p per kWh at night. Timer switches installed on washing machines, tumble dryers, dishwashers and immersion heaters may be set so that they switch on only when the rate is lower.

Commercial Usage

Large commercial and industrial premises may use electronic meters which record power usage in blocks of half an hour or less. This is because most electricity grids have demand surges throughout the day, and the power company may wish to give price incentives to large customers to reduce demand at these times. These demand surges often correspond to meal times or, famously, to advertisements in popular television programmes.

Appliance Energy Meters

Plug in electricity meters (or "Plug load" meters) measure energy used by individual appliances. There are a variety of models available on the market today but they all work on the same basic principle. The meter is plugged into an outlet, and the appliance to be measured is plugged into the meter. Such meters can help in energy conservation by identifying major energy users, or devices that consume excessive standby power. Web resources can also be used, if an estimate of the power consumption is enough for the research purposes. A power meter can often be borrowed from the local power authorities or a local public library.

In-home Energy Use Displays

A potentially powerful means to reduce household energy consumption is to provide convenient real-time feedback to users so they can change their energy using behaviour. Recently, low-cost energy feedback displays have become available. A study using a consumer-readable meter in 500 Ontario homes by *Hydro One* showed an average 6.5% drop in total electricity use when compared with a similarly sized control group. *Hydro One* subsequently offered free power monitors to 30,000 customers based on the success of the pilot. Projects such as Google Power Meter, take information from a smart meter and make it more readily available to users to help encourage conservation.

Smart Meters

Smart meters go a step further than simple AMR (automatic meter reading). They offer additional functionality including a real-time or near real-time reads, power outage notification, and power quality monitoring. They allow price setting agencies to introduce different prices for consumption based on the time of day and the season.

These price differences can be used to reduce peaks in demand (load shifting or peak lopping), reducing the need for additional power plants and in particular the higher polluting and costly to operate natural gas powered peaker plants. The feedback they provide to consumers has also been shown to cut overall energy consumption.

Another type of smart meter uses non-intrusive load monitoring to automatically determine the number and type of appliances in a residence, how much energy each uses and when. This meter is used by electric utilities to do surveys of energy use. It eliminates the need to put timers on all of the appliances in a house to determine how much energy each uses.

Prepayment Meters

The standard business model of electricity retailing involves the electricity company billing the customer for the amount of energy used in the previous month or quarter. In some countries, if the retailer believes that the customer may not pay the bill, a prepayment meter may be installed. This requires the customer to make advance payment before electricity can be used.If the available credit is exhausted then the supply of electricity is cut off by a relay.

In the UK, mechanical prepayment meters used to be common in rented accommodation. Disadvantages of these included the need for regular visits to remove cash, and risk of theft of the cash in the meter.

Modern solid-state electricity meters, in conjunction with smart cards, have removed these disadvantages and such meters are commonly used for customers considered to be a poor credit risk. In the UK, one system is the Pay Point network, where rechargeable tokens (Quantum cards for natural gas, or plastic "keys" for electricity) can be loaded with whatever money the customer has available.

Recently smartcards are introduced as much reliable tokens that allows two way data exchange between meter and the utility.

In South Africa, Sudan and Northern Ireland prepaid meters are recharged by entering a unique, encoded twenty digit number using a keypad. This makes the tokens, essentially a slip of paper, very cheap to produce.

Around the world, experiments are going on, especially in developing countries, to test pre-payment systems. In some cases, prepayment meters have not been accepted by customers. There are various groups, such as the Standard Transfer Specification (STS) association, which promote common standards for prepayment metering systems across manufacturers. Prepaid meters using the STS standard are used in many countries.

Time of Day Metering

Time of Day metering (TOD), also known as Time of Usage (TOU) or Seasonal Time of Day (SToD), metering involves dividing the day, month and year into tariff slots and with higher rates at peak load periods and low tariff rates at off-peak load periods. While this can be used to automatically control usage on the part of the customer (resulting in automatic load control), it is often simply the customers responsibility to control his own usage, or pay accordingly (voluntary load control). This also allows the utilities to plan their transmission infrastructure appropriately.

TOD metering normally splits rates into an arrangement of multiple segments including on-peak, off-peak, mid-peak or shoulder, and critical peak. A typical arrangement is a peak occurring during the day (non-holiday days only), such as from 1 pm to 9 pm Monday through Friday during the summer and from 6:30 am to 12 noon and 5 pm to 9 pm during the winter. More complex arrangements include the use of critical peaks which occur during high demand periods. The times of peak demand/cost will vary in different markets around the world.

Large commercial users can purchase power by the hour using either forecast pricing or real time pricing. Prices range from we pay you to take it (negative) to $1000/MWh (100 cents/kWh).

Some utilities allow residential customers to pay hourly rates, such as Illinois, which uses day ahead pricing.

Power Export Metering

Many electricity customers are installing their own electricity generating equipment, whether for reasons of economy, redundancy or environmental reasons.

When a customer is generating more electricity than required for his own use, the surplus may be exported back to the power grid. Customers that generate back into the "grid" usually must have special equipment and safety devices to protect the grid components (as well as the customer's own) in case of faults (electrical short circuits) or maintenance of the grid (say voltage on a downed line coming from an exporting customers facility).

This exported energy may be accounted for in the simplest case by the meter running backwards during periods of net export, thus reducing the customer's recorded energy usage by the amount exported. This in effect results in the customer being paid for his/her exports at the full retail price of electricity. Unless equipped with a detent or equivalent, a standard meter will accurately record power flow in each direction by simply running backwards when power is exported. Where allowed by law, utilities maintain a profitable margin between the price of energy delivered to the consumer and the rate credited for consumer-generated energy that flows back to the grid. Lately, upload sources typically originate from renewable sources (*e.g.*, wind turbines, photovoltaic cells), or gas or steam turbines, which are often found in cogeneration systems. Another potential upload source that has been proposed is plug-in hybrid car batteries (vehicle-to-grid power systems). This requires a "smart grid," which includes meters that measure electricity via communication networks that require remote control and give customers timing and pricing options. Vehicle-to-grid systems could be installed at workplace parking lots and garages and at park and rides and could help drivers charge their batteries at home at night when off-peak power prices are cheaper, and receive bill crediting for selling excess electricity back to the grid during high-demand hours.

Ownership

Following the deregulation of electricity supply markets in many countries (*e.g.*, UK), the company responsible for an electricity meter may not be obvious. Depending on the arrangements in place, the meter may be the property of the meter Operator, electricity distributor, the retailer or for some large users of electricity the meter may belong to the customer.

The company responsible for reading the meter may not always be the company which owns it. Meter reading is now sometimes sub-contracted and in some areas the same person may read gas, water and electricity meters at the same time.

Communication Methods

Remote meter reading is a practical example of telemetry. It saves the cost of a human meter reader and the resulting mistakes, but it also allows more measurements, and remote provisioning. Many smart meters now include a switch to interrupt or restore service.

Historically, rotating meters could report their metered information remotely, using a pair of electrical contacts attached to a *KYZ* line.

A KYZ interface is a Form C contact supplied from the meter. In a KYZ interface, the Y and Z wires are switch contacts, shorted to K for a measured amount of energy. When one contact closes the other contact opens to provide count accuracy security. Each contact change of state is considered one pulse. The frequency of pulses indicates the power demand. The number of pulses indicates energy metered.

KYZ outputs were historically attached to "totaliser relays" feeding a "totaliser" so that many meters could be read all at once in one place.

KYZ outputs are also the classic way of attaching electricity meters to programmable logic controllers, HVACs or other control systems. Some modern meters also supply a contact closure that warns when the meter detects a demand near a higher electricity tariff, to improve demand side management.

Some meters have an open collector or IR LED output that give 32-100 ms pulses for each metered amount of electrical energy, usually 1000-10000 pulses per kWh. Output is limited to max 27 V DC and 27 mA DC. These outputs usually follow the DIN 43864 standard.

Often, meters designed for semi-automated reading have a serial port on that communicates by infrared LED through the faceplate of the meter. In some multi-unit buildings, a similar protocol is used, but in a wired bus using a serial current loop to connect all the meters to a single plug. The plug is often near a more easily accessible point. In the European Union, the most common infrared and protocol is "FLAG", a simplified subset of mode C of IEC 61107. In the U.S. and Canada, the favoured infrared protocol is ANSI C12.18. Some industrial meters use a protocol for programmable logic controllers (Modbus or DNP3).

One protocol proposed for this purpose is DLMS/COSEM which can operate over any medium, including serial ports. The data can be transmitted by Zigbee, WiFi, telephone lines or over the power lines themselves. Some meters can be read over the internet. Other more modern protocols are also becoming widely used.

Electronic meters now use low-power radio, GSM, GPRS, Bluetooth, IrDA, as well as RS-485 wired link. The meters can now store the entire usage profiles with time stamps and relay them at a click of a button. The demand readings stored with the profiles accurately indicate the load requirements of the customer. This load profile data is processed at the utilities for billing and planning purposes.

AMR (Automatic Meter Reading) and *RMR* (Remote Meter Reading) describe various systems that allow meters to be checked without the need to send a meter reader out. An electronic meter can transmit its readings by telephone line or radio to a central billing office. Automatic meter reading can be done with GSM (Global System for Mobile Communications) modems, one is attached to each meter and the other is placed at the central utility office.

Location

The location of an electricity meter varies with each installation. Possible locations include on a utility pole serving the property, in a street-side cabinet (meter box) or inside the premises adjacent to the consumer unit / distribution board. Electricity companies may prefer external locations as the meter can be read without gaining access to the premises but external meters may be more prone to vandalism.

Current transformers permit the meter to be located remotely from the current-carrying conductors. This is common in large installations. For example, a sub-station serving a single large customer may have metering equipment installed in a cabinet, without bringing heavy cables into the cabinet.

Customer Drop and Metering Equation

Since electrical standards vary in different regions, "customer drops" from the grid to the customer also vary depending on the standards and the type of installation. There are several common types of connections between a grid and a customer. Each type has a different *metering equation*. Blondel's theorem states that for any system with N current-carrying conductors, that N-1 measuring elements are sufficient to measure electrical energy. This indicates that different metering is needed, for example, for a three-phase three-wire system than for a three-phase four-wire (with neutral) system.

In Europe, Asia, Africa and most other locations, single phase is common for residential and small commercial customers. Single phase distribution is less-expensive, because one set of transformers in a substation normally serve a large area with relatively high voltages (usually 220V) and no local transformers. These have a simple metering equation : Watts = Volts x Amps, with Volts measured from the neutral to the phase wire. In the United States, Canada, and parts of Latin and South America similar customers are normally served by three-wire single phase. Three-wire single-phase requires local transformers, as few as one per ten residences, but provides lower, safer voltages at the socket (usually 120V), and provides two voltages to customers : neutral to phase (usually 120V), and phase to phase (usually 240v). Additionally, three-wire customers normally have neutral wired to the zero side of the generator's windings, which gives earthing that can be easily measured to be safe. These meters have a metering equation of Watts = 0.5 x Volts x (Amps of phase A - Amps of phase B), with Volts measured between the phase wires.

Industrial power is normally supplied as three phase power. There are two forms : three wire, or four wire with a system neutral. In "three wire" or "three wire delta," the generator is wired as a triangle (or "delta"), and an earth ground is the safety ground. The three phases have voltage only relative to each other. This distribution method has one fewer wire, is less expensive, and is common in Asia, Africa, and many parts of Europe. In regions that mix residences and light industry, it is common for this to be the only distribution method. A meter for this type normally measures two of the windings relative to the third winding, and adds the watts. One disadvantage of this system is that if the safety earth fails, it is difficult to discover this by direct measurement, because no phase has a voltage relative to earth.

In the four-wire three-phase system, sometimes called "four-wire wye", the safety ground is connected to a neutral wire that is physically connected to the zero-voltage side of the three windings of the generator. Since all power phases are relative to the neutral in this system, if the neutral is disconnected, it can be directly measured. In the U.S., the National Electrical Code requires neutrals to be of this type. In this system, power meters measure and sum all three phases relative to the neutral.

In North America, it is common for electricity meters to plug into a standardised socket outdoors, on the side of a building. This allows the meter to be replaced without disturbing the wires to the socket, or the occupant of the building. Some sockets may have a bypass while the meter is removed for service. The amount of electricity used without being recorded during this small time is considered insignificant when compared to the inconvenience which might be caused to the customer by cutting off the electricity supply. Most electronic meters in North America use a serial protocol, ANSI C12.18.

In many other countries the supply and load terminals are in the meter housing itself. Cables are connected directly to the meter. In some areas the meter is outside, often on a utility pole. In others, it is inside the building in a niche. If inside, it may share a data connection with other meters. If it exists, the shared connection is often a small plug near the post box. The connection is often EIA-485 or infra-red with a serial protocol such as IEC 62056.

In 2014, networking to meters is rapidly changing. The most common schemes seem to combine an existing national standard for data (*e.g.* ANSI C12.19 or IEC 62056) operating via the internet protocol with a small circuit board for powerline communication, or a digital radio for a mobile phone network, or an ISM band.

Tampering and Security

Meters can be manipulated to make them under-register, effectively allowing power use without paying for it. This theft or fraud can be dangerous as well as dishonest.

Power companies often install remote-reporting meters specifically to enable remote detection of tampering, and specifically to discover energy theft. The change to smart power meters is useful to stop energy theft.

When tampering is detected, the normal tactic, legal in most areas of the USA, is to switch the subscriber to a "tampering" tariff charged at the meter's maximum designed current. At US$ 0.095/kWh, a standard residential 50 A meter causes a legally collectible charge of about US$ 5,000.00 per month. Meter readers are trained to spot signs of tampering, and with crude mechanical meters, the maximum rate may be charged each billing period until the tamper is removed, or the service is disconnected.

A common method of tampering on mechanical disk meters is to attach magnets to the outside of the meter. Strong magnets saturate the magnetic fields in the meter so that the motor portion of a mchanical meter does not operate. Lower power magnets can add to the drag resistance of the internal disk resistance magnets. Magnets can also saturate current transformers or power-supply transformers in electronic meters, though countermeasures are common.

Rectified DC loads cause mechanical (but not electronic) meters to under-register. DC current does not cause the coils to make eddy currents in the disk, so this causes reduced rotation and a lower bill.

Some combinations of capacitive and inductive load can interact with the coils and mass of a rotor and cause reduced or reverse motion.

All of these effects can be detected by the electric company, and many modern meters can detect or compensate for them.

The owner of the meter normally secures the meter against tampering. Revenue meters' mechanisms and connections are sealed. Meters may also measure VAR-hours (the reflected load), neutral and DC currents (elevated by most electrical tampering), ambient magnetic fields, etc. Even simple mechanical meters can have mechanical flags that are dropped by magnetic tampering or large DC currents.

Newer computerised meters usually have counter-measures against tampering. AMR (Automated Meter Reading) meters often have sensors that can report opening of the meter cover, magnetic anomalies, extra clock setting, glued buttons, inverted installation, reversed or switched phases etc.

Some tampers bypass the meter, wholly or in part. Safe tampers of this type normally increase the neutral current at the meter. Most split-phase residential meters in the United States are unable to detect neutral currents. However, modern tamper-resistant meters can detect and bill it at standard rates.

Disconnecting a meter's neutral connector is unsafe because shorts can then pass through people or equipment rather than a metallic ground to the generator or earth.

A phantom loop connection via an earth ground is often much higher resistance than the metallic neutral connector. Even if an earth ground is safe, metering at the substation can alert the operator to tampering. Substations, inter-ties, and transformers normally have a high-accuracy meter for the area served. Power companies normally investigate discrepancies between the total billed and the total generated, in order to find and fix power distribution problems. These investigations are an effective method to discover tampering.

Power thefts in the U.S. are often connected with indoor marijuana grow operations. Narcotics detectives associate abnormally high power usage with the lighting such operations require. Indoor marijuana growers aware of this are particularly motivated to steal electricity simply to conceal their usage of it.

Privacy Issues

The introduction of advanced meters in residential areas has produced additional privacy issues that may affect ordinary customers. These meters are often capable of recording energy usage every 15, 30 or 60 minutes. In some meters real time usage is transmitted on an IR light, that can be viewed with Night Vison viewers. These can be used for surveillance, revealing information about people's possessions and behaviour. For instance, it can show when the customer is away for extended periods. Non-intrusive load monitoring gives even more detail about what appliances people have and their living and use patterns.

A more detailed and recent analysis of this issue was performed by the Illinois Security Lab.

INDUCTANCE

In electromagnetism and electronics, inductance is the property of a conductor by which a change in current flowing through it "induces" (creates) a voltage (electromotive force) in both the conductor itself (self-inductance) and in any nearby conductors (mutual inductance).

Explanation

These effects are derived from two fundamental observations of physics : First, that a steady current creates a steady magnetic field (Oersted's law), and second, that a time-varying magnetic field induces voltage in nearby conductors (Faraday's law of induction). According to Lenz's law, a changing electric current through a circuit that contains inductance, induces a proportional voltage, which opposes the change in current (self-inductance). The varying field in this circuit may also induce an e.m.f. in neighbouring circuits (mutual inductance).

Origin of Term

The term 'inductance' was coined by Oliver Heaviside in February 1886. It is customary to use the symbol L for inductance, in honour of the physicist Heinrich Lenz. In the SI system the measurement unit for inductance is the henry, H, named in honour of the scientist who discovered inductance, Joseph Henry.

Circuit Analysis

To add inductance to a circuit, electrical or electronic components called inductors are used. Inductors are typically manufactured out of coils of wire, with this design delivering two circumstances, one, a concentration of the magnetic field, and two, a linking of the magnetic field into the circuit more than once.

The relationship between the self-inductance L of an electrical circuit (in henries), voltage, and current is

$$v(t) = L\frac{di}{dt}$$

Where $v(t)$ denotes the voltage in volts across the circuit, and $i(t)$ the current in amperes through the circuit. The formula implicitly states that a voltage is induced across an inductor, equal to the product of the inductor's inductance, and current's rate of change through the inductor.

All practical circuits have some inductance, which may provide beneficial or detrimental effects. For a tuned circuit, inductance is used to provide a frequency selective circuit. Practical inductors may be used to provide filtering, or energy storage, in a given network. The inductance of a transmission line is one of the properties that determines its characteristic impedance; balancing the inductance and capacitance of cables is important for distortion-free telegraphy and telephony. The inductance of lengthy power transmission lines effectively results in a lessened delivery of AC power, due to the combination of inductance, coupled with transmission lines being spread across great distances. Sensitive circuits, such as microphone and computer network cables, may utilize special cabling construction, limiting the mutual inductance between signal circuits.

The generalization to the case of K electrical circuits with currents i_m and voltages v_m reads

$$u_m = \sum_{n=1}^{K} L_{m,n} \frac{di_n}{dt}.$$

Here, inductance L is a symmetric matrix. The diagonal coefficients $L_{m,m}$ are called coefficients of self-inductance, the off-diagonal elements are called coefficients of mutual inductance. The coefficients of inductance are constant, as long as no magnetizable material with non-linear characteristics are involved. This is a direct consequence of the linearity of Maxwell's equations in the fields and the current density.

Derivation from Faraday's Law of Inductance

The inductance equations above are a consequence of Maxwell's equations. There is a straightforward derivation in the important case of electrical circuits consisting of thin wires.

Consider a system of K wire loops, each with one or several wire turns. The flux linkage of loop m is given by

$$N_m \Phi_m = \sum_{n=1}^{K} L_{m,n} i_n.$$

Here N_m denotes the number of turns in loop m, Φ_m the magnetic flux through this loop, and $L_{m,n}$ are some constants. This equation follows from Ampere's law - magnetic fields and fluxes are linear functions of the currents. By Faraday's law of induction we have

$$v_m = N_m \frac{d\Phi_m}{dt} = \sum_{n=1}^{K} L_{m,n} \frac{di_n}{dt},$$

where v_m denotes the voltage induced in circuit m. This agrees with the definition of inductance above if the coefficients $L_{m,n}$ are identified with the coefficients of inductance. Because the total currents $N_n i_n$ contribute to Φ_m it also follows that $L_{m,n}$ is proportional to the product of turns $N_m N_n$.

Inductance and Magnetic Field Energy

Multiplying the equation for v_m above with $i_m dt$ and summing over m gives the energy transferred to the system in the time interval dt,

$$\sum_{m}^{K} i_m v_m dt = \sum_{n,n=1}^{K} i_m L_{m,n} di_n \stackrel{!}{=} \sum_{n=1}^{K} \frac{\partial W(i)}{\partial i_n} di_n.$$

This must agree with the change of the magnetic field energy W caused by the currents. The integrability condition

$$\partial^2 W/\partial i_m \partial i_n = \partial^2 W/\partial i_n \partial i_m$$

requires $L_{m,n}=L_{n,m}$. The inductance matrix $L_{m,n}$ thus is symmetric. The integral of the energy transfer is the magnetic field energy as a function of the currents,

$$W(i) = \frac{1}{2} \sum_{m,n=1}^{K} i_m L_{m,n} i_n.$$

This equation also is a direct consequence of the linearity of Maxwell's equations. It is helpful to associate changing electric currents with a build-up or decrease of magnet field energy. The corresponding energy transfer requires or generates a voltage. A mechanical analogy in the $K=1$ case with magnetic field energy $(1/2)Li^2$ is a body with mass M, velocity u and kinetic energy $(1/2)Mu^2$. The rate of change of velocity (current) multiplied with mass (inductance) requires or generates a force (an electrical voltage).

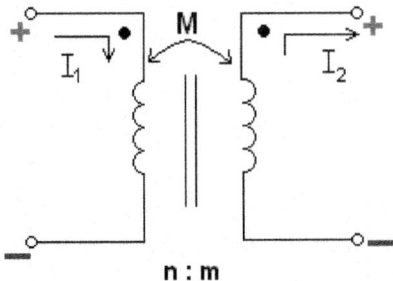

Fig. : The circuit diagram representation of mutually coupled inductors. The two vertical lines between the inductors indicate a *solid core* that the wires of the inductor are wrapped around. "n : m" shows the ratio between the number of windings of the left inductor to windings of the right inductor. This picture also shows the dot convention.

Coupled Inductors and Mutual Inductance

Mutual inductance occurs when the change in current in one inductor induces a voltage in another nearby inductor. It is important as the mechanism by which transformers work, but it can also cause unwanted coupling between conductors in a circuit.

The mutual inductance, M, is also a measure of the coupling between two inductors. The mutual inductance by circuit i on circuit j is given by the double integral *Neumann formula*.

The mutual inductance also has the relationship :

$$M_{21} = N_1 N_2 P_{21}$$

where :

M_{21} is the mutual inductance, and the subscript specifies the relationship of the voltage induced in coil 2 due to the current in coil 1.

N_1 is the number of turns in coil 1,
N_2 is the number of turns in coil 2,
P_{21} is the permeance of the space occupied by the flux.

The mutual inductance also has a relationship with the coupling coefficient. The coupling coefficient is always between 1 and 0, and is a convenient way to specify the relationship between a certain orientation of inductors with arbitrary inductance:

$$M = k\sqrt{L_1 L_2}$$

where:

k is the *coupling coefficient* and $0 \leq k \leq 1$,
L_1 is the inductance of the first coil, and
L_2 is the inductance of the second coil.

Once the mutual inductance, M, is determined from this factor, it can be used to predict the behaviour of a circuit:

$$v_1 = L_1 \frac{di_1}{dt} - M \frac{di_2}{dt}$$

where:

v_1 is the voltage across the inductor of interest,
L_1 is the inductance of the inductor of interest,
di_1/dt is the derivative, with respect to time, of the current through the inductor of interest,
di_2/dt is the derivative, with respect to time, of the current through the inductor that is coupled to the first inductor, and
M is the mutual inductance.

The minus sign arises because of the sense the current i_2 has been defined in the diagram. With both currents defined going into the dots the sign of M will be positive.

When one inductor is closely coupled to another inductor through mutual inductance, such as in a transformer, the voltages, currents, and number of turns can be related in the following way:

$$V_s = \frac{N_s}{N_p} V_p$$

where:

V_s is the voltage across the secondary inductor,
V_p is the voltage across the primary inductor (the one connected to a power source),
N_s is the number of turns in the secondary inductor, and
N_p is the number of turns in the primary inductor.

Conversely the current :

$$I_s = \frac{N_p}{N_s} I_p$$

where :

I_s is the current through the secondary inductor,

I_p is the current through the primary inductor (the one connected to a power source),

N_s is the number of turns in the secondary inductor, and

N_p is the number of turns in the primary inductor.

Note that the power through one inductor is the same as the power through the other. Also note that these equations don't work if both inductors are forced (with power sources).

When either side of the transformer is a tuned circuit, the amount of mutual inductance between the two windings determines the shape of the frequency response curve. Although no boundaries are defined, this is often referred to as loose-, critical-, and over-coupling. When two tuned circuits are loosely coupled through mutual inductance, the bandwidth will be narrow. As the amount of mutual inductance increases, the bandwidth continues to grow. When the mutual inductance is increased beyond a critical point, the peak in the response curve begins to drop, and the center frequency will be attenuated more strongly than its direct sidebands. This is known as overcoupling.

Calculation Techniques

In the most general case, inductance can be calculated from Maxwell's equations. Many important cases can be solved using simplifications. Where high frequency currents are considered, with skin effect, the surface current densities and magnetic field may be obtained by solving the Laplace equation. Where the conductors are thin wires, self-inductance still depends on the wire radius and the distribution of the current in the wire. This current distribution is approximately constant (on the surface or in the volume of the wire) for a wire radius much smaller than other length scales.

Mutual Inductance of Two Wire Loops

The mutual inductance by a filamentary circuit m on a filamentary circuit n is given by the double integral *Neumann formula*

$$L_{m,n} = \frac{\mu_0}{4\pi} \oint_{C_m} \oint_{C_n} \frac{dx_m \cdot dx_n}{|X_m - X_n|}$$

The symbol μ_0 denotes the magnetic constant ($4\pi \times 10^{-7}$ H/m), C_m and C_n are the curves spanned by the wires.

Self-inductance of a Wire Loop

Formally the self-inductance of a wire loop would be given by the above equation with $m = n$. The problem, however, is that $1/|x-x'|$ now becomes infinite, making it necessary to take the finite wire radius a and the distribution of the current in the wire into account. There remain the contribution from the integral over all points with $|x-x'| > a/2$ and a correction term,

$$L = \left(\frac{\mu_0}{4\pi} \oint_C \oint_{C'} \frac{dx \cdot dx'}{|x-x'|} \right)_{|x-x'|>a/2} + \frac{\mu_0}{4\pi} lY + O(\mu_0 a).$$

Here a and l denote radius and length of the wire, and Y is a constant that depends on the distribution of the current in the wire : $Y = 0$ when the current flows in the surface of the wire (skin effect), $Y = 1/2$ when the current is homogeneous across the wire. This approximation is accurate when the wires are long compared to their cross-sectional dimensions.

Method of Images

In some cases different current distributions generate the same magnetic field in some section of space. This fact may be used to relate self inductances (method of images). As an example consider the two systems :

- A wire at distance $d/2$ in front of a perfectly conducting wall (which is the return)
- Two parallel wires at distance d, with opposite current.

The magnetic field of the two systems coincides (in a half space). The magnetic field energy and the inductance of the second system thus are twice as large as that of the first system.

Relation between Inductance and Capacitance

Inductance per length L' and capacitance per length C' are related to each other in the special case of transmission lines consisting of two parallel perfect conductors of arbitrary but constant cross-section,

$$L'C' = \varepsilon \mu.$$

Here ε and μ denote dielectric constant and magnetic permeability of the medium the conductors are embedded in. There is no electric and no magnetic field inside the conductors (complete skin effect, high frequency). Current flows down on one line and returns on the other. Signals will propagate along the transmission line at the speed of electromagnetic radiation in the non-conductive medium enveloping the conductors.

Self-inductance of Simple Electrical Circuits in Air

The self-inductance of many types of electrical circuits can be given in closed form.

Inductance of simple electrical circuits in air		
Type	Inductance	Comment
Single layer solenoid	$\frac{\mu_0 r^2 N^2}{3l}\left[-8w + 4\frac{\sqrt{1+m}}{m}\left(K\left(\sqrt{\frac{m}{1+m}}\right) - (1-m)E\left(\sqrt{\frac{m}{1+m}}\right)\right)\right]$ $= \frac{\mu_0 r^2 N^2 \pi}{l}\left[1 - \frac{8w}{3\pi} + \sum_{n=1}^{\infty}\frac{(2n)!^2}{n!^4(n+1)(2n-1)2^{2n}}(-1)^{n+1}w^{2n}\right]$ $= \frac{\mu_0 r^2 N^2 \pi}{l}\left(1 - \frac{8w}{3\pi} + \frac{w^2}{2} - \frac{w^4}{4} + \frac{5w^6}{16} - \frac{35w^8}{64} + \ldots\right)$ for $w \ll 1$ $= \mu_0 r N^2\left[\left(1 + \frac{1}{32w^2} + O\left(\frac{1}{w^4}\right)\right)\ln(8w) - 1/2 + \frac{1}{128w^2} + O\left(\frac{1}{w^4}\right)\right]$ for $w \gg 1$	N : Number of turns r : Radius l : Length w = r/l m = 4w2 E,K : Elliptic integrals
Coaxial cable, high frequency	$\frac{\mu_0 l}{2\pi}\ln\left(\frac{a_1}{a}\right)$	a1 : Outer radius a : Inner radius l : Length
Circular loop	$\mu_0 r \cdot \left(\ln\left(\frac{8r}{a}\right) - 2 + \frac{Y}{2} + O\left(a^2/r^2\right)\right)$	r : Loop radius a : Wire radius
Rectangle	$\frac{\mu_0}{\pi}\left(b\ln\left(\frac{2b}{a}\right) + d\ln\left(\frac{2d}{a}\right) - (b+d)\left(2 - \frac{Y}{2}\right) + 2\sqrt{b^2 + d^2}\right)$ $-\frac{\mu_0}{\pi}\left(b \cdot \operatorname{arsinh}\left(\frac{b}{d}\right) + d \cdot \operatorname{arsinh}\left(\frac{d}{b}\right) + O(a)\right)$	b, d : Border length d >> a, b >> a a : Wire radius
Pair of parallel wires	$\frac{\mu_0 l}{\pi}\left(\ln\left(\frac{d}{a}\right) + \frac{Y}{2}\right)$	a : Wire radius d : Distance, d ≥ 2a l : Length of pair
Pair of parallel wires, high frequency	$\frac{\mu_0 l}{\pi}\operatorname{arcosh}\left(\frac{d}{2a}\right) = \frac{\mu_0 l}{\pi}\ln\left(\frac{d}{2a} + \sqrt{\frac{d^2}{4a^2} - 1}\right)$	a : Wire radius d : Distance, d ≥ 2a l : Length of pair
Wire parallel to perfectly conducting wall	$\frac{\mu_0 l}{2\pi}\left(\ln\left(\frac{2d}{a}\right) + \frac{Y}{2}\right)$	a : Wire radius d : Distance, d ≥ a l : Length

Wire parallel to conducting wall, high frequency	$\dfrac{\mu_0 l}{2\pi} \operatorname{arcosh}\left(\dfrac{d}{a}\right) = \dfrac{\mu_0 l}{2\pi} \ln\left(\dfrac{d}{a} + \sqrt{\dfrac{d^2}{a^2} - 1}\right)$	a : Wire radius d : Distance, $d \geq a$ l : Length

The symbol μ_0 denotes the magnetic constant ($4\pi \times 10^{-7}$ H/m). For high frequencies the electric current flows in the conductor surface (skin effect), and depending on the geometry it sometimes is necessary to distinguish low and high frequency inductances. This is the purpose of the constant Y : $Y = 0$ when the current is uniformly distributed over the surface of the wire (skin effect), $Y = 1/2$ when the current is uniformly distributed over the cross-section of the wire. In the high frequency case, if conductors approach each other, an additional screening current flows in their surface, and expressions containing Y become invalid.

Inductance with Physical Symmetry

Inductance of a Solenoid

A solenoid is a long, thin coil, i.e. a coil whose length is much greater than the diameter. Under these conditions, and without any magnetic material used, the magnetic flux density B within the coil is practically constant and is given by

$$B = \mu_0 N i / l$$

where μ_0 is the magnetic constant, N the number of turns, i the current and l the length of the coil. Ignoring end effects, the total magnetic flux through the coil is obtained by multiplying the flux density B by the cross-section area A :

$$\Phi = \mu_0 N i A / l,$$

When this is combined with the definition of inductance,

$$L = N\Phi / i$$

it follows that the inductance of a solenoid is given by :

$$L = \mu_0 N^2 A / l.$$

A table of inductance for short solenoids of various diameter to length ratios has been calculated by Dellinger, Whittmore, and Ould.

This, and the inductance of more complicated shapes, can be derived from Maxwell's equations. For rigid air-core coils, inductance is a function of coil geometry and number of turns, and is independent of current.

Similar analysis applies to a solenoid with a magnetic core, but only if the length of the coil is much greater than the product of the relative permeability of the magnetic core and the diameter. That limits the simple analysis to low-

Electrical Power Systems Technology

permeability cores, or extremely long thin solenoids. Although rarely useful, the equations are,

$$B = \mu_0 \mu_r Ni/l$$

where μ_r the relative permeability of the material within the solenoid,

$$\Phi = \mu_0 \mu_r NiA/l$$

from which it follows that the inductance of a solenoid is given by :

$$L = \mu_0 \mu_r N^2 A/l.$$

where N is squared because of the definition of inductance.

Note that since the permeability of ferromagnetic materials changes with applied magnetic flux, the inductance of a coil with a ferromagnetic core will generally vary with current.

Inductance of a Coaxial Line

Let the inner conductor have radius r_i and permeability μ_i, let the dielectric between the inner and outer conductor have permeability μ_d, and let the outer conductor have inner radius r_{01}, outer radius r_{02}, and permeability μ_0. Assume that a DC current I flows in opposite directions in the two conductors, with uniform current density. The magnetic field generated by these currents points in the azimuthal direction and is a function of radius r; it can be computed using Ampère's law :

$$0 \le r \le r_i : B(r) = \frac{\mu_i I r}{2\pi r_i^2}$$

$$r_i \le r \le r_{01} : B(r) = \frac{\mu_d I}{2\pi r}$$

$$r_{01} \le r \le r_{02} : B(r) = \frac{\mu_0 I}{2\pi r} \left(\frac{r_{02}^2 - r^2}{r_{02}^2 - r_{01}^1} \right)$$

The flux per length l in the region between the conductors can be computed by drawing a surface containing the axis :

$$\frac{d\phi_d}{dl} = \int_{r_i}^{r_{01}} B(r) dr = \frac{\mu_d I}{2\pi} \ln \frac{r_{01}}{r_i}$$

Inside the conductors, L can be computed by equating the energy stored in an inductor, $\frac{1}{2} LI^2$, with the energy stored in the magnetic field :

$$\frac{1}{2} LI^2 = \int_V \frac{B^2}{2\mu} dV$$

For a cylindrical geometry with no l dependence, the energy per unit length is

$$\frac{1}{2} L'I^2 = \int_{r1}^{r2} \frac{B^2}{2\mu} 2\pi r \, dr$$

where L' is the inductance per unit length. For the inner conductor, the integral on the right-hand-side is $\dfrac{\mu_i I^2}{16\pi}$; for the outer conductor it is

$$\frac{\mu_o I^2}{4\pi}\left(\frac{r_{o2}^2}{r_{o2}^2-r_{o1}^2}\right)^2 \ln\frac{r_{o2}}{r_{o1}} - \frac{\mu_o I^2}{8\pi}\left(\frac{r_{o2}^2}{r_{o2}^2-r_{o1}^2}\right) - \frac{\mu_o I^2}{16\pi}$$

Solving for L' and summing the terms for each region together gives a total inductance per unit length of :

$$L' = \frac{\mu_i}{8\pi} + \frac{\mu_d}{2\pi}\ln\frac{r_{o1}}{r_i} + \frac{\mu_o}{2\pi}\left(\frac{r_{o2}^2}{r_{o2}^1-r_{o1}^2}\right)^2 \ln\frac{r_{o2}}{r_{o1}} - \frac{\mu_o}{4\pi}\left(\frac{r_{o2}^2}{r_{o2}^2-r_{o1}^2}\right) - \frac{\mu_o}{8\pi}$$

However, for a typical coaxial line application we are interested in passing (non-DC) signals at frequencies for which the resistive skin effect cannot be neglected. In most cases, the inner and outer conductor terms are negligible, in which case one may approximate

$$L' = \frac{dL}{dl} \approx \frac{\mu_d}{2\pi}\ln\frac{r_{o1}}{r_i}$$

Phasor Circuit Analysis and Impedance

Using phasors, the equivalent impedance of an inductance is given by :

$$Z_L = V/I = j\omega L$$

where :

    j is the imaginary unit,
    L is the inductance,
    $\omega = 2\pi f$ is the angular frequency,
    f is the frequency and
    $\omega L = X_L$ is the inductive reactance.

Non-linear Inductance

Many inductors make use of magnetic materials. These materials over a large enough range exhibit a non-linear permeability with such effects as saturation. This in-turn makes the resulting inductance a function of the applied current. Faraday's Law still holds but inductance is ambiguous and is different whether you are calculating circuit parameters or magnetic fluxes.

The secant or large-signal inductance is used in flux calculations. It is defined as :

$$L_s(i) \stackrel{\text{def}}{=} \frac{N\Phi}{i} = \frac{\Lambda}{i}$$

Electrical Power Systems Technology

The differential or small-signal inductance, on the other hand, is used in calculating voltage. It is defined as :

$$L_d(i) \stackrel{\text{def}}{=} \frac{d(N\Phi)}{di} = \frac{d\Lambda}{di}$$

The circuit voltage for a non-linear inductor is obtained via the differential inductance as shown by Faraday's Law and the chain rule of calculus.

$$v(t) = \frac{d\Lambda}{dt} = \frac{d\Lambda}{di}\frac{di}{dt} = L_d(i)\frac{di}{dt}$$

There are similar definitions for non-linear mutual inductances.

Electrical Impedance

Electrical impedance is the measure of the opposition that a circuit presents to a current when a voltage is applied.

In quantitative terms, it is the complex ratio of the voltage to the current in an alternating current (AC) circuit. Impedance extends the concept of resistance to AC circuits, and possesses both magnitude and phase, unlike resistance, which has only magnitude. When a circuit is driven with direct current (DC), there is no distinction between impedance and resistance; the latter can be thought of as impedance with zero phase angle.

It is necessary to introduce the concept of impedance in AC circuits because there are two additional impeding mechanisms to be taken into account besides the normal resistance of DC circuits : the induction of voltages in conductors self-induced by the magnetic fields of currents (inductance), and the electrostatic storage of charge induced by voltages between conductors (capacitance). The impedance caused by these two effects is collectively referred to as reactance and forms the imaginary part of complex impedance whereas resistance forms the real part.

The symbol for impedance is usually Z and it may be represented by writing its magnitude and phase in the form $|Z|\angle\theta$. However, complex number representation is often more powerful for circuit analysis purposes. The term *impedance* was coined by Oliver Heaviside in July 1886. Arthur Kennelly was the first to represent impedance with complex numbers in 1893.

Impedance is defined as the frequency domain ratio of the voltage to the current. In other words, it is the voltage–current ratio for a single complex exponential at a particular frequency ω. In general, impedance will be a complex number, with the same units as resistance, for which the SI unit is the ohm (Ω). For a sinusoidal current or voltage input, the polar form of the complex impedance relates the amplitude and phase of the voltage and current. In particular,

- The magnitude of the complex impedance is the ratio of the voltage amplitude to the current amplitude.
- The phase of the complex impedance is the phase shift by which the current lags the voltage.

The reciprocal of impedance is admittance (*i.e.,* admittance is the current-to-voltage ratio, and it conventionally carries units of siemens, formerly called mhos).

Complex Impedance

Impedance is represented as a complex quantity Z and the term *complex impedance* may be used interchangeably; the polar form conveniently captures both magnitude and phase characteristics,

$$Z = |Z| e^{j \arg(Z)}$$

where the magnitude $|Z|$ represents the ratio of the voltage difference amplitude to the current amplitude, while the argument $\arg(Z)$ commonly given the symbol θ) gives the phase difference between voltage and current. j is the imaginary unit, and is used instead of i in this context to avoid confusion with the symbol for electric current. In Cartesian form,

$$Z = R + jX$$

where the real part of impedance is the resistance R and the imaginary part is the reactance X.

Where it is required to add or subtract impedances the cartesian form is more convenient, but when quantities are multiplied or divided the calculation becomes simpler if the polar form is used. A circuit calculation, such as finding the total impedance of two impedances in parallel, may require conversion between forms several times during the calculation. Conversion between the forms follows the normal conversion rules of complex numbers.

Ohm's Law

The meaning of electrical impedance can be understood by substituting it into Ohm's law.

$$V = IZ = I|Z| e^{j \arg(Z)}$$

The magnitude of the impedance $|Z|$ acts just like resistance, giving the drop in voltage amplitude across an impedance Z for a given current I. The phase factor tells us that the current lags the voltage by a phase of $\theta = \arg(Z)$ (*i.e.,* in the time domain, the current signal is shifted $\frac{\theta}{2\pi}T$ later with respect to the voltage signal).

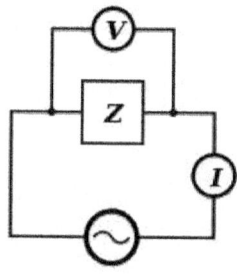

Fig. : An AC supply applying a voltage V, across a load Z, driving a current I.

Electrical Power Systems Technology

Just as impedance extends Ohm's law to cover AC circuits, other results from DC circuit analysis such as voltage division, current division, Thévenin's theorem, and Norton's theorem can also be extended to AC circuits by replacing resistance with impedance.

Complex Voltage and Current

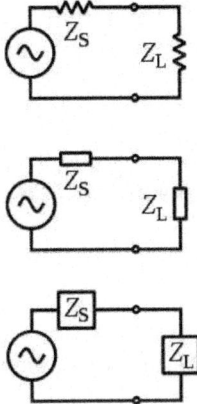

Fig. : Generalized impedances in a circuit can be drawn with the same symbol as a resistor (US ANSI or DIN Euro) or with a labeled box.

In order to simplify calculations, sinusoidal voltage and current waves are commonly represented as complex-valued functions of time denoted as V and I.

$$V = |V| e^{j(\omega t + \phi_v)}$$
$$I = |I| e^{j(\omega t + \phi_I)}$$

Impedance is defined as the ratio of these quantities.

$$Z = \frac{V}{I}$$

Substituting these into Ohm's law we have

$$|V| e^{j(\omega t + \phi_v)} = |I| e^{j(\omega t + \phi_I)} |Z| e^{j\theta}$$
$$= |I| |Z| e^{j(\omega t + \phi_I + \theta)}$$

Noting that this must hold for all t, we may equate the magnitudes and phases to obtain

$$|V| = |I| |Z|$$
$$\phi_v = \phi_I + \theta$$

The magnitude equation is the familiar Ohm's law applied to the voltage and current amplitudes, while the second equation defines the phase relationship.

Validity of Complex Representation

This representation using complex exponentials may be justified by noting that (by Euler's formula) :

$$\cos(\omega t + \phi) = \frac{1}{2}\left[e^{j(\omega t + \phi)} + e^{-j(\omega t + \phi)} \right]$$

The real-valued sinusoidal function representing either voltage or current may be broken into two complex-valued functions. By the principle of superposition, we may analyse the behaviour of the sinusoid on the left-hand side by analysing the behaviour of the two complex terms on the right-hand side. Given the symmetry, we only need to perform the analysis for one right-hand term; the results will be identical for the other. At the end of any calculation, we may return to real-valued sinusoids by further noting that

$$\cos(\omega t + \phi) = \Re\{e^{j(\omega t + \phi)}\}$$

Phasors

A phasor is a constant complex number, usually expressed in exponential form, representing the complex amplitude (magnitude and phase) of a sinusoidal function of time. Phasors are used by electrical engineers to simplify computations involving sinusoids, where they can often reduce a differential equation problem to an algebraic one.

The impedance of a circuit element can be defined as the ratio of the phasor voltage across the element to the phasor current through the element, as determined by the relative amplitudes and phases of the voltage and current. This is identical to the definition from Ohm's law given above, recognising that the factors of $e^{j\omega t}$ cancel.

Device Examples

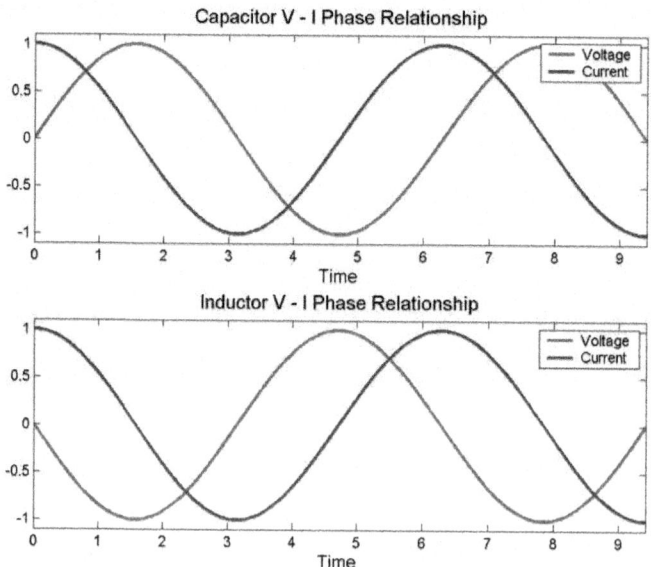

Fig. : The phase angles in the equations for the impedance of inductors and capacitors indicate that the voltage across a capacitor lags the current through it by a phase of $\pi/2$, while the voltage across an inductor leads the current through it by $\pi/2$. The identical voltage and current amplitudes indicate that the magnitude of the impedance is equal to one.

The impedance of an ideal resistor is purely real and is referred to as a *resistive impedance* :

$$Z_R = R$$

In this case, the voltage and current waveforms are proportional and in phase. Ideal inductors and capacitors have a purely imaginary *reactive impedance* : the impedance of inductors increases as frequency increases;

$$Z_L = j\omega L$$

the impedance of capacitors decreases as frequency increases;

$$Z_C = \frac{1}{j\omega C}$$

In both cases, for an applied sinusoidal voltage, the resulting current is also sinusoidal, but in quadrature, 90 degrees out of phase with the voltage. However, the phases have opposite signs : in an inductor, the current is *lagging*; in a capacitor the current is *leading*.

Note the following identities for the imaginary unit and its reciprocal :

$$j \equiv \cos\left(\frac{\pi}{2}\right) + j\sin\left(\frac{\pi}{2}\right) \equiv e^{j\frac{\pi}{2}}$$

$$\frac{1}{j} \equiv -j \equiv \cos\left(-\frac{\pi}{2}\right) + j\sin\left(-\frac{\pi}{2}\right) \equiv e^{j(-\frac{\pi}{2})}$$

Thus the inductor and capacitor impedance equations can be rewritten in polar form :

$$Z_L = \omega L e^{j\frac{\pi}{2}}$$

$$Z_C = \frac{1}{\omega C} e^{j(-\frac{\pi}{2})}$$

The magnitude gives the change in voltage amplitude for a given current amplitude through the impedance, while the exponential factors give the phase relationship.

Deriving the Device-specific Impedances

What follows below is a derivation of impedance for each of the three basic circuit elements : the resistor, the capacitor, and the inductor. Although the idea can be extended to define the relationship between the voltage and current of any arbitrary signal, these derivations will assume sinusoidal signals, since any arbitrary signal can be approximated as a sum of sinusoids through Fourier analysis.

Resistor

For a resistor, there is the relation :

$$v_R(t) = i_R(t)\,R$$

This is Ohm's law.

Considering the voltage signal to be

$$v_R(t) = V_p \sin(\omega t)$$

it follows that

$$\frac{v_R(t)}{i_R(t)} = \frac{V_p \sin(\omega t)}{I_p \sin(\omega t)} = R$$

This says that the ratio of AC voltage amplitude to alternating current (AC) amplitude across a resistor is R, and that the AC voltage leads the current across a resistor by 0 degrees.

This result is commonly expressed as

$$Z_{resistor} = R$$

Capacitor

For a capacitor, there is the relation:

$$i_C(t) = C \frac{d v_c(t)}{dt}$$

Considering the voltage signal to be

$$v_c(t) = V_p \sin(\omega t)$$

it follows that

$$\frac{d v_C(t)}{dt} = \omega V_p \cos(\omega t)$$

And thus

$$\frac{v_C(t)}{i_C(t)} = \frac{V_p \sin(\omega t)}{\omega V_p C \cos(\omega t)} = \frac{\sin(\omega t)}{\omega_C \sin(\omega t + \frac{\pi}{2})}$$

This says that the ratio of AC voltage amplitude to AC current amplitude across a capacitor is $\frac{1}{\omega C}$, and that the AC voltage lags the AC current across a capacitor by 90 degrees (or the AC current leads the AC voltage across a capacitor by 90 degrees).

This result is commonly expressed in polar form, as

$$Z_{capacitor} = \frac{1}{\omega C} e^{-j\frac{\pi}{2}}$$

or, by applying Euler's formula, as

$$Z_{capacitor} = -j \frac{1}{\omega C} = \frac{1}{j\omega C}$$

ELECTRICAL POWER SYSTEMS TECHNOLOGY

Inductor

For the inductor, we have the relation :

$$v_L(t) = L \frac{d\, i_L(t)}{d\, t}$$

This time, considering the current signal to be

$$i_L(t) = I_p \sin(\omega t)$$

it follows that

$$\frac{d\, i_L(t)}{d\, t} = \omega I_p \cos(\omega t)$$

And thus

$$\frac{v_L(t)}{i_L(t)} = \frac{\omega I_p L \cos(\omega t)}{I_p \sin(\omega t)} = \frac{\omega L \sin\left(\omega t + \frac{\pi}{2}\right)}{\sin(\omega t)}$$

This says that the ratio of AC voltage amplitude to AC current amplitude across an inductor is ω_L, and that the AC voltage leads the AC current across an inductor by 90 degrees.

This result is commonly expressed in polar form, as

$$Z_{inductor} = \omega L e^{j\frac{\pi}{2}}$$

or, using Euler's formula, as

$$Z_{inductor} = j\omega L$$

Generalised s-plane Impedance

Impedance defined in terms of $j\omega$ can strictly only be applied to circuits which are driven with a steady-state AC signal. The concept of impedance can be extended to a circuit energised with any arbitrary signal by using complex frequency instead of $j\omega$. Complex frequency is given the symbol s and is, in general, a complex number. Signals are expressed in terms of complex frequency by taking the Laplace transform of the time domain expression of the signal. The impedance of the basic circuit elements in this more general notation is as follows :

Element	Impedance expression
Resistor	R
Inductor	sL
Capacitor	$\dfrac{1}{sC}$

For a DC circuit this simplifies to $s = 0$. For a steady-state sinusoidal AC signal $s = j\omega$.

Resistance vs Reactance

Resistance and reactance together determine the magnitude and phase of the impedance through the following relations:

$$|Z| = \sqrt{ZZ^*} = \sqrt{R^2 + X^2}$$

$$\theta = \arctan\left(\frac{X}{R}\right)$$

In many applications the relative phase of the voltage and current is not critical so only the magnitude of the impedance is significant.

Resistance

Resistance R is the real part of impedance; a device with a purely resistive impedance exhibits no phase shift between the voltage and current.

$$R = |Z|\cos\theta$$

Reactance

Reactance X is the imaginary part of the impedance; a component with a finite reactance induces a phase shift θ between the voltage across it and the current through it.

$$X = |Z|\sin\theta$$

A purely reactive component is distinguished by the sinusoidal voltage across the component being in quadrature with the sinusoidal current through the component. This implies that the component alternately absorbs energy from the circuit and then returns energy to the circuit. A pure reactance will not dissipate any power.

Capacitive Reactance

A capacitor has a purely reactive impedance which is inversely proportional to the signal frequency. A capacitor consists of two conductors separated by an insulator, also known as a dielectric.

$$X_C = (\omega_C)^{-1} = (2\pi fC)^{-1}$$

At low frequencies a capacitor is open circuit, as no charge flows in the dielectric. A DC voltage applied across a capacitor causes charge to accumulate on one side; the electric field due to the accumulated charge is the source of the opposition to the current. When the potential associated with the charge exactly balances the applied voltage, the current goes to zero.

Driven by an AC supply, a capacitor will only accumulate a limited amount of charge before the potential difference changes sign and the charge dissipates. The higher the frequency, the less charge will accumulate and the smaller the opposition to the current.

Electrical Power Systems Technology

Inductive Reactance

Inductive reactance X_L is proportional to the signal frequency f and the inductance L.

$$X_L = \omega L = 2\pi f L$$

An inductor consists of a coiled conductor. Faraday's law of electromagnetic induction gives the back emf ε (voltage opposing current) due to a rate-of-change of magnetic flux density B through a current loop.

$$\varepsilon = -\frac{d\Phi_B}{dt}$$

For an inductor consisting of a coil with N loops this gives.

$$\varepsilon = -N\frac{d\Phi_B}{dt}$$

The back-emf is the source of the opposition to current flow. A constant direct current has a zero rate-of-change, and sees an inductor as a short-circuit (it is typically made from a material with a low resistivity). An alternating current has a time-averaged rate-of-change that is proportional to frequency, this causes the increase in inductive reactance with frequency.

Total Reactance

The total reactance is given by

$$X = X_L - X_C$$

so that the total impedance is

$$Z = R + jX$$

Combining Impedances

The total impedance of many simple networks of components can be calculated using the rules for combining impedances in series and parallel. The rules are identical to those used for combining resistances, except that the numbers in general will be complex numbers. In the general case however, equivalent impedance transforms in addition to series and parallel will be required.

Series Combination

For components connected in series, the current through each circuit element is the same; the total impedance is the sum of the component impedances.

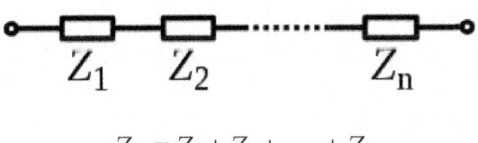

$$Z_{eq} = Z_1 + Z_2 + \ldots + Z_n$$

Or explicitly in real and imaginary terms :
$$Z_{eq} = R + jX = (R_1 + R_2 + \cdots + R_n) + j(X_1 + X_2 + \cdots + X_n)$$

Parallel Combination

For components connected in parallel, the voltage across each circuit element is the same; the ratio of currents through any two elements is the inverse ratio of their impedances.

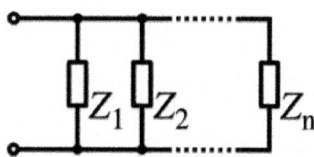

Hence the inverse total impedance is the sum of the inverses of the component impedances :
$$\frac{1}{Z_{eq}} = \frac{1}{Z_1} + \frac{1}{Z_2} + \cdots + \frac{1}{Z_n}$$

or, when n = 2 :
$$\frac{1}{Z_{eq}} = \frac{1}{Z_1} + \frac{1}{Z_2} = \frac{Z_1 + Z_2}{Z_1 Z_2}$$

$$Z_{eq} = \frac{Z_1 Z_2}{Z_1 + Z_2}$$

The equivalent impedance Z_{eq} can be calculated in terms of the equivalent series resistance R_{eq} and reactance X_{eq}.
$$Z_{eq} = R_{eq} + jX_{eq}$$
$$R_{eq} = \frac{(X_1 R_2 + X_2 R_1)(X_1 + X_2) + (R_1 R_2 - X_1 X_2)(R_1 + R_2)}{(R_1 + R_2)^2 + (X_1 + X_2)^2}$$
$$X_{eq} = \frac{(X_1 R_2 + X_2 R_1)(R_1 + R_2) - (R_1 R_2 - X_1 X_2)(X_1 + X_2)}{(R_1 + R_2)^2 + (X_1 + X_2)^2}$$

Measurement

The measurement of the impedance of devices and transmission lines is a practical problem in radio technology and others. Measurements of impedance may be carried out at one frequency, or the variation of device impedance over a range of frequencies may be of interest. The impedance may be measured or displayed directly in ohms, or other values related to impedance may be displayed; for example in a radio antenna the standing wave ratio or reflection coefficient may be more useful than the impedance alone. Measurement of impedance requires measurement of the magnitude of voltage and current, and the phase difference between them. Impedance is often measured by "bridge" methods, similar to the direct-current Wheatstone bridge; a calibrated reference impedance is adjusted to balance off the effect of the impedance of the device under test. Impedance measurement in power electronic devices may require simultaneous measurement and provision of power to the operating device.

The impedance of a device can be calculated by complex division of the voltage and current. The impedance of the device can be calculated by applying a sinusoidal voltage to the device in series with a resistor, and measuring the voltage across the resistor and across the device. Performing this measurement by sweeping the frequencies of the applied signal provides the impedance phase and magnitude.

The use of an impulse response may be used in combination with the fast Fourier transform (FFT) to rapidly measure the electrical impedance of various electrical devices.

The LCR meter (Inductance (L), Capacitance (C), and Resistance (R)) is a device commonly used to measure the inductance, resistance and capacitance of a component; from these values the impedance at any frequency can be calculated.

Variable Impedance

In general, neither impedance nor admittance can be time varying as they are defined for complex exponentials for $-\infty < t < +\infty$. If the complex exponential voltage–current ratio changes over time or amplitude, the circuit element cannot be described using the frequency domain. However, many systems (*e.g.*, varicaps that are used in radio tuners) may exhibit non-linear or time-varying voltage–current ratios that appear to be linear time-invariant (LTI) for small signals over small observation windows; hence, they can be roughly described as having a time-varying impedance. That is, this description is an approximation; over large signal swings or observation windows, the voltage–current relationship is non-LTI and cannot be described by impedance.

ADMITTANCE

In electrical engineering, **admittance** is a measure of how easily a circuit or device will allow a current to flow. It is defined as the inverse of impedance. The SI unit of admittance is the siemens (symbol S). Oliver Heaviside coined the term *admittance* in December 1887.

Admittance is defined as

$$Y \equiv \frac{1}{Z}$$

where :
    Y is the admittance, measured in siemens
    Z is the impedance, measured in ohms.

The synonymous unit mho, and the symbol ℧ (an upside-down uppercase omega Ω), are also in common use.

Resistance is a measure of the opposition of a circuit to the flow of a steady current, while impedance takes into account not only the resistance but also dynamic effects (known as reactance). Likewise, admittance is not only a measure of the ease with which a steady current can flow, but also the dynamic effects of the material's susceptance to polarization :

$$Y = G + jB$$

where :

- Y is the admittance, measured in siemens.
- G is the conductance, measured in siemens.
- B is the susceptance, measured in siemens.
- $j^2 = -1$.

Conversion from Impedance to Admittance

The impedance, Z, is composed of real and imaginary parts,

$$Z = R + jX$$

where :

- R is the resistance, measured in ohms
- X is the reactance, measured in ohms.

$$Y = Z^{-1} = \frac{1}{R + jX} = \left(\frac{1}{R^2 + X^2}\right)(R - jX)$$

Admittance, just like impedance, is a complex number, made up of a real part (the conductance, G), and an imaginary part (the susceptance, B), thus :

$$Y = G + jB$$

where G (conductance) and B (susceptance) are given by :

$$G = \Re(Y) = \frac{R}{R^2 + X^2}$$
$$B = \Im(Y) = -\frac{X}{R^2 + X^2}$$

The magnitude and phase of the admittance are given by :

$$|Y| = \sqrt{G^2 + B^2} = \frac{1}{\sqrt{R^2 + X^2}}$$
$$\angle Y = \arctan\left(\frac{B}{G}\right) = \arctan\left(-\frac{X}{R}\right)$$

where :

- G is the conductance, measured in siemens
- B is the susceptance, also measured in siemens

Note that the signs of reactances become reversed in the admittance domain; *i.e.* capacitive susceptance is positive and inductive suceptance is negative.

Chapter 4

POWER SYSTEM FUNDAMENTALS

An electric power system is a network of electrical components used to supply, transmit and use electric power. An example of an electric power system is the network that supplies a region's homes and industry with power - for sizable regions, this power system is known as the grid and can be broadly divided into the generators that supply the power, the transmission system that carries the power from the generating centres to the load centres and the distribution system that feeds the power to nearby homes and industries. Smaller power systems are also found in industry, hospitals, commercial buildings and homes. The majority of these systems rely upon three-phase AC power - the standard for large-scale power transmission and distribution across the modern world. Specialised power systems that do not always rely upon three-phase AC power are found in aircraft, electric rail systems, ocean liners and automobiles.

HISTORY

In 1881 two electricians built the world's first power system at Godalming in England. It was powered by a power station consisting of two waterwheels that produced an alternating current that in turn supplied seven Siemens arc lamps at 250 volts and 34 incandescent lamps at 40 volts. However supply to the lamps was intermittent and in 1882 Thomas Edison and his company, The Edison Electric Light Company, developed the first steam powered electric power station on Pearl Street in New York City. The Pearl Street Station initially powered around 3,000 lamps for 59 customers. The power station used direct current and operated at a single voltage. Direct current power could not be easily transformed to the higher voltages necessary to minimise power loss during long-distance transmission, so the maximum economic distance between the generators and load was limited to around half-a-mile (800 m).

That same year in London Lucien Gaulard and John Dixon Gibbs demonstrated the first transformer suitable for use in a real power system. The practical value of Gaulard and Gibbs' transformer was demonstrated in 1884 at Turin where the transformer was used to light up forty kilometres (25 miles) of railway from a single alternating current generator. Despite the success of the system, the pair made some fundamental mistakes. Perhaps the most serious was connecting the primaries of the transformers in series so that active lamps would affect the brightness of other lamps further down the line. Following the demonstration George Westinghouse, an American entrepreneur, imported a number of the transformers along with a Siemens generator and set his engineers to experimenting with them in the hopes of improving them for use in a commercial power system. In July 1888, Westinghouse also licensed Nikola Tesla's US patents for a polyphase AC induction motor and transformer designs and hired Tesla for one year to be a consultant at the Westinghouse Electric & Manufacturing Company's Pittsburgh labs.

One of Westinghouse's engineers, William Stanley, recognised the problem with connecting transformers in series as opposed to parallel and also realised that making the iron core of a transformer a fully enclosed loop would improve the voltage regulation of the secondary winding. Using this knowledge he built a much improved alternating current power system at Great Barrington, Massachusetts in 1886.

By 1890 the electric power industry was flourishing, and power companies had built thousands of power systems (both direct and alternating current) in the United States and Europe. These networks were effectively dedicated to providing electric lighting. During this time a fierce rivalry known as the "War of Currents" emerged between Thomas Edison and George Westinghouse over which form of transmission (direct or alternating current) was superior. In 1891, Westinghouse installed the first major power system that was designed to drive a 100 horsepower (75 kW) synchronous electric motor, not just provide electric lighting, at Telluride, Colorado. On the other side of the Atlantic, Oskar von Miller built a 20 kV 176 km three-phase transmission line from Lauffen am Neckar to Frankfurt am Main for the Electrical Engineering Exhibition in Frankfurt. In 1895, after a protracted decision-making process, the Adams No. 1 generating station at Niagara Falls began transferring three-phase alternating current power to Buffalo at 11 kV. Following completion of the Niagara Falls project, new power systems increasingly chose alternating current as opposed to direct current for electrical transmission.

Developments in power systems continued beyond the nineteenth century. In 1936 the first experimental HVDC (high voltage direct current) line using mercury arc valves was built between Schenectady and Mechanicville, New York. HVDC had previously been achieved by series-connected direct current generators and motors (the Thury system) although this suffered from serious reliability issues. In 1957 Siemens demonstrated the first solid-state rectifier, but it was not until the early 1970s that solid-state devices became the standard in HVDC. In recent times, many important developments have come from extending innovations in the ICT field to the power engineering field. For example, the development of computers meant load flow studies could be run more efficiently allowing for much better planning of power systems. Advances in information technology and telecommunication also allowed for remote control of a power system's switchgear and generators.

BASICS OF ELECTRIC POWER

Electric power is the product of two quantities : current and voltage. These two quantities can vary with respect to time (AC power) or can be kept at constant levels (DC power).

Most refrigerators, air conditioners, pumps and industrial machinery use AC power whereas most computers and digital equipment use DC power (the digital devices you plug into the mains typically have an internal or external power adapter to convert from AC to DC power). AC power has the advantage of being easy to transform between voltages and is able to be generated and utilised by brushless machinery. DC power remains the only practical choice in digital systems and can be more economical to transmit over long distances at very high voltages.

The ability to easily transform the voltage of AC power is important for two reasons : Firstly, power can be transmitted over long distances with less loss at higher voltages. So in power systems where generation is distant from the load, it is desirable to step-up (increase) the voltage of power at the generation point and then step-down (decrease) the voltage near the load. Secondly, it is often more economical to install turbines that produce higher voltages than would be used by most appliances, so the ability to easily transform voltages means this mismatch between voltages can be easily managed.

Solid state devices, which are products of the semi-conductor revolution, make it possible to transform DC power to different voltages, build brushless DC machines and convert between AC and DC power. Nevertheless devices utilising solid state technology are often more expensive than their traditional counterparts, so AC power remains in widespread use.

BALANCING THE GRID

One of the main difficulties in power systems is that the amount of active power consumed plus losses should always equal the active power produced. If more power would be produced than consumed the frequency would rise and vice versa. Even small deviations from the nominal frequency value would damage synchronous machines and other appliances. Making sure the frequency is constant is usually the task of a transmission system operator. In some countries (for example in the European Union) this is achieved through a balancing market using ancillary services.

COMPONENTS OF POWER SYSTEMS

Supplies

All power systems have one or more sources of power. For some power systems, the source of power is external to the system but for others it is part of the system itself. Direct current power can be supplied by batteries, fuel cells or photovoltaic cells. Alternating current power is typically supplied by a rotor that spins in a magnetic field in a device known as a turbo generator. There have been a wide range of techniques used to spin a turbine's rotor, from steam heated using fossil fuel (including coal, gas and oil) or nuclear energy, falling water (hydroelectric power) and wind (wind power).

The speed at which the rotor spins in combination with the number of generator poles determines the frequency of the alternating current produced by the generator. All generators on a single synchronous system, for example the national grid, rotate at sub-multiples of the same speed and so generate electrical current at the same frequency. If the load on the system increases, the generators will require more torque to spin at that speed and, in a typical power station, more steam must be supplied to the turbines driving them. Thus the steam used and the fuel expended are directly dependent on the quantity of electrical energy supplied. An exception exists for generators incorporating power electronics such as gearless wind turbines or linked to a grid through an asynchronous tie such as a HVDC link – these can operate at frequencies independent of the power system frequency.

Depending on how the poles are fed, alternating current generators can produce a variable number of phases of power. A higher number of phases leads to more efficient power system operation but also increases the infrastructure requirements of the system.

Electricity grid systems connect multiple generators and loads operating at the same frequency and number of phases, the commonest being three-phase at 50 or 60 Hz. However there are other considerations. These range from the obvious: How much power should the generator be able to supply? What is an acceptable length of time for starting the generator (some generators can take hours to start)? Is the availability of the power source acceptable (some renewables are only available when the sun is shining or the wind is blowing)? To the more technical: How should the generator start (some turbines act like a motor to bring themselves up to speed in which case they need an appropriate starting circuit)? What is the mechanical speed of operation for the turbine and consequently what are the number of poles required? What type of generator is suitable (synchronous or asynchronous) and what type of rotor (squirrel-cage rotor, wound rotor, salient pole rotor or cylindrical rotor)?

Loads

Power systems deliver energy to loads that perform a function. These loads range from household appliances to industrial machinery. Most loads expect a certain voltage and, for alternating current devices, a certain frequency and number of phases. The appliances found in your home, for example, will typically be single-phase operating at 50 or 60 Hz with a voltage between 110 and 260 volts (depending on national standards). An exception exists for centralized air conditioning systems as these are now typically three-phase because this allows them to operate more efficiently. All devices in your house will also have a wattage, this specifies the amount of power the device consumes. At any one time, the net amount of power consumed by the loads on a power system must equal the net amount of power produced by the supplies less the power lost in transmission.

Making sure that the voltage, frequency and amount of power supplied to the loads is in line with expectations is one of the great challenges of power system engineering. However it is not the only challenge, in addition to the power used by a load to do useful work (termed real power) many alternating current devices also use an additional amount of power because they cause the alternating voltage and alternating current to become slightly out-of-sync (termed reactive power). The reactive power like the real power must balance (that is the reactive power produced on a system must equal the reactive power consumed) and can be supplied from the generators, however it is often more economical to supply such power from capacitors.

A final consideration with loads is to do with power quality. In addition to sustained overvoltages and undervoltages (voltage regulation issues) as well as sustained deviations from the system frequency (frequency regulation issues), power system loads can be adversely affected by a range of temporal issues. These include voltage sags, dips and swells, transient overvoltages, flicker, high frequency noise, phase imbalance and poor power factor. Power quality issues occur when the power supply to a load deviates from the ideal: For an AC supply, the ideal is the current and voltage in-sync fluctuating as a perfect sine wave at a prescribed frequency with the voltage at a prescribed amplitude. For DC supply, the ideal is the voltage not varying from a prescribed level. Power quality issues can be especially important when it comes to specialist industrial machinary or hospital equipment.

Conductors

Conductors carry power from the generators to the load. In a grid, conductors may be classified as belonging to the transmission system, which carries large amounts of power at high voltages (typically more than 50 kV) from the generating centres to the load centres, or the distribution system, which feeds smaller amounts of power at lower voltages (typically less than 50 kV) from the load centres to nearby homes and industry.

Choice of conductors is based upon considerations such as cost, transmission losses and other desirable characteristics of the metal like tensile strength. Copper, with lower resistivity than aluminium, was the conductor of choice for most power systems. However, aluminum has lower cost for the same current carrying capacity and is the primary metal used for transmission line conductors. Overhead line conductors may be reinforced with steel or aluminum alloys.

Conductors in exterior power systems may be placed overhead or underground. Overhead conductors are usually air insulated and supported on porcelain, glass or polymer insulators. Cables used for underground transmission or building wiring are insulated with cross-linked polyethylene or other flexible insulation. Large conductors are stranded for ease of handling; small conductors used for building wiring are often solid, especially in light commercial or residential construction.

Conductors are typically rated for the maximum current that they can carry at a given temperature rise over ambient conditions. As current flow increases through a conductor it heats up. For insulated conductors, the rating is determined by the insulation. For overhead conductors, the rating is determined by the point at which the sag of the conductors would become unacceptable.

Capacitors and Reactors

The majority of the load in a typical AC power system is inductive; the current lags behind the voltage. Since the voltage and current are out-of-sync, this leads to the emergence of a "useless" form of power known as reactive power. Reactive power does no measurable work but is transmitted back and forth between the reactive power source and load every cycle. This reactive power can be provided by the generators themselves but it is often cheaper to provide it through capacitors, hence capacitors are often placed near inductive loads to reduce current demand on the power system. Power factor correction may be applied at a central substation or adjacent to large loads.

Reactors consume reactive power and are used to regulate voltage on long transmission lines. In light load conditions, where the loading on transmission lines is well below the surge impedance loading, the efficiency of the power system may actually be improved by switching in reactors. Reactors installed in series in a power system also limit rushes of current flow, small reactors are therefore almost always installed in series with capacitors to limit the current rush associated with switching in a capacitor. Series reactors can also be used to limit fault currents.

Capacitors and reactors are switched by circuit breakers, which results in moderately large steps in reactive power. A solution comes in the form of static VAR compensators and static synchronous compensators. Briefly, static VAR compensators work by switching in capacitors using thyristors as opposed to circuit breakers allowing capacitors to be switched-in and switched-out within a single cycle. This provides a far more refined response than circuit breaker switched capacitors. Static synchronous compensators take a step further by achieving reactive power adjustments using only power electronics.

Power Electronics

Power electronics are semi-conductor based devices that are able to switch quantities of power ranging from a few hundred watts to several hundred megawatts. Despite their relatively simple function, their speed of operation (typically in the order of nanoseconds) means they are capable of a wide range of tasks that would be difficult or impossible with conventional technology. The classic function of power electronics is rectification, or the conversion of AC-to-DC power, power electronics are therefore found in almost every digital device that is supplied from an AC source either as an adapter that plugs into the wall or as component internal to the device. High-powered power electronics can also be used to convert AC power to DC power for long distance transmission in a system known as HVDC. HVDC is used because it proves to be more economical than similar high voltage AC systems for very long distances (hundreds to thousands of kilometres). HVDC is also desirable for interconnects because it allows frequency independence thus improving system stability. Power electronics are also essential for any power source that is required to produce an AC output but that by its nature produces a DC output. They are therefore used by many photovoltaic installations both industrial and residential.

Power electronics also feature in a wide range of more exotic uses. They are at the heart of all modern electric and hybrid vehicles - where they are used for both motor control and as part of the brushless DC motor. Power electronics are also found in practically all modern petrol-powered vehicles, this is because the power provided by the car's batteries alone is insufficient to provide ignition, air-conditioning, internal lighting, radio and dashboard displays for the life of the car. So the batteries must be recharged while driving using DC power from the engine - a feat that is typically accomplished using power electronics. Whereas conventional technology would be unsuitable for a modern electric car, commutators can and have been used in petrol-powered cars, the switch to alternators in combination with power electronics has occurred because of the improved durability of brushless machinery.

Some electric railway systems also use DC power and thus make use of power electronics to feed grid power to the locomotives and often for speed control of the locomotive's motor. In the middle twentieth century, rectifier locomotives were popular, these used power electronics to convert AC power from the railway network for use by a DC motor. Today most electric locomotives are supplied with AC power and run using AC motors, but still use power electronics to provide suitable motor control. The use of power electronics to assist with motor control and with starter circuits cannot be underestimated and, in addition to rectification, is responsible for power electronics appearing in a wide range of industrial machinery. Power electronics even appear in modern residential air conditioners.

Power electronics are also at the heart of the variable-speed wind turbine. Conventional wind turbines require significant engineering to ensure they operate at some ratio of the system frequency, however by using power electronics this requirement can be eliminated leading to quieter, more flexible and (at the moment) more costly wind turbines.

Protective Devices

Power systems contain protective devices to prevent injury or damage during failures. The quintessential protective device is the fuse. When the current through a fuse exceeds a certain threshold, the fuse element melts, producing an arc across the resulting gap that is then extinguished, interrupting the circuit. Given that fuses can be built as the weak point of a system, fuses are ideal for protecting circuitry from damage. Fuses however have two problems : First, after they have functioned, fuses must be replaced as they cannot be reset. This can prove inconvenient if the fuse is at a remote site or a spare fuse is not on hand. And second, fuses are typically inadequate as the sole safety device in most power systems as they allow current flows well in excess of that that would prove lethal to a human or animal.

The first problem is resolved by the use of circuit breakers - devices that can be reset after they have broken current flow. In modern systems that use less than about 10 kW, miniature circuit breakers are typically used. These devices combine the mechanism that initiates the trip (by sensing excess current) as well as the mechanism that breaks the current flow in a single unit. Some miniature circuit breakers operate solely on the basis of electromagnetism. In these miniature circuit breakers, the current is run through a solenoid, and, in the event of excess current flow, the magnetic pull of the solenoid is sufficient to force open the circuit breaker's contacts (often indirectly through a tripping mechanism). A better design however arises by inserting a bimetallic strip before the solenoid - this means that instead of always producing a magnetic force, the solenoid only produces a magnetic force when the current is strong enough to deform the bimetallic strip and complete the solenoid's circuit.

In higher powered applications, the protective relays that detect a fault and initiate a trip are separate from the circuit breaker. Early relays worked based upon electromagnetic principles similar to those mentioned in the previous paragraph, modern relays are application-specific computers that determine whether to trip based upon readings from the power system. Different relays will initiate trips depending upon different protection schemes. For example, an overcurrent relay might initiate a trip if the current on any phase exceeds a certain threshold whereas a set of differential relays might initiate a trip if the sum of currents between them indicates there may be current leaking to earth. The circuit breakers in higher powered applications are different too. Air is typically no longer sufficient to quell the arc that forms when the contacts are forced open so a variety of techniques are used. One of the most popular techniques is to keep the chamber enclosing the contacts flooded with sulfur hexafluoride (SF_6) - a non-toxic gas that has sound arc-quelling properties. Other techniques are discussed in the reference.

The second problem, the inadequacy of fuses to act as the sole safety device in most power systems, is probably best resolved by the use of residual current devices (RCDs). In any properly functioning electrical appliance the current flowing into the appliance on the active line should equal the current flowing out of the appliance on the neutral line. A residual current device works by monitoring the active and neutral lines and tripping the active line if it notices a difference. Residual current devices require a separate neutral line for each phase and to be able to trip within a time frame before harm occurs. This is typically not a problem in most residential applications where standard wiring provides an active and neutral line for each appliance (that's why your power plugs always have at least two tongs) and the voltages are relatively low however these issues do limit the effectiveness of RCDs in other applications such as industry. Even with the installation of an RCD, exposure to electricity can still prove lethal.

SCADA Systems

In large electric power systems, Supervisory Control And Data Acquisition (SCADA) is used for tasks such as switching on generators, controlling generator output and switching in or out system elements for maintenance. The first supervisory control systems implemented consisted of a panel of lamps and switches at a central console near the controlled plant. The lamps provided feedback on the state of plant (the data acquisition function) and the switches allowed adjustments to the plant to be made (the supervisory control function). Today, SCADA systems are much more sophisticated and, due to advances in communication systems, the consoles controlling the plant no longer need to be near the plant itself. Instead it is now common for plant to be controlled from a with equipment similar to (if not identical to) a desktop computer. The ability to control such plant through computers has increased the need for security and already there have been reports of cyber-attacks on such systems causing significant disruptions to power systems.

POWER SYSTEMS IN PRACTICE

Despite their common components, power systems vary widely both with respect to their design and how they operate.

Residential Power Systems

Residential dwellings almost always take supply from the low voltage distribution lines or cables that run past the dwelling. These operate at voltages of between 110 and 260 volts (phase-to-earth) depending upon national standards. A few decades ago small dwellings would be fed a single phase using a dedicated two-core service cable (one core for the active phase and one core for the neutral return). The active line would then be run through a main isolating switch in the fuse box and then split into one or more circuits to feed lighting and appliances inside the house. By convention, the lighting and appliance circuits are kept separate so the failure of an appliance does not leave the dwelling's occupants in the dark. All circuits would be fused with an appropriate fuse based upon the wire size used for that circuit. Circuits would have both an active and neutral wire with both the lighting and power sockets being connected in parallel. Sockets would also be provided with a protective earth. This would be made available to appliances to connect to any metallic casing. If this casing were to become live, the theory is the connection to earth would cause an RCD or fuse to trip - thus preventing the future electrocution of an occupant handling the appliance. Earthing systems vary between regions, but in countries such as the United Kingdom and Australia both the protective earth and neutral line would be earthed together near the fuse box before the main isolating switch and the neutral earthed once again back at the distribution transformer.

There have been a number of minor changes over the year to practice of residential wiring. Some of the most significant ways modern residential power systems tend to vary from older ones include :

- For convenience, miniature circuit breakers are now almost always used in the fuse box instead of fuses as these can easily be reset by occupants.
- For safety reasons, RCDs are now installed on appliance circuits and, increasingly, even on lighting circuits.
- Dwellings are typically connected to all three-phases of the distribution system with the phases being arbitrarily allocated to the house's single-phase circuits.
- Whereas air conditioners of the past might have been fed from a dedicated circuit attached to a single phase, centralised air conditioners that require three-phase power are now becoming common.
- Protective earths are now run with lighting circuits to allow for metallic lamp holders to be earthed.
- Increasingly residential power systems are incorporating micro-generators, most notably, photovoltaic cells.

Commercial Power Systems

Commercial power systems such as shopping centers or high-rise buildings are larger in scale than residential systems. Electrical designs for larger commercial systems are usually studied for load flow, short-circuit fault levels, and voltage drop for steady-state loads and during starting of large motors. The objectives of the studies are to assure proper equipment and conductor sizing, and to co-ordinate protective devices so that minimal disruption is cause when a fault is cleared. Large commercial installations will have an orderly system of sub-panels, separate from the main distribution board to allow for better system protection and more efficient electrical installation.

Typically one of the largest appliances connected to a commercial power system is the HVAC unit, and ensuring this unit is adequately supplied is an important consideration in commercial power systems. Regulations for commercial establishments place other requirements on commercial systems that are not placed on residential systems. For example, in Australia, commercial systems must comply with AS 2293, the standard for emergency lighting, which requires emergency lighting be maintained for at least 90 minutes in the event of loss of mains supply. In the United States, the National Electrical Code requires commercial systems to be built with at least one 20A sign outlet in order to light outdoor signage. Building code regulations may place special requirements on the electrical system for emergency lighting, evacuation, emergency power, smoke control and fire protection.

ELECTRICAL NETWORK

An **electrical network** is an interconnection of electrical elements such as resistors, inductors, capacitors, voltage sources, current sources and switches. An **electrical circuit** is a network consisting of a closed loop, giving a return path for the current. Linear electrical networks, a special type consisting only of sources (voltage or current), linear lumped elements (resistors, capacitors, inductors), and linear distributed elements (transmission lines), have the property that signals are linearly superimposable. They are thus more easily analyzed, using powerful frequency domain methods such as Laplace transforms, to determine DC response, AC response, and transient response.

A **resistive circuit** is a circuit containing only resistors and ideal current and voltage sources. Analysis of resistive circuits is less complicated than analysis of circuits containing capacitors and inductors. If the sources are constant (DC) sources, the result is a DC circuit.

A network that contains active electronic components is known as an *electronic circuit*. Such networks are generally non-linear and require more complex design and analysis tools.

Classification

By Passivity

An active network is a network that consists of at least one active source like a voltage source or current source

A passive network is a network which does not contain any active device.

By Linearity

A network is linear if its signals obey the principle of superposition; otherwise it is non-linear. A linear network will be composed entirely of independent sources, linear dependent sources and linear passive elements.

Classification of Sources

Sources can be classified as independent sources and dependent sources

Independent Sources

Ideal Independent Source maintains same voltage or current regardless of the other elements present in the circuit. Its value is either constant (DC) or sinusoidal (AC). The strength of voltage or current is not changed by any variation in connected network.

Dependent Sources

Dependent Sources depend upon a particular element of the circuit for delivering the power or voltage or current depending upon the type of source it is.

Design Methods

To design any electrical circuit, either analog or digital, electrical engineers need to be able to predict the voltages and currents at all places within the circuit. Linear circuits, that is, circuits with the same input and output frequency, can be analyzed by hand using complex number theory. Other circuits can only be analyzed with specialized software programs or estimation techniques such as the piecewise-linear model.

Circuit simulation software, such as HSPICE, and languages such as VHDL-AMS and verilog-AMS allow engineers to design circuits without the time, cost and risk of error involved in building circuit prototypes.

Other more complex laws may be needed if the network contains non-linear or reactive components. Non-linear self-regenerative heterodyning systems can be approximated. Applying these laws results in a set of simultaneous equations that can be solved either algebraically or numerically.

Network Simulation Software

More complex circuits can be analyzed numerically with software such as SPICE or GNUCAP, or symbolically using software such as SapWin.

Linearization Around Operating Point

When faced with a new circuit, the software first tries to find a steady state solution, that is, one where all nodes conform to Kirchhoff's Current Law *and* the voltages across and through each element of the circuit conform to the voltage/current equations governing that element.

Once the steady state solution is found, the **operating points** of each element in the circuit are known. For a small signal analysis, every non-linear element can be linearized around its operation point to obtain the small-signal estimate of the voltages and currents. This is an application of Ohm's Law. The resulting linear circuit matrix can be solved with Gaussian elimination.

Piecewise-linear Approximation

Software such as the PLECS interface to Simulink uses piecewise-linear approximation of the equations governing the elements of a circuit. The circuit is treated as a completely linear network of ideal diodes. Every time a diode switches from on to off or *vice versa*, the configuration of the linear network changes. Adding more detail to the approximation of equations increases the accuracy of the simulation, but also increases its running time.

ELECTRICAL LAWS

Kirchhoff's Circuit Laws

Kirchhoff's circuit laws are two equalities that deal with the current and potential difference (commonly known as voltage) in the lumped element model of electrical circuits. They were first described in 1845 by Gustav Kirchhoff. This generalized the work of Georg Ohm and preceded the work of Maxwell. Widely used in electrical engineering, they are also called Kirchhoff's *rules* or simply Kirchhoff's *laws*.

Both of Kirchhoff's laws can be understood as corollaries of the Maxwell equations in the low-frequency limit. They are accurate for DC circuits, and for AC circuits at frequencies where the wavelengths of electromagnetic radiation are very large compared to the circuits.

Kirchhoff's Current Law (KCL)

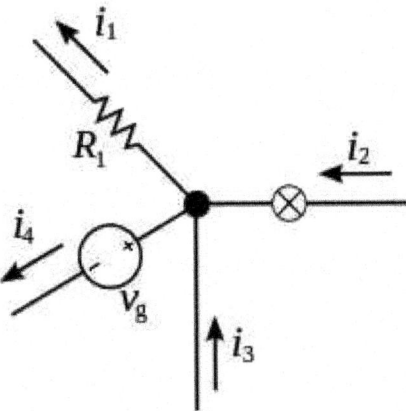

Fig. : The current entering any junction is equal to the current leaving that junction.
$$i_2 + i_3 = i_1 + i_4.$$

This law is also called **Kirchhoff's first law, Kirchhoff's point rule,** or **Kirchhoff's junction rule** (or nodal rule).

The principle of conservation of electric charge implies that :

At any node (junction) in an electrical circuit, the sum of currents flowing into that node is equal to the sum of currents flowing out of that node, or :

The algebraic sum of currents in a network of conductors meeting at a point is zero.

Recalling that current is a signed (positive or negative) quantity reflecting direction towards or away from a node, this principle can be stated as :

$$\sum_{k=1}^{n} I_k = 0$$

n is the total number of branches with currents flowing towards or away from the node.

This formula is valid for complex currents :

$$\sum_{k=1}^{n} \tilde{I}_k = 0$$

The law is based on the conservation of charge whereby the charge (measured in coulombs) is the product of the current (in amperes) and the time (in seconds).

Uses

A matrix version of Kirchhoff's current law is the basis of most circuit simulation software, such as SPICE. Kirchhoff's current law combined with Ohm's Law is used in nodal analysis.

Kirchhoff's Voltage Law (KVL)

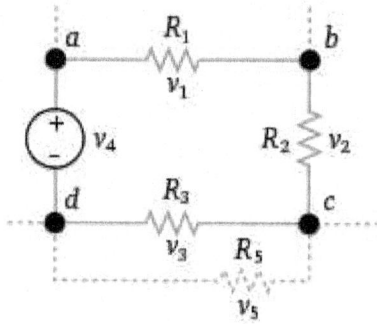

Fig. : The sum of all the voltages around the loop is equal to zero. $v_1 + v_2 + v_3 - v_4 = 0$.

This law is also called **Kirchhoff's second law, Kirchhoff's loop (or mesh) rule,** and **Kirchhoff's second rule.**

The principle of conservation of energy implies that

The directed sum of the electrical potential differences (voltage) around any closed network is zero, or :

More simply, the sum of the emfs in any closed loop is equivalent to the sum of the potential drops in that loop, or :

The algebraic sum of the products of the resistances of the conductors and the currents in them in a closed loop is equal to the total emf available in that loop.

Power System Fundamentals

Similarly to KCL, it can be stated as :

$$\sum_{k=1}^{n} V_k = 0$$

Here, n is the total number of voltages measured. The voltages may also be complex :

$$\sum_{k=1}^{n} \tilde{V}_k = 0$$

This law is based on the conservation of energy whereby voltage is defined as the energy per unit charge. The total amount of energy gained per unit charge must equal the amount of energy lost per unit charge, as energy and charge are both conserved.

Generalization

In the low-frequency limit, the voltage drop around any loop is zero. This includes imaginary loops arranged arbitrarily in space – not limited to the loops delineated by the circuit elements and conductors. In the low-frequency limit, this is a corollary of Faraday's law of induction (which is one of the Maxwell equations).

This has practical application in situations involving "static electricity".

Limitations

KCL and KVL both depend on the lumped element model being applicable to the circuit in question. When the model is not applicable, the laws do not apply.

KCL, in its usual form, is dependent on the assumption that current flows only in conductors, and that whenever current flows into one end of a conductor it immediately flows out the other end. This is not a safe assumption for high-frequency AC circuits, where the lumped element model is no longer applicable. It is often possible to improve the applicability of KCL by considering "parasitic capacitances" distributed along the conductors. Significant violations of KCL can occur even at 60Hz, which is not a very high frequency.

In other words, KCL is valid only if the total electric charge, Q, remains constant in the region being considered. In practical cases this is always so when KCL is applied at a geometric point. When investigating a finite region, however, it is possible that the charge density within the region may change. Since charge is conserved, this can only come about by a flow of charge across the region boundary. This flow represents a net current, and KCL is violated.

KVL is based on the assumption that there is no fluctuating magnetic field linking the closed loop. This is not a safe assumption for high-frequency (short-wavelength) AC circuits. In the presence of a changing magnetic field the electric field is not a conservative vector field. Therefore the electric field can not be the gradient of any potential. That is to say, the line integral of the electric field around the loop is not zero, directly contradicting KVL.

It is often possible to improve the applicability of KVL by considering "parasitic inductances" (including mutual inductances) distributed along the conductors. These are treated as imaginary circuit elements that produce a voltage drop equal to the rate-of-change of the flux.

Example

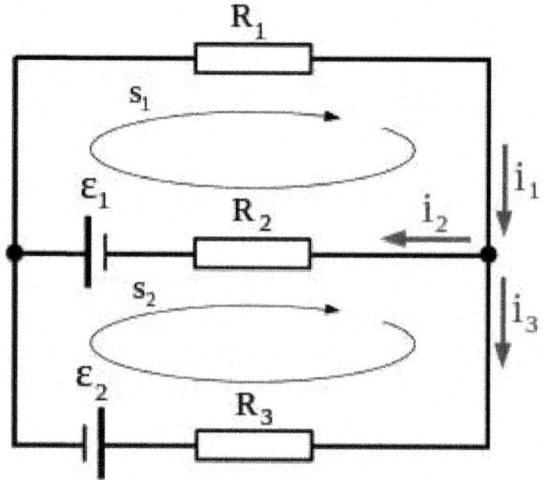

Assume an electric network consisting of two voltage sources and three resistors.

According to the first law we have

$$i_1 - i_2 - i_3 = 0$$

The second law applied to the closed circuit s_1 gives

$$-R_2 i_2 + \epsilon_1 - R_1 i_1 = 0$$

The second law applied to the closed circuit s_2 gives

$$- R_3 i_3 - \epsilon_2 - \epsilon_1 + R_2 i_2 = 0$$

Thus we get a linear system of equations in i_1, i_2, i_3:

$$\begin{cases} i_1 - i_2 - i_3 & = 0 \\ -R_2 i_2 + \epsilon_1 - R_1 i_1 & = 0 \\ -R_3 i_3 - \epsilon_2 - \epsilon_1 + R_2 i_2 & = 0 \end{cases}$$

Assuming

$R_1 = 100, R_2 = 200, R_3 = 300$ (ohms); $\epsilon_1 = 3, \epsilon_2 = 4$ (volts)

the solution is

Power System Fundamentals

$$\begin{cases} i_1 = \dfrac{1}{1100} \\ i_2 = \dfrac{4}{275} \\ i_3 = -\dfrac{3}{220} \end{cases}$$

i_3 has a negative sign, which means that the direction of i_3 is opposite to the assumed direction (the direction defined in the picture).

Norton's Theorem

Known in Europe as the **Mayer–Norton theorem**, **Norton's theorem** holds, to illustrate in DC circuit theory terms, that :

- Any linear electrical network with voltage and current sources and only resistances can be replaced at terminals A-B by an equivalent current source I_{NO} in parallel connection with an equivalent resistance R_{NO}.
- This equivalent current I_{NO} is the current obtained at terminals A-B of the network with terminals A-B short circuited.
- This equivalent resistance R_{NO} is the resistance obtained at terminals A-B of the network with all its voltage sources short circuited and all its current sources open circuited.

For AC systems the theorem can be applied to reactive impedances as well as resistances.

The **Norton equivalent** circuit is used to represent any network of linear sources and impedances at a given frequency.

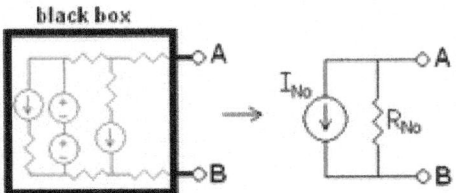

Fig. : Any black box containing resistances only and voltage and current sources can be replaced by an equivalent circuit consisting of an equivalent current source in parallel connection with an equivalent resistance.

Norton's theorem and its dual, Thévenin's theorem, are widely used for circuit analysis simplification and to study circuit's initial-condition and steady-state response.

Norton's theorem was independently derived in 1926 by Siemens & Halske researcher Hans Ferdinand Mayer (1895–1980) and Bell Labs engineer Edward Lawry Norton (1898–1983).

To find the equivalent,

1. Find the Norton current I_{No}. Calculate the output current, $I_{AB'}$ with a short circuit as the load (meaning 0 resistance between A and B). This is I_{No}.
2. Find the Norton resistance R_{No}. When there are no dependent sources (all current and voltage sources are independent), there are two methods of determining the Norton impedance R_{No}.
   - Calculate the output voltage, $V_{AB'}$ when in open circuit condition (i.e., no load resistor – meaning infinite load resistance). R_{No} equals this V_{AB} divided by I_{No}.

   or

   - Replace independent voltage sources with short circuits and independent current sources with open circuits. The total resistance across the output port is the Norton impedance R_{No}.

   This is equivalent to calculating the Thevenin resistance.

   However, when there are dependent sources, the more general method must be used.

   - Connect a constant current source at the output terminals of the circuit with a value of 1 Ampere and calculate the voltage at its terminals. This voltage divided by the 1 A current is the Norton impedance R_{No}. This method must be used if the circuit contains dependent sources, but it can be used in all cases even when there are no dependent sources.

Example of a Norton Equivalent Circuit

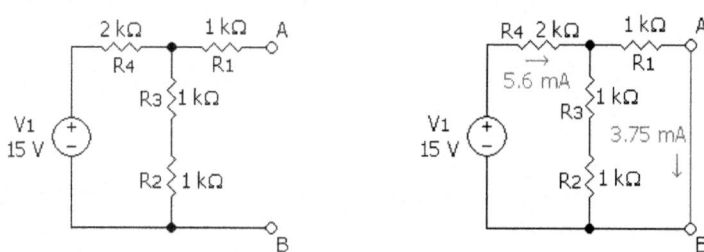

| Step 0 : The original circuit | Step 1 : Calculating the equivalent output current |

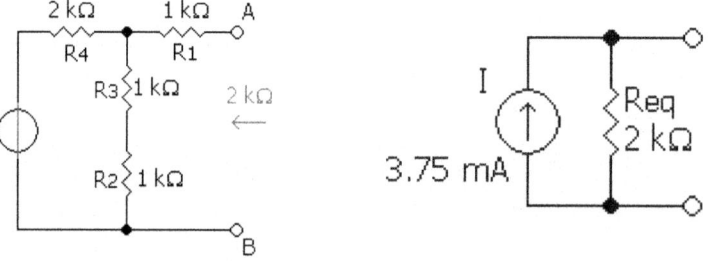

| Step 2 : Calculating the equivalent resistance | Step 3 : Design the equivalent circuit |

Power System Fundamentals

In the example, the total current I_{total} is given by :

$$I_{total} = \frac{15\,V}{2k\Omega + (1\,k\Omega\,||+1k\Omega+1k\Omega))} = 5.625\,mA.$$

The current through the load is then, using the current divider rule :

$$I_{No} = \frac{1k\Omega+1k\Omega}{(1k\Omega+1k\Omega+1k\Omega)} \cdot I_{total}$$

$$= 2/3.5.625\,mA = 3.75mA.$$

And the equivalent resistance looking back into the circuit is :

$$R_{eq} = 1k\Omega + (2k\Omega\,||+1k\Omega+1k\Omega)) = 2k\Omega.$$

So the equivalent circuit is a 3.70 mA current source in parallel with a 2 kΩ resistor.

Conversion to a Thévenin Equivalent

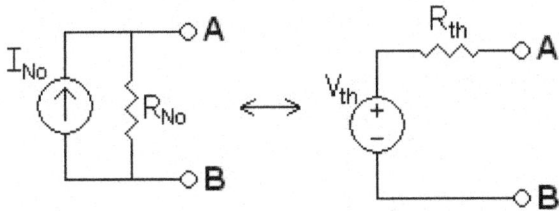

A Norton equivalent circuit is related to the Thévenin equivalent by the following equations :

$$R_{Th} = R_{No}$$
$$V_{Th} = I_{No}\,R_{No}$$
$$\frac{V_{Th}}{R_{Th}} = I_{No}.$$

Queueing Theory

The passive circuit equivalent of "Norton's theorem" in queuing theory is called the Chandy Herzog Woo theorem. In a reversible queueing system, it is often possible to replace an uninteresting subset of queues by a single (FCFS or PS) queue with an appropriately chosen service rate.

Thévenin's Theorem

Thévenin's theorem is an equivalence principle in circuit theory. For DC systems it states :

- Any linear electrical network with voltage and current sources and only resistances can be replaced at terminals A-B by an equivalent voltage source V_{th} in series connection with an equivalent resistance R_{th}.

- This equivalent voltage V_{th} is the voltage obtained at terminals A-B of the network with terminals A-B open circuited.
- This equivalent resistance R_{th} is the resistance obtained at terminals A-B of the network with all its independent current sources open circuited and all its independent voltage sources short circuited.

For AC systems, the statement of the theorem allows for reactive impedances as well as resistances.

The theorem was independently derived in 1853 by the German scientist Hermann von Helmholtz and in 1883 by Léon Charles Thévenin (1857–1926), an electrical engineer with France's national Postes et Télégraphes telecommunications organization.

Thévenin's theorem and its dual, Norton's theorem, are widely used for circuit analysis simplification and to study circuit's initial-condition and steady-state response. Thévenin's theorem can be used to convert any circuit's sources and impedances to a **Thévenin equivalent**; use of the theorem may in some cases be more convenient than use of Kirchhoff's circuit laws.

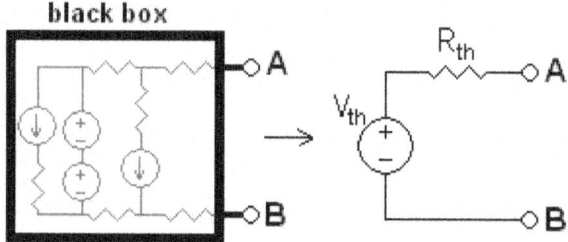

Fig.: Any black box containing resistances only and voltage and current sources can be replaced to a Thévenin equivalent circuit consisting of an equivalent voltage source in series connection with an equivalent resistance.

Calculating the Thévenin Equivalent

To calculate the equivalent circuit, the resistance and voltage are needed, so two equations are required. These two equations are usually obtained by using the following steps, but any conditions placed on the terminals of the circuit should also work:

1. Calculate the output voltage, V_{AB}, when in open circuit condition (no load resistor — meaning infinite resistance). This is V_{Th}.
2. Calculate the output current, I_{AB}, when the output terminals are short circuited (load resistance is 0). R_{Th} equals V_{Th} divided by this I_{AB}.

The equivalent circuit is a voltage source with voltage V_{Th} in series with a resistance R_{Th}.

Step 2 could also be thought of as:

2a. Replace the independent voltage sources with short circuits, and independent current sources with open circuits.

2b. Calculate the resistance between terminals A and B. This is R_{Th}.

The Thévenin-equivalent voltage is the voltage at the output terminals of the original circuit. When calculating a Thévenin-equivalent voltage, the voltage divider principle is often useful, by declaring one terminal to be V_{out} and the other terminal to be at the ground point.

The Thévenin-equivalent resistance is the resistance measured across points A and B "looking back" into the circuit. It is important to first replace all voltage- and current-sources with their internal resistances. For an ideal voltage source, this means replace the voltage source with a short circuit. For an ideal current source, this means replace the current source with an open circuit. Resistance can then be calculated across the terminals using the formulae for series and parallel circuits. This method is valid only for circuits with independent sources. If there are dependent sources in the circuit, another method must be used such as connecting a test source across A and B and calculating the voltage across or current through the test source.

Example

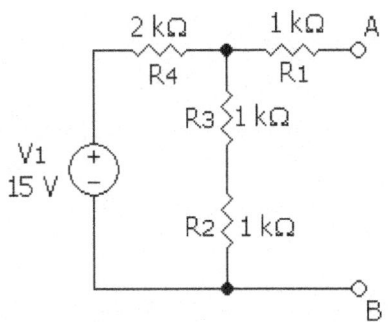

Step 0 : The original circuit

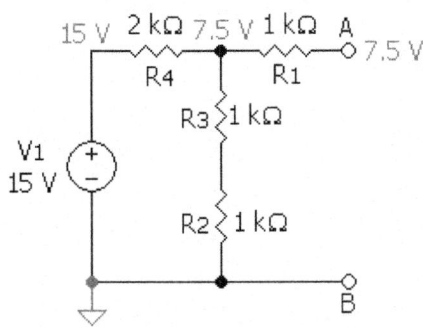

Step 1 : Calculating the equivalent output voltage

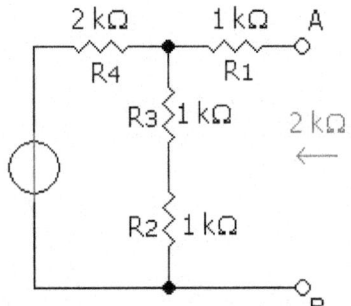

Step 2 : Calculating the equivalent resistance

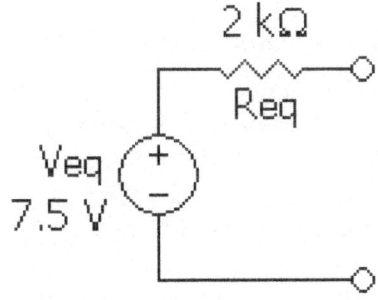

Step 3 : The equivalent circuit

In the example, calculating the equivalent voltage:

$$V_{Th} = \frac{R_2 + R_3}{(R_2 + R_3) + R_4} \cdot V_1$$

$$= \frac{1k\Omega + 1k\Omega}{(1k\Omega + 1k\Omega) + 2k\Omega} \cdot 15V$$

$$= \frac{1}{2} \cdot 15V = 7.5 \text{ V}$$

(notice that R_1 is not taken into consideration, as above calculations are done in an open circuit condition between A and B, therefore no current flows through this part, which means there is no current through R_1 and therefore no voltage drop along this part)

Calculating equivalent resistance:

$$R_{Th} = R_1 + [(R_2 + R_3) || R_4]$$
$$= 1k\Omega + [(1k\Omega + 1k\Omega) || 2k\Omega]$$
$$= 1k\Omega + \left(\frac{1}{1k\Omega + 1k\Omega} + \frac{1}{(2k\Omega)}\right)^{-1} \; 2k\Omega.$$

Conversion to a Norton Equivalent

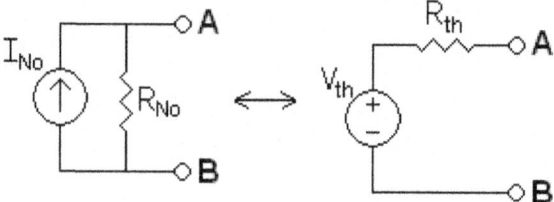

A Norton equivalent circuit is related to the Thévenin equivalent by the following:

$$R_{Th} = R_{No}$$
$$V_{Th} = I_{No} R_{No}$$
$$I_{No} = V_{Th}/R_{Th}.$$

Practical Limitations

- Many, if not most circuits are only linear over a certain range of values, thus the Thévenin equivalent is valid only within this linear range and may not be valid outside the range.
- The Thévenin equivalent has an equivalent I-V characteristic only from the point of view of the load.
- The power dissipation of the Thévenin equivalent is not necessarily identical to the power dissipation of the real system. However, the power dissipated by

an external resistor between the two output terminals is the same regardless of how the internal circuit is represented.

A Proof of the Theorem

The proof involves two steps. First use superposition theorem to construct a solution, and then use uniqueness theorem to show the solution is unique. The second step is usually implied. Firstly, using the superposition theorem, in general for any linear "black box" circuit which contains voltage sources and resistors, one can always write down its voltage as a linear function of the corresponding current as follows :

$$V = V_{Eq} - Z_{Eq} I$$

where the first term reflects the linear summation of contributions from each voltage source, while the second term measures the contribution from all the resistors. The above argument is due to the fact that the voltage of the black box for a given current I is identical to the linear superposition of the solutions of the following problems : (1) to leave the black box open circuited but activate individual voltage source one at a time and, (2) to short circuit all the voltage sources but feed the circuit with a certain ideal voltage source so that the resulting current exactly reads I (or an ideal current source of current I). Once the above expression is established, it is straightforward to show that V_{Eq} and Z_{Eq} are the single voltage source and the single series resistor in question.

Superposition Theorem

The **superposition theorem** for electrical circuits states that for a linear system the response (voltage or current) in any branch of a bilateral linear circuit having more than one independent source equals the algebraic sum of the responses caused by each independent source acting alone, where all the other independent sources are replaced by their internal impedances.

To ascertain the contribution of each individual source, all of the other sources first must be "turned off" (set to zero) by :

1. Replacing all other independent voltage sources with a short circuit (thereby eliminating difference of potential i.e. V=0; internal impedance of ideal voltage source is zero (short circuit)).
2. Replacing all other independent current sources with an open circuit (thereby eliminating current i.e. I=0; internal impedance of ideal current source is infinite (open circuit)).

This procedure is followed for each source in turn, then the resultant responses are added to determine the true operation of the circuit. The resultant circuit operation is the superposition of the various voltage and current sources.

The superposition theorem is very important in circuit analysis. It is used in converting any circuit into its Norton equivalent or Thevenin equivalent.

The theorem is applicable to linear networks (time varying or time invariant) consisting of independent sources, linear dependent sources, linear passive elements (resistors, inductors, capacitors) and linear transformers.

Another point that should be considered is that superposition only works for voltage and current but not power. In other words the sum of the powers of each source with the other sources turned off is not the real consumed power. To calculate power we should first use superposition to find both current and voltage of each linear element and then calculate the sum of the multiplied voltages and currents.

Maximum Power Transfer Theorem

In electrical engineering, the **maximum power transfer theorem** states that, to obtain *maximum* external power from a source with a finite internal resistance, the resistance of the load must equal the resistance of the source as viewed from its output terminals. Moritz von Jacobi published the maximum power (transfer) theorem around 1840; it is also referred to as "Jacobi's law".

The theorem results in maximum *power* transfer, and not maximum *efficiency*. If the resistance of the load is made larger than the resistance of the source, then efficiency is higher, since a higher percentage of the source power is transferred to the load, but the *magnitude* of the load power is lower since the total circuit resistance goes up.

If the load resistance is smaller than the source resistance, then most of the power ends up being dissipated in the source, and although the total power dissipated is higher, due to a lower total resistance, it turns out that the amount dissipated in the load is reduced.

The theorem states how to choose (so as to maximize power transfer) the load resistance, once the source resistance is given. It is a common mis-conception to apply the theorem in the opposite scenario. It does *not* say how to choose the source resistance for a given load resistance. In fact, the source resistance that maximizes power transfer is always zero, regardless of the value of the load resistance.

The theorem can be extended to AC circuits that include reactance, and states that maximum power transfer occurs when the load impedance is equal to the complex conjugate of the source impedance.

Maximizing Power Transfer versus Power Efficiency

The theorem was originally misunderstood (notably by Joule) to imply that a system consisting of an electric motor driven by a battery could not be more than 50% efficient since, when the impedances were matched, the power lost as heat in the battery would always be equal to the power delivered to the motor. In 1880 this assumption was shown to be false by either Edison or his colleague Francis Robbins Upton, who realized that maximum efficiency was not the same as maximum power transfer. To achieve maximum efficiency, the resistance of

Power System Fundamentals

the source (whether a battery or a dynamo) could be made close to zero. Using this new understanding, they obtained an efficiency of about 90%, and proved that the electric motor was a practical alternative to the heat engine.

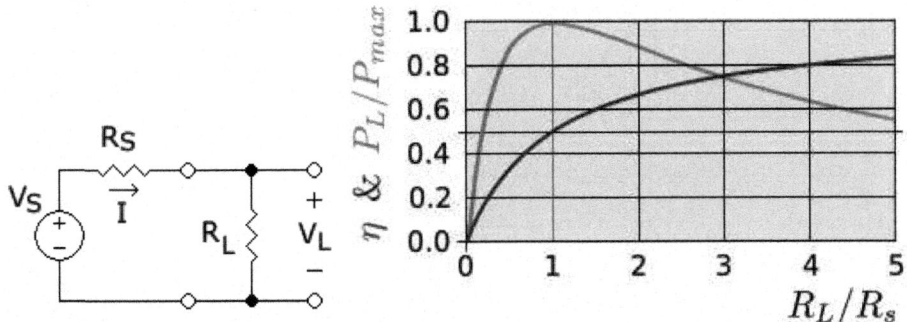

The condition of maximum power transfer does not result in maximum efficiency. If we define the efficiency η as the ratio of power dissipated by the load to power developed by the source, then it is straightforward to calculate from the above circuit diagram that

$$\eta = \frac{R_{load}}{R_{load} + R_{source}} = \frac{1}{1 + \frac{R_{source}}{R_{load}}}.$$

Consider three particular cases:

- If $R_{load} = R_{source}$, then η = 0.5.
- If $R_{load} = \infty$ or $R_{source} = 0$ then η = 1.
- If $R_{load} = 0$, then η = 0.

The efficiency is only 50% when maximum power transfer is achieved, but approaches 100% as the load resistance approaches infinity, though the total power level tends towards zero. Efficiency also approaches 100% if the source resistance approaches zero, and 0% if the load resistance approaches zero. In the latter case, all the power is consumed inside the source (unless the source also has no resistance), so the power dissipated in a short circuit is zero.

Impedance Matching

A related concept is reflectionless impedance matching. In radio, transmission lines, and other electronics, there is often a requirement to match the source impedance (such as a transmitter) to the load impedance (such as an antenna) to avoid reflections in the transmission line.

Calculus-based Proof for Purely Resistive Circuits

In the diagram opposite, power is being transferred from the source, with voltage V and fixed source resistance R_s, to a load with resistance R_L, resulting in a current I. By Ohm's law, I is simply the source voltage divided by the total circuit resistance:

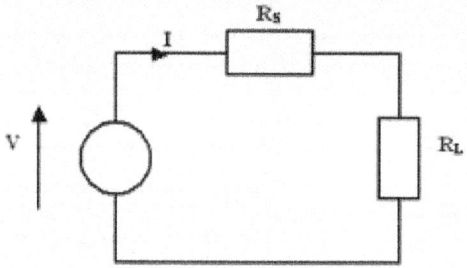

$$I = \frac{V}{R_S + R_L}.$$

The power P_L dissipated in the load is the square of the current multiplied by the resistance:

$$P_L = I^2 R_L = \left(\frac{V}{R_S + R_L}\right)^2 R_L = \frac{V^2}{R_S^2/R_L + 2R_S + R_L}$$

The value of R_L for which this expression is a maximum could be calculated by differentiating it, but it is easier to calculate the value of R_L for which the denominator

$$R_S^2/R_L + 2R_S + R_L$$

is a minimum. The result will be the same in either case. Differentiating the denominator with respect to R_L:

$$\frac{d}{dR_L}(R_S^2/R_L + 2R_S + R_L) = -R_S^2/R_L^2 + 1.$$

For a maximum or minimum, the first derivative is zero, so

$$R_S^2/R_L^2 = 1$$

or

$$R_L = \pm R_S.$$

In practical resistive circuits, R_S and R_L are both positive, so the positive sign in the above is the correct solution. To find out whether this solution is a minimum or a maximum, the denominator expression is differentiated again:

$$\frac{d^2}{dR_L^2}(R_S^2/R_L + 2R_S + R_L) = R_S^2/R_L^2.$$

This is always positive for positive values of R_S and R_L, showing that the denominator is a minimum, and the power is therefore a maximum, when $R_S = R_L$.

A note of caution is in order here. This last statement, as written, implies to many people that for a given load, the source resistance must be set equal to the load resistance for maximum power transfer. However, this equation only applies

Power System Fundamentals

if the source resistance cannot be adjusted, *e.g.*, with antennas. For any given load resistance a source resistance of zero is the way to transfer maximum power to the load. As an example, a 100 volt source with an internal resistance of 10 ohms connected to a 10 ohm load will deliver 250 watts to that load. Make the source resistance zero ohms and the load power jumps to 1000 watts.

In Reactive Circuits

The theorem also applies where the source and/or load are not totally resistive. This invokes a refinement of the maximum power theorem, which says that any reactive components of source and load should be of equal magnitude but opposite phase. This means that the source and load impedances should be *complex conjugates* of each other. In the case of purely resistive circuits, the two concepts are identical. However, physically realizable sources and loads are not usually totally resistive, having some inductive or capacitive components, and so practical applications of this theorem, under the name of complex conjugate impedance matching, do, in fact, exist.

If the source is totally inductive (capacitive), then a totally capacitive (inductive) load, in the absence of resistive losses, would receive 100% of the energy from the source but send it back after a quarter cycle. The resultant circuit is nothing other than a resonant LC circuit in which the energy continues to oscillate to and fro. This is called reactive power. Power factor correction (where an inductive reactance is used to "balance out" a capacitive one), is essentially the same idea as complex conjugate impedance matching although it is done for entirely different reasons.

For a fixed reactive *source*, the maximum power theorem maximizes the real power (P) delivered to the load by complex conjugate matching the load to the source.

For a fixed reactive *load*, power factor correction minimizes the apparent power (S) (and unnecessary current) conducted by the transmission lines, while maintaining the same amount of real power transfer. This is done by adding a reactance to the load to balance out the load's own reactance, changing the reactive load impedance into a resistive load impedance.

Proof

In this diagram, AC power is being transferred from the source, with phasor magnitude voltage $|V_S|$ (peak voltage) and fixed source impedance Z_S, to a load with impedance Z_L, resulting in a phasor magnitude current $|I|$. $|I|$ is simply the source voltage divided by the total circuit impedance :

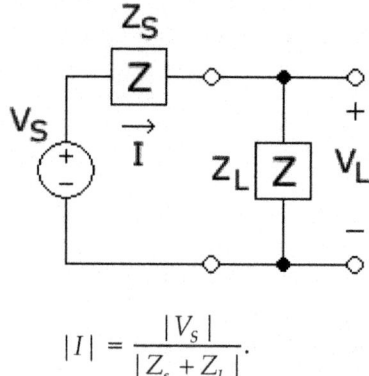

$$|I| = \frac{|V_S|}{|Z_s + Z_L|}.$$

The average power P_L dissipated in the load is the square of the current multiplied by the resistive portion (the real part) R_L of the load impedance:

$$P_L = I^2_{rms} R_L = \frac{1}{2}|I|^2 R_L = \frac{1}{2}\left(\frac{|V_S|}{|Z_S Z_L|}\right)^2 R_L$$

$$= \frac{1}{2}\frac{|V_S|^2 R_L}{(R_S + R_L)^2 + (X_S + X_L)^2},$$

where the resistance R_S and reactance X_S are the real and imaginary parts of Z_S, and X_L is the imaginary part of Z_L.

To determine the values of R_L and X_L (since V_S, R_S, and X_S are fixed) for which this expression is a maximum, we first find, for each fixed positive value of R_L, the value of the reactive term X_L for which the denominator

$$(R_S + R_L)^2 + (X_S + X_L)^2$$

is a minimum. Since reactances can be negative, this denominator is easily minimized by making

$$X_L = -X_S.$$

The power equation is now reduced to:

$$P_L = \frac{1}{2}\frac{|V_S|^2 R_L}{2(R_S + R_L)^2}$$

and it remains to find the value of R_L which maximizes this expression. However, this maximization problem has exactly the same form as in the purely resistive case, and the maximizing condition $R_L = R_S$ can be found in the same way.

The combination of conditions

- $R_L = R_S$
- $X_L = -X_S$

can be concisely written with a complex conjugate (the *) as:

$$Z_L = Z^*_S.$$

PHASOR

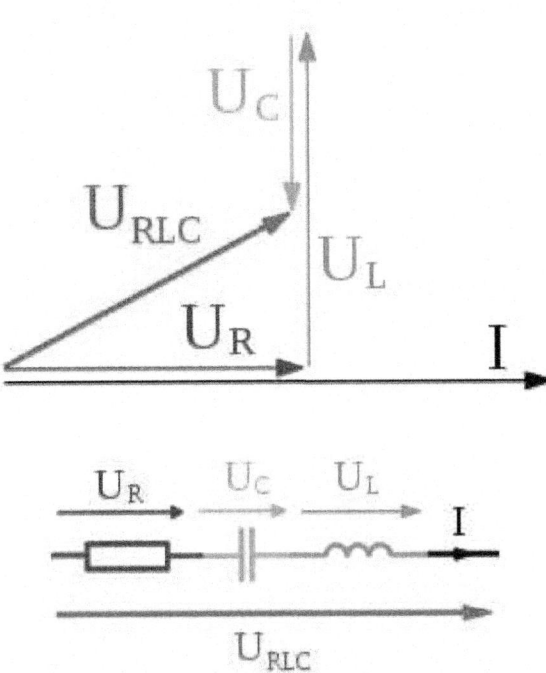

Fig.: An example of series RLC circuit and respective phasor diagram for a specific ω.

In physics and engineering, a **phase vector**, or **phasor**, is a representation of a sinusoidal function whose amplitude (**A**), frequency (ω), and phase (θ) are time-invariant. It is a subset of a more general concept called analytic representation. Phasors separate the dependencies on **A**, ω, and θ into three independent factors. This can be particularly useful because the frequency factor (which includes the time-dependence of the sinusoid) is often common to all the components of a linear combination of sinusoids. In those situations, phasors allow this common feature to be factored out, leaving just the **A** and θ features. The result is that trigonometry reduces to algebra, and linear differential equations become algebraic ones. The term *phasor* therefore often refers to just those two factors. In older texts, a **phasor** is also referred to as a **sinor**.

Definition

Euler's formula indicates that sinusoids can be represented mathematically as the sum of two complex-valued functions:

$$A.\cos(\omega t + \theta) = A. \frac{e^{i(\omega t + \theta)} + e^{-i(\omega t + \theta)}}{2},$$

or as the real part of one of the functions:

$$A \cdot \cos(\omega t + \theta) = Re\{A \cdot e^{i(\omega t + \theta)}\}$$
$$= Re\{Ae^{i\theta} \cdot e^{i\omega t}\}$$

The term *phasor* can refer to either $Ae^{i\theta} \cdot e^{i\omega t}$ or just the complex constant, $Ae^{i\theta}$. In the latter case, it is understood to be a shorthand notation, encoding the amplitude and phase of an underlying sinusoid.

An even more compact shorthand is angle notation : $A \angle \theta$.

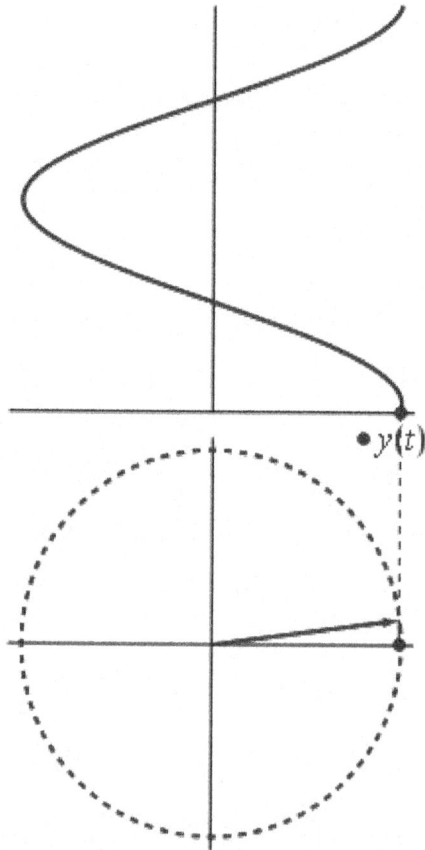

Fig. : A phasor can be considered a vector rotating about the origin in a complex plane. The cosine function is the projection of the vector onto the real axis. Its amplitude is the modulus of the vector, and its argument is the total phase $\omega t + \theta$. The phase constant θ represents the angle that the vector forms with the real axis at $t = 0$.

Phasor Arithmetic

Multiplication by a Constant (Scalar)

Multiplication of the phasor $Ae^{i\theta}e^{i\omega t}$ by a complex constant, $Be^{i\phi}$, produces another phasor. That means its only effect is to change the amplitude and phase of the underlying sinusoid :

Power System Fundamentals

$$Re\{(Ae^{i\theta} \cdot Be^{i\phi}) \cdot e^{i\omega t}\} = Re\{(ABe^{i(\theta+\phi)}) \cdot e^{i\omega t}\}$$
$$= AB\cos(\omega t + (\theta + \phi))$$

In electronics, $Be^{i\phi}$ would represent an impedance, which is independent of time. In particular it is *not* the shorthand notation for another phasor. Multiplying a phasor current by an impedance produces a phasor voltage. But the product of two phasors (or squaring a phasor) would represent the product of two sinusoids, which is a non-linear operation that produces new frequency components. Phasor notation can only represent systems with one frequency, such as a linear system stimulated by a sinusoid.

Differentiation and Integration

The time derivative or integral of a phasor produces another phasor. For example:

$$Re\left\{\frac{d}{dt}(Ae^{i\theta} \cdot e^{i\omega t})\right\} = Re\{Ae^{i\theta} \cdot i\omega e^{i\omega t}\}$$
$$= Re\{Ae^{i\theta} \cdot e^{i\pi/2} \omega e^{i\omega t}\}$$
$$= Re\{\omega A e^{i(\theta + \pi/2)} \cdot e^{i\omega t}\}$$
$$= \omega A \cdot \cos\{\omega t + \theta + \pi/2\}$$

Therefore, in phasor representation, the time derivative of a sinusoid becomes just multiplication by the constant, $i\omega = (e^{i\pi/2} \cdot \omega)$.

Similarly, integrating a phasor corresponds to multiplication by $\frac{1}{i\omega} = \frac{e^{-i\pi/2}}{\omega}$.

The time-dependent factor, $e^{i\omega t}$, is unaffected.

When we solve a linear differential equation with phasor arithmetic, we are merely factoring $e^{i\omega t}$ out of all terms of the equation, and reinserting it into the answer. For example, consider the following differential equation for the voltage across the capacitor in an RC circuit:

$$\frac{dv_C(t)}{dt} + \frac{1}{R_C}v_C(t) = \frac{1}{R_C}v_s(t)$$

When the voltage source in this circuit is sinusoidal:

$$v_s(t) = V_p \cdot \cos(\omega t + \theta),$$

we may substitute:

$$v_s(t) = Re\{V_s \cdot e^{i\omega t}\}$$
$$v_s(t) = Re\{V_C \cdot e^{i\omega t}\}$$

where phasor $V_s = V_p e^{i\theta}$ and phasor V_C is the unknown quantity to be determined.

In the phasor shorthand notation, the differential equation reduces to:

$$i\omega V_c + \frac{1}{R_C}V_c = \frac{1}{R_C}V_s$$

Solving for the phasor capacitor voltage gives:

$$V_C = \frac{1}{1+i\omega R_C} \cdot (V_S) = \frac{1-i\omega RC}{1+(\omega RC)^2} \cdot (V_p e^{i\theta})$$

As we have seen, the factor multiplying V_S represents differences of the amplitude and phase of $V_c(t)$ relative to V_p and θ.

In polar co-ordinate form, it is:

$$\frac{1}{\sqrt{1+(\omega RC)^2}} \cdot e^{-i\phi(\omega)}, \text{ where } \phi(\omega) = \arctan(\omega RC).$$

Therefore:

$$v_c(t) = \frac{1}{\sqrt{1+(\omega RC)^2}} \cdot V_p \cos(\omega t + \theta - \phi(\omega))$$

Addition

The sum of multiple phasors produces another phasor. That is because the sum of sinusoids with the same frequency is also a sinusoid with that frequency:

$$A_1 \cos(\omega t + \theta_1) + A_2 \cos(\omega t + \theta_2) = \text{Re}\{A_1 e^{i\theta_1} e^{i\omega t}\} + \text{Re}\{A_2 e^{i\theta_2} e^{i\omega t}\}$$
$$= \text{Re}\{A_1 e^{i\theta_1} e^{i\omega t} + A_2 e^{i\theta_2} e^{i\omega t}\}$$
$$= \text{Re}\{(A_1 e^{i\theta_1} + A_2 e^{i\theta_2}) e^{i\omega t}\}$$
$$= \text{Re}\{(A_3 e^{i\theta_3}) e^{i\omega t}\}$$
$$= A_3 \cos(\omega t + \theta_3),$$

where:

$$A^2_3 = (A_1 \cos\theta_1 + A_2 \cos\theta_2)^2 + (A_1 \sin\theta_1 + A_2 \sin\theta_2)$$

$$\theta_3 = \arctan\left(\frac{A_1 \sin\theta_1 + A_2 \sin\theta_2}{A_1 \cos\theta_1 + A_2 \cos\theta_2}\right)$$

or, via the law of cosines on the complex plane (or the trigonometric identity for angle differences):

$$A^2_3 = A^2_1 + A^2_2 - 2A_1 A_2 \cos(180° - \Delta\theta), = A^2_1 + A^2_2 + 2A_1 A_2 \cos(\Delta\theta),$$

where $\Delta\theta = \theta_1 - \theta_2$. A key point is that A_3 and θ_3 do not depend on ω or t, which is what makes phasor notation possible. The time and frequency dependence can be suppressed and re-inserted into the outcome as long as the only operations used in between are ones that produce another phasor. In angle notation, the operation shown above is written:

$$A_1 \angle \theta_1 + A_2 \angle \theta_2 = A_3 \angle \theta_3.$$

Another way to view addition is that two **vectors** with co-ordinates [A1 cos(ωt + θ1), A1 sin(ωt + θ1)] and [A2 cos(ωt + θ2), A2 sin(ωt + θ2)] are added vectorially to produce a resultant vector with co-ordinates [A3 cos(ωt + θ3), A3 sin(ωt + θ3)].

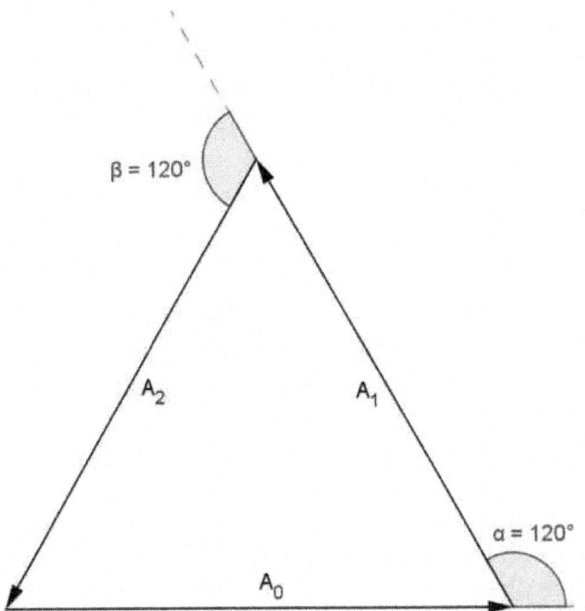

Fig. : Phasor diagram of three waves in perfect destructive interference.

In physics, this sort of addition occurs when sinusoids interfere with each other, constructively or destructively. The static vector concept provides useful insight into questions like this : "What phase difference would be required between three identical sinusoids for perfect cancellation?" In this case, simply imagine taking three vectors of equal length and placing them head to tail such that the last head matches up with the first tail. Clearly, the shape which satisfies these conditions is an equilateral triangle, so the angle between each phasor to the next is 120° ($2\pi/3$ radians), or one third of a wavelength $\lambda/_3$. So the phase difference between each wave must also be 120°, as is the case in three-phase power

In other words, what this shows is :

$$\cos(\omega t) + \cos(\omega t + 2\pi/3) + \cos(\omega t - 2\pi/3) = 0.$$

In the example of three waves, the phase difference between the first and the last wave was 240 degrees, while for two waves destructive interference happens at 180 degrees. In the limit of many waves, the phasors must form a circle for destructive interference, so that the first phasor is nearly parallel with the last. This means that for many sources, destructive interference happens when the first and last wave differ by 360 degrees, a full wavelength λ. This is why in single slit diffraction, the minima occurs when light from the far edge travels a full wavelength further than the light from the near edge.

Phasor Diagrams

Electrical engineers, electronics engineers, electronic engineering technicians and aircraft engineers all use phasor diagrams to visualize complex constants and variables (phasors). Like vectors, arrows drawn on graph paper or computer displays represent phasors. Cartesian and polar representations each have advantages, with the Cartesian co-ordinates showing the real and imaginary parts of the phasor and the polar co-ordinates showing its magnitude and phase.

Circuit Laws

With phasors, the techniques for solving DC circuits can be applied to solve AC circuits.

- **Ohm's law for resistors**: a resistor has no time delays and therefore doesn't change the phase of a signal therefore $V=IR$ remains valid.
- **Ohm's law for resistors, inductors, and capacitors**: $V = IZ$ where Z is the complex impedance.
- In an AC circuit we have real power (P) which is a representation of the average power into the circuit and reactive power (Q) which indicates power flowing back and forward. We can also define the complex power $S = P + jQ$ and the apparent power which is the magnitude of S. The power law for an AC circuit expressed in phasors is then $S = VI^*$ (where I^* is the complex conjugate of I).
- Kirchhoff's circuit laws work with phasors in complex form.

Given this we can apply the techniques of analysis of resistive circuits with phasors to analyze single frequency AC circuits containing resistors, capacitors, and inductors. Multiple frequency linear AC circuits and AC circuits with different waveforms can be analyzed to find voltages and currents by transforming all waveforms to sine wave components with magnitude and phase then analyzing each frequency separately, as allowed by the superposition theorem.

Power Engineering

In analysis of three phase AC power systems, usually a set of phasors is defined as the three complex cube roots of unity, graphically represented as unit magnitudes at angles of 0, 120 and 240 degrees. By treating polyphase AC circuit quantities as phasors, balanced circuits can be simplified and unbalanced circuits can be treated as an algebraic combination of symmetrical circuits. This approach greatly simplifies the work required in electrical calculations of voltage drop, power flow, and short-circuit currents. In the context of power systems analysis, the phase angle is often given in degrees, and the magnitude in rms value rather than the peak amplitude of the sinusoid.

The technique of synchrophasors uses digital instruments to measure the phasors representing transmission system voltages at widespread points in a transmission network. Small changes in the phasors are sensitive indicators of power flow and system stability.

Chapter 5

ELECTRICAL POWER PRODUCTION SYSTEMS

POWER STATION

A power station (also referred to as a generating station, power plant, powerhouse or generating plant) is an industrial facility for the generation of electric power. At the center of nearly all power stations is a generator, a rotating machine that converts mechanical power into electrical power by creating relative motion between a magnetic field and a conductor. The energy source harnessed to turn the generator varies widely. It depends chiefly on which fuels are easily available, cheap enough and on the types of technology that the power company has access to. Most power stations in the world burn fossil fuels such as coal, oil, and natural gas to generate electricity, and some use nuclear power, but there is an increasing use of cleaner renewable sources such as solar, wind, wave and hydroelectric.

History

The world's first power station was designed and built by Lord Armstrong at Cragside, England in 1868. Water from one of the lakes was used to power Siemens dynamos. The electricity supplied power to lights, heating, produced hot water, ran an elevator as well as labour-saving devices and farm buildings.

The first public power station was the *Edison Electric Light Station*, built in London at 57, Holborn Viaduct, which started operation in January 1882. This was an initiative of Thomas Edison that was organized and managed by his partner, Edward Johnson. A Babcock and Wilcox boiler powered a 125 horsepower steam engine that drove a 27 ton generator called Jumbo, after the celebrated elephant. This supplied electricity to premises in the area that could be reached through the culverts of the viaduct without digging up the road, which was the monopoly of the gas companies. The customers included the City Temple and the Old Bailey.

Another important customer was the Telegraph Office of the General Post Office, but this could not be reached though the culverts. Johnson arranged for the supply cable to be run overhead, via Holborn Tavern and Newgate.

In September 1882 in New York, the Pearl Street Station was established by Edison to provide electric lighting in the lower Manhattan Island area. The station ran until destroyed by fire in 1890. The station used reciprocating steam engines to turn direct-current generators. Because of the DC distribution, the service area was small, limited by voltage drop in the feeders. The War of Currents eventually resolved in favour of AC distribution and utilization, although some DC systems persisted to the end of the 20th century. DC systems with a service radius of a mile (kilometer) or so were necessarily smaller, less efficient of fuel consumption, and more labour-intensive to operate than much larger central AC generating stations.

AC systems used a wide range of frequencies depending on the type of load; lighting load using higher frequencies, and traction systems and heavy motor load systems preferring lower frequencies. The economics of central station generation improved greatly when unified light and power systems, operating at a common frequency, were developed. The same generating plant that fed large industrial loads during the day, could feed commuter railway systems during rush hour and then serve lighting load in the evening, thus improving the system load factor and reducing the cost of electrical energy overall. Many exceptions existed, generating stations were dedicated to power or light by the choice of frequency, and rotating frequency changers and rotating converters were particularly common to feed electric railway systems from the general lighting and power network.

Throughout the first few decades of the 20th century central stations became larger, using higher steam pressures to provide greater efficiency, and relying on interconnections of multiple generating stations to improve reliability and cost. High-voltage AC transmission allowed hydroelectric power to be conveniently moved from distant waterfalls to city markets. The advent of the steam turbine in central station service, around 1906, allowed great expansion of generating capacity. Generators were no longer limited by the power transmission of belts or the relatively slow speed of reciprocating engines, and could grow to enormous sizes. For example, Sebastian Ziani de Ferranti planned what would have been the largest reciprocating steam engine ever built for a proposed new central station, but scrapped the plans when turbines became available in the necessary size. Building power systems out of central stations required combinations of engineering skill and financial acumen in equal measure. Pioneers of central station generation include George Westinghouse and Samuel Insull in the United States, Ferranti and Charles Hesterman Merz in UK, and many others.

Thermal Power Station

A thermal power station is a power plant in which the prime mover is steam driven. Water is heated, turns into steam and spins a steam turbine which drives an electrical generator. After it passes through the turbine, the steam is condensed

Electrical Power Production Systems

in a condenser and recycled to where it was heated; this is known as a Rankine cycle. The greatest variation in the design of thermal power stations is due to the different fossil fuel resources generally used to heat the water. Some prefer to use the term energy center because such facilities convert forms of heat energy into electrical energy. Certain thermal power plants also are designed to produce heat energy for industrial purposes of district heating, or desalination of water, in addition to generating electrical power. Globally, fossil fueled thermal power plants produce a large part of man-made CO_2 emissions to the atmosphere, and efforts to reduce these are varied and widespread.

Introductory Overview

.coal, nuclear, geothermal, solar thermal electric, and waste incineration plants, as well as many natural gas power plants are thermal. Natural gas is frequently combusted in gas turbines as well as boilers. The waste heat from a gas turbine can be used to raise steam, in a combined cycle plant that improves overall efficiency. Power plants burning coal, fuel oil, or natural gas are often called *fossil-fuel power plants*. Some biomass-fueled thermal power plants have appeared also. Non-nuclear thermal power plants, particularly fossil-fueled plants, which do not use co-generation are sometimes referred to as *conventional power plants*.

Commercial electric utility power stations are usually constructed on a large scale and designed for continuous operation. Electric power plants typically use three-phase electrical generators to produce alternating current (AC) electric power at a frequency of 50 Hz or 60 Hz. Large companies or institutions may have their own power plants to supply heating or electricity to their facilities, especially if steam is created anyway for other purposes. Steam-driven power plants have been used in various large ships, but are now usually used in large naval ships. Shipboard power plants usually directly couple the turbine to the ship's propellers through gearboxes. Power plants in such ships also provide steam to smaller turbines driving electric generators to supply electricity. Shipboard steam power plants can be either fossil fuel or nuclear. Nuclear marine propulsion is, with few exceptions, used only in naval vessels. There have been perhaps about a dozen turbo-electric ships in which a steam-driven turbine drives an electric generator which powers an electric motor for propulsion.

Combined heat and power plants (CH & P plants), often called *co-generation plants*, produce both electric power and heat for process heat or space heating. Steam and hot water lose energy when piped over substantial distance, so carrying heat energy by steam or hot water is often only worthwhile within a local area, such as a ship, industrial plant, or district heating of nearby buildings.

History

The initially developed reciprocating steam engine has been used to produce mechanical power since the 18th Century, with notable improvements being made by James Watt. When the first commercially developed central electrical power stations were established in 1882 at Pearl Street Station in New York and Holborn Viaduct power station in London, reciprocating steam engines were used. The development of the steam turbine in 1884 provided larger and more efficient machine designs for central generating stations. By 1892 the turbine was considered a better alternative to reciprocating engines; turbines offered higher speeds, more compact machinery, and stable speed regulation allowing for parallel synchronous operation of generators on a common bus. After about 1905, turbines entirely replaced reciprocating engines in large central power stations.

The largest reciprocating engine-generator sets ever built were completed in 1901 for the Manhattan Elevated Railway. Each of seventeen units weighed about 500 tons and was rated 6000 kilowatts; a contemporary turbine set of similar rating would have weighed about 20% as much.

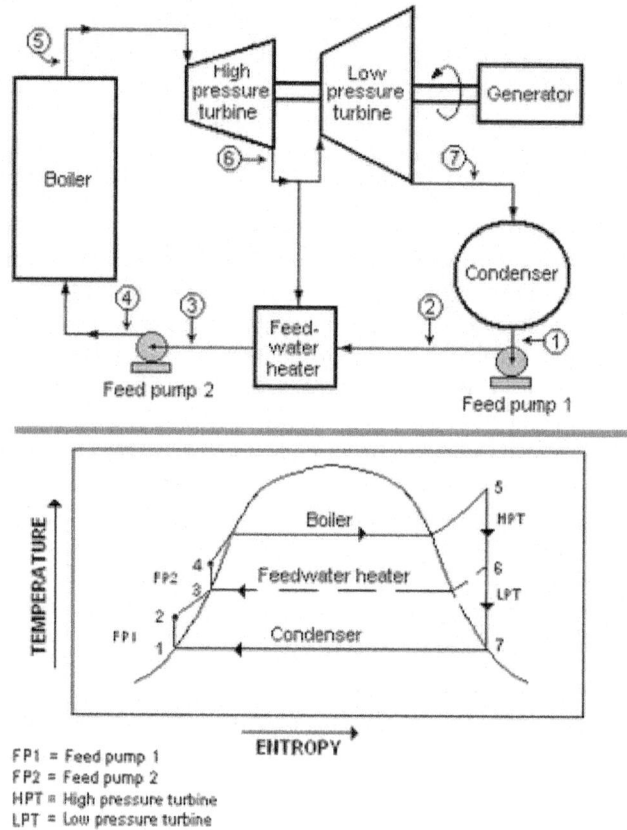

Fig. : A Rankine cycle with a two-stage steam turbine and a single feed water heater.

Efficiency

The energy efficiency of a conventional thermal power station, considered salable energy produced as a percent of the heating value of the fuel consumed, is typically 33% to 48%. As with all heat engines, their efficiency is limited, and governed by the laws of thermodynamics. By comparison, most hydropower stations in the United States are about 90 per cent efficient in converting the energy of falling water into electricity.

The energy of a thermal not utilized in power production must leave the plant in the form of heat to the environment. This waste heat can go through a condenser and be disposed of with cooling water or in cooling towers. If the waste heat is instead utilized for district heating, it is called co-generation. An important class of thermal power station are associated with desalination facilities; these are typically found in desert countries with large supplies of natural gas and in these plants, freshwater production and electricity are equally important co-products.

The Carnot efficiency dictates that higher efficiencies can be attained by increasing the temperature of the steam. Sub-critical fossil fuel power plants can achieve 36–40% efficiency. Super critical designs have efficiencies in the low to mid 40% range, with new "ultra critical" designs using pressures of 4400 psi (30.3 MPa) and multiple stage reheat reaching about 48% efficiency. Above the critical point for water of 705 °F (374 °C) and 3212 psi (22.06 MPa), there is no phase transition from water to steam, but only a gradual decrease in density.

Currently most of the nuclear power plants must operate below the temperatures and pressures that coal-fired plants do, since the pressurized vessel is very large and contains the entire bundle of nuclear fuel rods. The size of the reactor limits the pressure that can be reached. This, in turn, limits their thermodynamic efficiency to 30–32%. Some advanced reactor designs being studied, such as the very high temperature reactor, advanced gas-cooled reactor and supercritical water reactor, would operate at temperatures and pressures similar to current coal plants, producing comparable thermodynamic efficiency.

Electricity Cost

The direct cost of electric energy produced by a thermal power station is the result of cost of fuel, capital cost for the plant, operator labour, maintenance, and such factors as ash handling and disposal. Indirect, social or environmental costs such as the economic value of environmental impacts, or environmental and health effects of the complete fuel cycle and plant decommissioning, are not usually assigned to generation costs for thermal stations in utility practice, but may form part of an environmental impact assessment.

Typical Steam Power Station

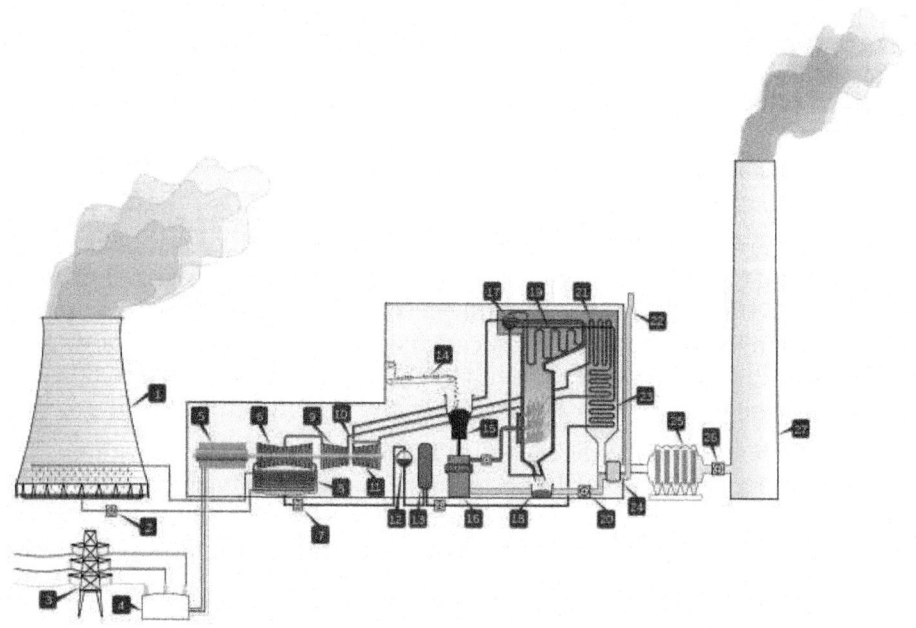

Fig. : Typical diagram of steam power station.

1. Cooling tower	10. Steam Control valve	19. Superheater
2. Cooling water pump	11. High pressure steam turbine	20. Forced draught (draft) fan
3. transmission line (3-phase)	12. Deaerator	21. Reheater
4. Step-up transformer (3-phase)	13. Feedwater heater	22. Combustion air intake
5. Electrical generator (3-phase)	23. Economiser	
6. Low pressure steam turbine	24. Air preheater	
7. Condensate pump	25. Precipitator	
8. Surface condenser	17. Boiler steam drum	26. Induced draught (draft) fan
9. Intermediate pressure steam turbine		

Boiler and Steam Cycle

In the nuclear plant field, *steam generator* refers to a specific type of large heat exchanger used in a pressurized water reactor (PWR) to thermally connect the primary (reactor plant) and secondary (steam plant) systems, which generates steam. In a nuclear reactor called a boiling water reactor (BWR), water is boiled to generate steam directly in the reactor itself and there are no units called steam generators.

In some industrial settings, there can also be steam-producing heat exchangers called *heat recovery steam generators* (HRSG) which utilize heat from some industrial process. The steam generating boiler has to produce steam at the high purity, pressure and temperature required for the steam turbine that drives the electrical generator.

Geothermal plants need no boiler since they use naturally occurring steam sources. Heat exchangers may be used where the geothermal steam is very corrosive or contains excessive suspended solids.

A fossil fuel steam generator includes an economizer, a steam drum, and the furnace with its steam generating tubes and superheater coils. Necessary safety valves are located at suitable points to avoid excessive boiler pressure. The air and flue gas path equipment include : forced draft (FD) fan, air preheater (AP), boiler furnace, induced draft (ID) fan, fly ash collectors (electrostatic precipitator or baghouse) and the flue gas stack.

Feed Water Heating and Deaeration

The boiler feedwater used in the steam boiler is a means of transferring heat energy from the burning fuel to the mechanical energy of the spinning steam turbine. The total feed water consists of recirculated *condensate* water and purified *makeup water*. Because the metallic materials it contacts are subject to corrosion at high temperatures and pressures, the makeup water is highly purified before use. A system of water softeners and ion exchange demineralizers produces water so pure that it coincidentally becomes an electrical insulator, with conductivity in the range of 0.3–1.0 microsiemens per centimeter. The makeup water in a 500 MWe plant amounts to perhaps 120 US gallons per minute (7.6 L/s) to replace water drawn off from the boiler drums for water purity management, and to also offset the small losses from steam leaks in the system.

The feed water cycle begins with condensate water being pumped out of the condenser after travelling through the steam turbines. The condensate flow rate at full load in a 500 MW plant is about 6,000 US gallons per minute (400 L/s).

The water is pressurized in two stages, and flows through a series of six or seven intermediate feed water heaters, heated up at each point with steam extracted from an appropriate duct on the turbines and gaining temperature at each stage. Typically, in the middle of this series of feedwater heaters, and before the second stage of pressurization, the condensate plus the makeup water flows through a deaerator that removes dissolved air from the water, further purifying and reducing its corrosiveness. The water may be dosed following this point with hydrazine, a chemical that removes the remaining oxygen in the water to below 5 parts per billion (ppb). It is also dosed with pH control agents such as ammonia or morpholine to keep the residual acidity low and thus non-corrosive.

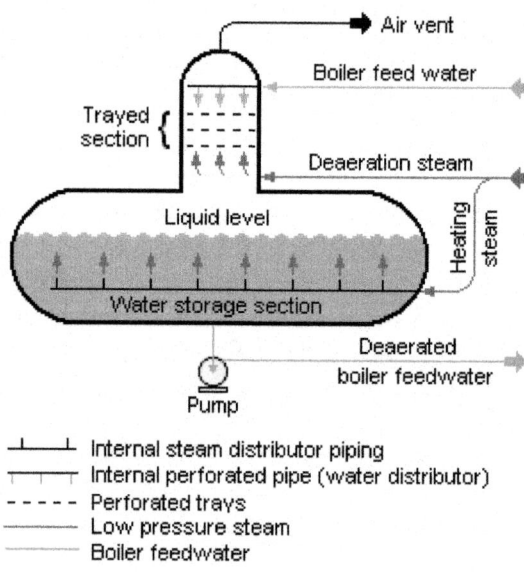

Fig.: Diagram of boiler feed water deaerator (with vertical, domed aeration section and horizontal water storage section).

Boiler Operation

The boiler is a rectangular furnace about 50 feet (15 m) on a side and 130 feet (40 m) tall. Its walls are made of a web of high pressure steel tubes about 2.3 inches (58 mm) in diameter.

Pulverized coal is air-blown into the furnace through burners located at the four corners, or along one wall, or two opposite walls, and it is ignited to rapidly burn, forming a large fireball at the center. The thermal radiation of the fireball heats the water that circulates through the boiler tubes near the boiler perimeter. The water circulation rate in the boiler is three to four times the throughput. As the water in the boiler circulates it absorbs heat and changes into steam. It is separated from the water inside a drum at the top of the furnace. The saturated steam is introduced into superheat pendant tubes that hang in the hottest part of the combustion gases as they exit the furnace. Here the steam is superheated to 1,000 °F (540 °C) to prepare it for the turbine.

Plants designed for lignite (brown coal) are increasingly used in locations as varied as Germany, Victoria, Australia and North Dakota. Lignite is a much younger form of coal than black coal. It has a lower energy density than black coal and requires a much larger furnace for equivalent heat output. Such coals may contain up to 70% water and ash, yielding lower furnace temperatures and requiring larger induced-draft fans. The firing systems also differ from black coal and typically draw hot gas from the furnace-exit level and mix it with the incoming coal in fan-type mills that inject the pulverized coal and hot gas mixture into the boiler.

Plants that use gas turbines to heat the water for conversion into steam use boilers known as heat recovery steam generators (HRSG). The exhaust heat from the gas turbines is used to make superheated steam that is then used in a conventional water-steam generation cycle.

Boiler Furnace and Steam Drum

The water enters the boiler through a section in the convection pass called the economizer. From the economizer it passes to the steam drum and from there it goes through downcomers to inlet headers at the bottom of the water walls. From these headers the water rises through the water walls of the furnace where some of it is turned into steam and the mixture of water and steam then re-enters the steam drum. This process may be driven purely by natural circulation (because the water is the downcomers is denser than the water/steam mixture in the water walls) or assisted by pumps. In the steam drum, the water is returned to the downcomers and the steam is passed through a series of steam separators and dryers that remove water droplets from the steam. The dry steam then flows into the superheater coils.

The boiler furnace auxiliary equipment includes coal feed nozzles and igniter guns, soot blowers, water lancing and observation ports (in the furnace walls) for observation of the furnace interior. Furnace explosions due to any accumulation of combustible gases after a trip-out are avoided by flushing out such gases from the combustion zone before igniting the coal.

The steam drum (as well as the super heater coils and headers) have air vents and drains needed for initial start up.

Superheater

Fossil fuel power plants often have a superheater section in the steam generating furnace. The steam passes through drying equipment inside the steam drum on to the superheater, a set of tubes in the furnace. Here the steam picks up more energy from hot flue gases outside the tubing and its temperature is now superheated above the saturation temperature. The superheated steam is then piped through the main steam lines to the valves before the high pressure turbine.

Nuclear-powered steam plants do not have such sections but produce steam at essentially saturated conditions. Experimental nuclear plants were equipped with fossil-fired super heaters in an attempt to improve overall plant operating cost.

Steam Condensing

The condenser condenses the steam from the exhaust of the turbine into liquid to allow it to be pumped. If the condenser can be made cooler, the pressure of the exhaust steam is reduced and efficiency of the cycle increases.

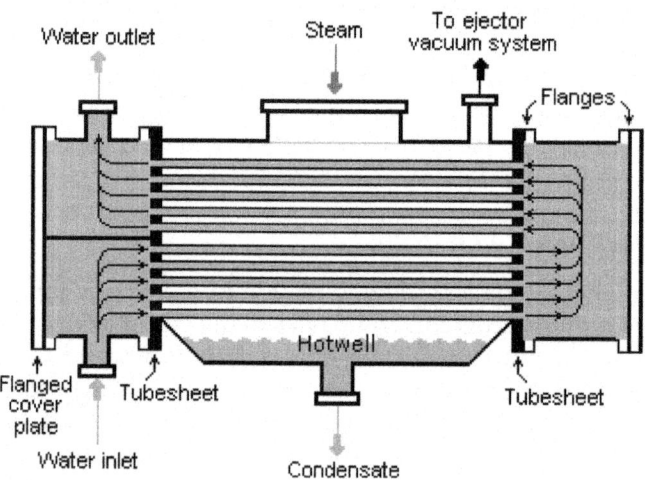

Fig. : Diagram of a typical water-cooled surface condenser.

The surface condenser is a shell and tube heat exchanger in which cooling water is circulated through the tubes. The exhaust steam from the low pressure turbine enters the shell where it is cooled and converted to condensate (water) by flowing over the tubes as shown in the adjacent diagram. Such condensers use steam ejectors or rotary motor-driven exhausters for continuous removal of air and gases from the steam side to maintain vacuum.

For best efficiency, the temperature in the condenser must be kept as low as practical in order to achieve the lowest possible pressure in the condensing steam. Since the condenser temperature can almost always be kept significantly below 100 °C where the vapour pressure of water is much less than atmospheric pressure, the condenser generally works under vacuum. Thus leaks of non-condensible air into the closed loop must be prevented.

Typically the cooling water causes the steam to condense at a temperature of about 35 °C (95 °F) and that creates an absolute pressure in the condenser of about 2-7 kPa (0.59-2.07 inHg), *i.e.* a vacuum of about −95 kPa (−28 inHg) relative to atmospheric pressure. The large decrease in volume that occurs when water vapour condenses to liquid creates the low vacuum that helps pull steam through and increase the efficiency of the turbines.

The limiting factor is the temperature of the cooling water and that, in turn, is limited by the prevailing average climatic conditions at the power plant's location (it may be possible to lower the temperature beyond the turbine limits during winter, causing excessive condensation in the turbine). Plants operating in hot climates may have to reduce output if their source of condenser cooling water becomes warmer; unfortunately this usually coincides with periods of high electrical demand for air conditioning.

Electrical Power Production Systems

The condenser generally uses either circulating cooling water from a cooling tower to reject waste heat to the atmosphere, or once-through water from a river, lake or ocean.

The heat absorbed by the circulating cooling water in the condenser tubes must also be removed to maintain the ability of the water to cool as it circulates. This is done by pumping the warm water from the condenser through either natural draft, forced draft or induced draft cooling towers (as seen in the image to the right) that reduce the temperature of the water by evaporation, by about 11 to 17 °C (20 to 30 °F) — expelling waste heat to the atmosphere. The circulation flow rate of the cooling water in a 500 MW unit is about 14.2 m^3/s (500 ft^3/s or 225,000 US gal/min) at full load.

The condenser tubes are made of brass or stainless steel to resist corrosion from either side. Nevertheless they may become internally fouled during operation by bacteria or algae in the cooling water or by mineral scaling, all of which inhibit heat transfer and reduce thermodynamic efficiency. Many plants include an automatic cleaning system that circulates sponge rubber balls through the tubes to scrub them clean without the need to take the system off-line.

The cooling water used to condense the steam in the condenser returns to its source without having been changed other than having been warmed. If the water returns to a local water body (rather than a circulating cooling tower), it is tempered with cool 'raw' water to prevent thermal shock when discharged into that body of water.

Another form of condensing system is the air-cooled condenser. The process is similar to that of a radiator and fan. Exhaust heat from the low pressure section of a steam turbine runs through the condensing tubes, the tubes are usually finned and ambient air is pushed through the fins with the help of a large fan. The steam condenses to water to be reused in the water-steam cycle. Air-cooled condensers typically operate at a higher temperature than water-cooled versions. While saving water, the efficiency of the cycle is reduced (resulting in more carbon dioxide per megawatt of electricity).

From the bottom of the condenser, powerful condensate pumps recycle the condensed steam (water) back to the water/steam cycle.

Reheater

Power plant furnaces may have a reheater section containing tubes heated by hot flue gases outside the tubes. Exhaust steam from the high pressure turbine is passed through these heated tubes to collect more energy before driving the intermediate and then low pressure turbines.

Air Path

External fans are provided to give sufficient air for combustion. The Primary air fan takes air from the atmosphere and, first warming it in the air preheater for better combustion, injects it via the air nozzles on the furnace wall.

The induced draft fan assists the FD fan by drawing out combustible gases from the furnace, maintaining a slightly negative pressure in the furnace to avoid backfiring through any closing.

Steam Turbine Generator

The turbine generator consists of a series of steam turbines interconnected to each other and a generator on a common shaft. There is a high pressure turbine at one end, followed by an intermediate pressure turbine, two low pressure turbines, and the generator. As steam moves through the system and loses pressure and thermal energy it expands in volume, requiring increasing diameter and longer blades at each succeeding stage to extract the remaining energy. The entire rotating mass may be over 200 metric tons and 100 feet (30 m) long. It is so heavy that it must be kept turning slowly even when shut down (at 3 rpm) so that the shaft will not bow even slightly and become unbalanced. This is so important that it is one of only five functions of blackout emergency power batteries on site. Other functions are emergency lighting, communication, station alarms and turbogenerator lube oil.

Superheated steam from the boiler is delivered through 14–16-inch (360–410 mm) diameter piping to the high pressure turbine where it falls in pressure to 600 psi (4.1 MPa) and to 600 °F (320 °C) in temperature through the stage. It exits via 24–26-inch (610–660 mm) diameter cold reheat lines and passes back into the boiler where the steam is reheated in special reheat pendant tubes back to 1,000 °F (540 °C). The hot reheat steam is conducted to the intermediate pressure turbine where it falls in both temperature and pressure and exits directly to the long-bladed low pressure turbines and finally exits to the condenser.

The generator, 30 feet (9 m) long and 12 feet (3.7 m) in diameter, contains a stationary stator and a spinning rotor, each containing miles of heavy copper conductor—no permanent magnets here. In operation it generates up to 21,000 amperes at 24,000 volts AC (504 MWe) as it spins at either 3,000 or 3,600 rpm, synchronized to the power grid. The rotor spins in a sealed chamber cooled with hydrogen gas, selected because it has the highest known heat transfer coefficient of any gas and for its low viscosity which reduces windage losses. This system requires special handling during startup, with air in the chamber first displaced by carbon dioxide before filling with hydrogen. This ensures that the highly explosive hydrogen–oxygen environment is not created.

The power grid frequency is 60 Hz across North America and 50 Hz in Europe, Oceania, Asia (Korea and parts of Japan are notable exceptions) and parts of Africa. The desired frequency affects the design of large turbines, since they are highly optimized for one particular speed.

The electricity flows to a distribution yard where transformers increase the voltage for transmission to its destination.

The steam turbine-driven generators have auxiliary systems enabling them to work satisfactorily and safely. The steam turbine generator being rotating equipment generally has a heavy, large diameter shaft. The shaft therefore requires not only supports but also has to be kept in position while running. To minimize the frictional resistance to the rotation, the shaft has a number of bearings. The bearing shells, in which the shaft rotates, are lined with a low friction material like Babbitt metal. Oil lubrication is provided to further reduce the friction between shaft and bearing surface and to limit the heat generated.

Stack Gas Path and Cleanup

As the combustion flue gas exits the boiler it is routed through a rotating flat basket of metal mesh which picks up heat and returns it to incoming fresh air as the basket rotates. This is called the air preheater. The gas exiting the boiler is laden with fly ash, which are tiny spherical ash particles. The flue gas contains nitrogen along with combustion products carbon dioxide, sulfur dioxide, and nitrogen oxides. The fly ash is removed by fabric bag filters or electrostatic precipitators. Once removed, the fly ash by product can sometimes be used in the manufacturing of concrete. This cleaning up of flue gases, however, only occurs in plants that are fitted with the appropriate technology. Still, the majority of coal-fired power plants in the world do not have these facilities. Legislation in Europe has been efficient to reduce flue gas pollution. Japan has been using flue gas cleaning technology for over 30 years and the US has been doing the same for over 25 years. China is now beginning to grapple with the pollution caused by coal-fired power plants.

Where required by law, the sulfur and nitrogen oxide pollutants are removed by stack gas scrubbers which use a pulverized limestone or other alkaline wet slurry to remove those pollutants from the exit stack gas. Other devices use catalysts to remove Nitrous Oxide compounds from the flue gas stream. The gas travelling up the flue gas stack may by this time have dropped to about 50 °C (120 °F). A typical flue gas stack may be 150–180 metres (490–590 ft) tall to disperse the remaining flue gas components in the atmosphere. The tallest flue gas stack in the world is 419.7 metres (1,377 ft) tall at the GRES-2 power plant in Ekibastuz, Kazakhstan.

In the United States and a number of other countries, atmospheric dispersion modelling studies are required to determine the flue gas stack height needed to comply with the local air pollution regulations. The United States also requires the height of a flue gas stack to comply with what is known as the "Good Engineering Practice (GEP)" stack height. In the case of existing flue gas stacks that exceed the GEP stack height, any air pollution dispersion modelling studies for such stacks must use the GEP stack height rather than the actual stack height.

Fly ash collection

Fly ash is captured and removed from the flue gas by electrostatic precipitators or fabric bag filters (or sometimes both) located at the outlet of the furnace and before the induced draft fan. The fly ash is periodically removed from the collection hoppers below the precipitators or bag filters. Generally, the fly ash is pneumatically transported to storage silos for subsequent transport by trucks or railroad cars .

Bottom Ash Collection and Disposal

At the bottom of the furnace, there is a hopper for collection of bottom ash. This hopper is always filled with water to quench the ash and clinkers falling down from the furnace. Some arrangement is included to crush the clinkers and for conveying the crushed clinkers and bottom ash to a storage site. Ash extractor is used to discharge ash from Municipal solid waste-fired boilers.

Auxiliary Systems

Boiler Make-up Water Treatment Plant and Storage

Since there is continuous withdrawal of steam and continuous return of condensate to the boiler, losses due to blowdown and leakages have to be made up to maintain a desired water level in the boiler steam drum. For this, continuous make-up water is added to the boiler water system. Impurities in the raw water input to the plant generally consist of calcium and magnesium salts which impart hardness to the water. Hardness in the make-up water to the boiler will form deposits on the tube water surfaces which will lead to overheating and failure of the tubes. Thus, the salts have to be removed from the water, and that is done by a water demineralising treatment plant (DM). A DM plant generally consists of cation, anion, and mixed bed exchangers. Any ions in the final water from this process consist essentially of hydrogen ions and hydroxide ions, which recombine to form pure water. Very pure DM water becomes highly corrosive once it absorbs oxygen from the atmosphere because of its very high affinity for oxygen.

The capacity of the DM plant is dictated by the type and quantity of salts in the raw water input. However, some storage is essential as the DM plant may be down for maintenance. For this purpose, a storage tank is installed from which DM water is continuously withdrawn for boiler make-up. The storage tank for DM water is made from materials not affected by corrosive water, such as PVC. The piping and valves are generally of stainless steel. Sometimes, a steam blanketing arrangement or stainless steel doughnut float is provided on top of the water in the tank to avoid contact with air. DM water make-up is generally added at the steam space of the surface condenser (*i.e.*, the vacuum side). This arrangement not only sprays the water but also DM water gets deaerated, with the dissolved gases being removed by a de-aerator through an ejector attached to the condenser.

Fuel Preparation System

In coal-fired power stations, the raw feed coal from the coal storage area is first crushed into small pieces and then conveyed to the coal feed hoppers at the boilers. The coal is next pulverized into a very fine powder. The pulverizers may be ball mills, rotating drum grinders, or other types of grinders.

Some power stations burn fuel oil rather than coal. The oil must kept warm (above its pour point) in the fuel oil storage tanks to prevent the oil from congealing and becoming unpumpable. The oil is usually heated to about 100 °C before being pumped through the furnace fuel oil spray nozzles.

Boilers in some power stations use processed natural gas as their main fuel. Other power stations may use processed natural gas as auxiliary fuel in the event that their main fuel supply (coal or oil) is interrupted. In such cases, separate gas burners are provided on the boiler furnaces.

Barring Gear

Barring gear (or "turning gear") is the mechanism provided to rotate the turbine generator shaft at a very low speed after unit stoppages. Once the unit is "tripped" (*i.e.*, the steam inlet valve is closed), the turbine coasts down towards standstill. When it stops completely, there is a tendency for the turbine shaft to deflect or bend if allowed to remain in one position too long. This is because the heat inside the turbine casing tends to concentrate in the top half of the casing, making the top half portion of the shaft hotter than the bottom half. The shaft therefore could warp or bend by millionths of inches.

This small shaft deflection, only detectable by eccentricity meters, would be enough to cause damaging vibrations to the entire steam turbine generator unit when it is restarted. The shaft is therefore automatically turned at low speed (about one percent rated speed) by the barring gear until it has cooled sufficiently to permit a complete stop.

Oil System

An auxiliary oil system pump is used to supply oil at the start-up of the steam turbine generator. It supplies the hydraulic oil system required for steam turbine's main inlet steam stop valve, the governing control valves, the bearing and seal oil systems, the relevant hydraulic relays and other mechanisms.

At a preset speed of the turbine during start-ups, a pump driven by the turbine main shaft takes over the functions of the auxiliary system.

Generator Cooling

While small generators may be cooled by air drawn through filters at the inlet, larger units generally require special cooling arrangements. Hydrogen gas cooling, in an oil-sealed casing, is used because it has the highest known heat transfer coefficient of any gas and for its low viscosity which reduces windage losses. This system requires special handling during start-up, with air in the generator enclosure first displaced by carbon dioxide before filling with hydrogen. This ensures that the highly flammable hydrogen does not mix with oxygen in the air.

The hydrogen pressure inside the casing is maintained slightly higher than atmospheric pressure to avoid outside air ingress. The hydrogen must be sealed against outward leakage where the shaft emerges from the casing. Mechanical seals around the shaft are installed with a very small annular gap to avoid rubbing between the shaft and the seals. Seal oil is used to prevent the hydrogen gas leakage to atmosphere.

The generator also uses water cooling. Since the generator coils are at a potential of about 22 kV, an insulating barrier such as Teflon is used to interconnect the water line and the generator high-voltage windings. Demineralized water of low conductivity is used.

Generator High-voltage System

The generator voltage for modern utility-connected generators ranges from 11 kV in smaller units to 22 kV in larger units. The generator high-voltage leads are normally large aluminium channels because of their high current as compared to the cables used in smaller machines. They are enclosed in well-grounded aluminium bus ducts and are supported on suitable insulators. The generator high-voltage leads are connected to step-up transformers for connecting to a high-voltage electrical substation (usually in the range of 115 kV to 765 kV) for further transmission by the local power grid.

The necessary protection and metering devices are included for the high-voltage leads. Thus, the steam turbine generator and the transformer form one unit. Smaller units may share a common generator step-up transformer with individual circuit breakers to connect the generators to a common bus.

Monitoring and Alarm System

Most of the power plant operational controls are automatic. However, at times, manual intervention may be required. Thus, the plant is provided with monitors and alarm systems that alert the plant operators when certain operating parameters are seriously deviating from their normal range.

Battery-supplied Emergency Lighting and Communication

A central battery system consisting of lead acid cell units is provided to supply emergency electric power, when needed, to essential items such as the power plant's control systems, communication systems, turbine lube oil pumps, and emergency lighting. This is essential for a safe, damage-free shutdown of the units in an emergency situation.

Transport of Coal Fuel to Site and to Storage

Most thermal stations use coal as the main fuel. Raw coal is transported from coal mines to a power station site by trucks, barges, bulk cargo ships or railway cars. Generally, when shipped by railways, the coal cars are sent as a full train of cars. The coal received at site may be of different sizes. The railway cars are unloaded at site by rotary dumpers or side tilt dumpers to tip over onto conveyor belts below. The coal is generally conveyed to crushers which crush the coal to about 3/4 inch (19 mm) size. The crushed coal is then sent by belt conveyors to a storage pile.

Normally, the crushed coal is compacted by bulldozers, as compacting of highly volatile coal avoids spontaneous ignition.

The crushed coal is conveyed from the storage pile to silos or hoppers at the boilers by another belt conveyor system.

Hydroelectricity

Hydroelectricity is the term referring to electricity generated by hydropower; the production of electrical power through the use of the gravitational force of falling or flowing water. It is the most widely used form of renewable energy, accounting for 16 per cent of global electricity generation – 3,427 terawatt-hours of electricity production in 2010, and is expected to increase about 3.1% each year for the next 25 years.

Hydropower is produced in 150 countries, with the Asia-Pacific region generating 32 per cent of global hydropower in 2010. China is the largest hydroelectricity producer, with 721 terawatt-hours of production in 2010, representing around 17 per cent of domestic electricity use. There are now three hydroelectricity plants larger than 10 GW : the Three Gorges Dam in China, Itaipu Dam across the Brazil/Paraguay border, and Guri Dam in Venezuela.

The cost of hydroelectricity is relatively low, making it a competitive source of renewable electricity. The average cost of electricity from a hydro plant larger than 10 megawatts is 3 to 5 U.S. cents per kilowatt-hour. Hydro is also a flexible source of electricity since plants can be ramped up and down very quickly to adapt to changing energy demands. However, damming interrupts the flow of rivers and can harm local ecosystems, and building large dams and reservoirs often involves displacing people and wildlife. Once a hydroelectric complex is constructed, the project produces no direct waste, and has a considerably lower output level of the greenhouse gas carbon dioxide (CO_2) than fossil fuel powered energy plants.

History

Hydropower has been used since ancient times to grind flour and perform other tasks. In the mid-1770s, French engineer Bernard Forest de Bélidor published *Architecture Hydraulique* which described vertical- and horizontal-axis hydraulic machines. By the late 19th century, the electrical generator was developed and could now be coupled with hydraulics. The growing demand for the Industrial Revolution would drive development as well. In 1878 the world's first hydroelectric power scheme was developed at Cragside in Northumberland, England by William George Armstrong. It was used to power a single arc lamp in his art gallery. The old Schoelkopf Power Station No. 1 near Niagara Falls in the U.S. side began to produce electricity in 1881. The first Edison hydroelectric power plant, the Vulcan Street Plant, began operating September 30, 1882, in Appleton, Wisconsin, with an output of about 12.5 kilowatts. By 1886 there were 45 hydroelectric power plants in the U.S. and Canada. By 1889 there were 200 in the U.S. alone.

At the beginning of the 20th century, many small hydroelectric power plants were being constructed by commercial companies in mountains near metropolitan areas. Grenoble, France held the International Exhibition of Hydropower and Tourism with over one million visitors. By 1920 as 40% of the power produced in the United States was hydroelectric, the Federal Power Act was enacted into law. The Act created the Federal Power Commission to regulate hydroelectric power plants on federal land and water. As the power plants became larger, their associated dams developed additional purposes to include flood control, irrigation and navigation. Federal funding became necessary for large-scale development and federally owned corporations, such as the Tennessee Valley Authority (1933) and the Bonneville Power Administration (1937) were created. Additionally, the Bureau of Reclamation which had began a series of western U.S. irrigation projects in the early 20th century was now constructing large hydroelectric projects such as the 1928 Hoover Dam. The U.S. Army Corps of Engineers was also involved in hydroelectric development, completing the Bonneville Dam in 1937 and being recognized by the Flood Control Act of 1936 as the premier federal flood control agency.

Hydroelectric power plants continued to become larger throughout the 20th century. Hydropower was referred to as *white coal* for its power and plenty. Hoover Dam's initial 1,345 MW power plant was the world's largest hydroelectric power plant in 1936; it was eclipsed by the 6809 MW Grand Coulee Dam in 1942. The Itaipu Dam opened in 1984 in South America as the largest, producing 14,000 MW but was surpassed in 2008 by the Three Gorges Dam in China at 22,500 MW. Hydroelectricity would eventually supply some countries, including Norway, Democratic Republic of the Congo, Paraguay and Brazil, with over 85% of their electricity. The United States currently has over 2,000 hydroelectric power plants that supply 6.4% of its total electrical production output, which is 49% of its renewable electricity.

Generating Methods

Conventional (Dams)

Most hydroelectric power comes from the potential energy of dammed water driving a water turbine and generator. The power extracted from the water depends on the volume and on the difference in height between the source and the water's outflow. This height difference is called the head. The amount of potential energy in water is proportional to the head. A large pipe (the "penstock") delivers water to the turbine.

Electrical Power Production Systems

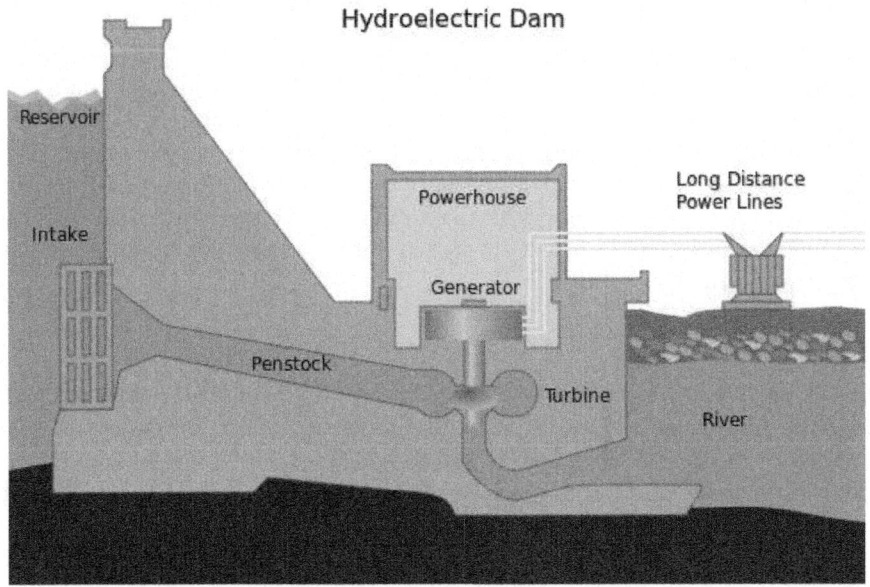

Fig. : Cross section of a conventional hydroelectric dam.

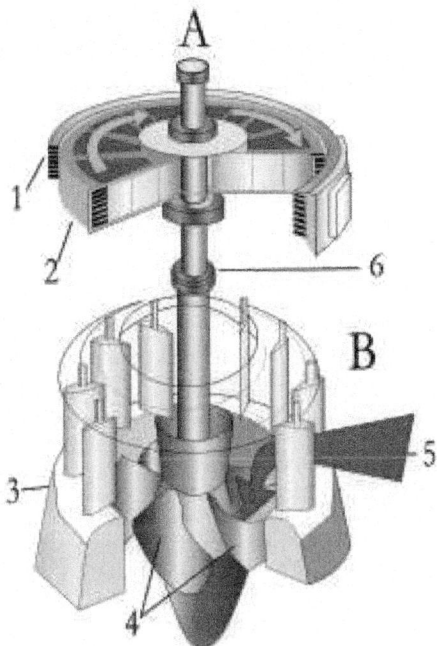

Fig. : A typical turbine and generator.

Pumped-storage

This method produces electricity to supply high peak demands by moving water between reservoirs at different elevations. At times of low electrical demand, excess generation capacity is used to pump water into the higher reservoir. When there is higher demand, water is released back into the lower reservoir through a turbine. Pumped-storage schemes currently provide the most commercially important means of large-scale grid energy storage and improve the daily capacity factor of the generation system. Pumped storage is not an energy source, and appears as a negative number in listings.

Run-of-the-river

Run-of-the-river hydroelectric stations are those with small or no reservoir capacity, so that the water coming from upstream must be used for generation at that moment, or must be allowed to bypass the dam. In the United States, run of the river hydropower could potentially provide 60,000 MW (about 13.7% of total use in 2011 if continuously available).

Tide

A tidal power plant makes use of the daily rise and fall of ocean water due to tides; such sources are highly predictable, and if conditions permit construction of reservoirs, can also be dispatchable to generate power during high demand periods. Less common types of hydro schemes use water's kinetic energy or undammed sources such as undershot waterwheels. Tidal power is viable in a relatively small number of locations around the world. In Great Britain, there are eight sites that could be developed, which have the potential to generate 20% of the electricity used in 2012.

Underground

An underground power station makes use of a large natural height difference between two waterways, such as a waterfall or mountain lake. An underground tunnel is constructed to take water from the high reservoir to the generating hall built in an underground cavern near the lowest point of the water tunnel and a horizontal tailrace taking water away to the lower outlet waterway.

Sizes and Capacities of Hydroelectric Facilities

Large Facilities

Although no official definition exists for the capacity range of large hydroelectric power stations, facilities from over a few hundred megawatts to more than 10 GW are generally considered large hydroelectric facilities. Currently, only three facilities over 10 GW (10,000 MW) are in operation worldwide; Three Gorges Dam at 22.5 GW, Itaipu Dam at 14 GW, and Guri Dam at 10.2 GW. Large-scale hydroelectric power stations are more commonly seen as the largest power producing facilities in the world, with some hydroelectric facilities capable of generating more than double the installed capacities of the current largest nuclear power stations.

Rank	Station	Country	Location	Capacity (MW)
1	Three Gorges Dam	China	30°49'15"N 111°00'08"E	20,300
2	Itaipu Dam	Brazil Paraguay	25°24'31"S 54°35'21"W	14,000
3	Guri Dam	Venezuela	07°45'59"N 62°59'57"W	10,200
4	Tucurui Dam	Brazil	03°49'53"S 49°38'36"W	8,370
5	Grand Coulee Dam	United States	47°57'23"N 118°58'56"W	6,809

Small

Small hydro is the development of hydroelectric power on a scale serving a small community or industrial plant. The definition of a small hydro project varies but a generating capacity of up to 10 megawatts (MW) is generally accepted as the upper limit of what can be termed small hydro. This may be stretched to 25 MW and 30 MW in Canada and the United States. Small-scale hydroelectricity production grew by 28% during 2008 from 2005, raising the total world small-hydro capacity to 85 GW. Over 70% of this was in China (65 GW), followed by Japan (3.5 GW), the United States (3 GW), and India (2 GW).

Small hydro plants may be connected to conventional electrical distribution networks as a source of low-cost renewable energy. Alternatively, small hydro projects may be built in isolated areas that would be uneconomic to serve from a network, or in areas where there is no national electrical distribution network. Since small hydro projects usually have minimal reservoirs and civil construction work, they are seen as having a relatively low environmental impact compared to large hydro. This decreased environmental impact depends strongly on the balance between stream flow and power production.

Micro

Micro hydro is a term used for hydroelectric power installations that typically produce up to 100 kW of power. These installations can provide power to an isolated home or small community, or are sometimes connected to electric power networks. There are many of these installations around the world, particularly in developing nations as they can provide an economical source of energy without purchase of fuel. Micro hydro systems complement photovoltaic solar energy systems because in many areas, water flow, and thus available hydro power, is highest in the winter when solar energy is at a minimum.

Pico

Pico hydro is a term used for hydroelectric power generation of under 5 kW. It is useful in small, remote communities that require only a small amount of electricity. For example, to power one or two fluorescent light bulbs and a TV or radio for a few homes. Even smaller turbines of 200-300W may power a single home in a developing country with a drop of only 1 m (3 ft). Pico-hydro setups typically are run-of-the-river, meaning that dams are not used, but rather pipes divert some of the flow, drop this down a gradient, and through the turbine before returning it to the stream.

Calculating Available Power

A simple formula for approximating electric power production at a hydroelectric plant is : $P = \rho hrgk$,

where :

- P is Power in watts,
- ρ is the density of water (~1000 kg/m^3),
- h is height in meters,
- r is flow rate in cubic meters per second,
- g is acceleration due to gravity of 9.8 m/s^2,
- k is a coefficient of efficiency ranging from 0 to 1. Efficiency is often higher (that is, closer to 1) with larger and more modern turbines.

Annual electric energy production depends on the available water supply. In some installations, the water flow rate can vary by a factor of 10 : 1 over the course of a year.

Advantages and Disadvantages

Advantages

Flexibility

Hydro is a flexible source of electricity since plants can be ramped up and down very quickly to adapt to changing energy demands. Hydro turbines have a start-up time of the order of few minutes. It takes around 60 to 90 seconds to bring a unit from cold start-up to full load; this is much shorter than for gas turbines or steam plants. Power generation can also be decreased quickly when there is a surplus power generation. Hence the limited capacity of hydropower units is not generally used to produce base power except for vacating the flood pool or meeting downstream needs. Instead, it serves as backup for non-hydro generators.

Low Power Costs

The major advantage of hydroelectricity is elimination of the cost of fuel. The cost of operating a hydroelectric plant is nearly immune to increases in the cost of fossil fuels such as oil, natural gas or coal, and no imports are needed. The average cost of electricity from a hydro plant larger than 10 megawatts is 3 to 5 U.S. cents per kilowatt-hour.

Hydroelectric plants have long economic lives, with some plants still in service after 50–100 years. Operating labour cost is also usually low, as plants are automated and have few personnel on site during normal operation.

Where a dam serves multiple purposes, a hydroelectric plant may be added with relatively low construction cost, providing a useful revenue stream to offset the costs of dam operation. It has been calculated that the sale of electricity from the Three Gorges Dam will cover the construction costs after 5 to 8 years of full generation. Additionally, some data shows that in most countries large hydropower dams will be too costly and take too long to build to deliver a positive risk adjusted return, unless appropriate risk management measures are put in place.

Suitability for Industrial Applications

While many hydroelectric projects supply public electricity networks, some are created to serve specific industrial enterprises. Dedicated hydroelectric projects are often built to provide the substantial amounts of electricity needed for aluminium electrolytic plants, for example. The Grand Coulee Dam switched to support Alcoa aluminium in Bellingham, Washington, United States for American World War II airplanes before it was allowed to provide irrigation and power to citizens (in addition to aluminium power) after the war. In Suriname, the Brokopondo Reservoir was constructed to provide electricity for the Alcoa aluminium industry. New Zealand's Manapouri Power Station was constructed to supply electricity to the aluminium smelter at Tiwai Point.

Reduced CO_2 Emissions

Since hydroelectric dams do not burn fossil fuels, they are claimed to not directly produce carbon dioxide. While some carbon dioxide is produced during manufacture and construction of the project, this is a tiny fraction of the operating emissions of equivalent fossil-fuel electricity generation. One measurement of greenhouse gas related and other externality comparison between energy sources can be found in the Extern E project by the Paul Scherrer Institut and the University of Stuttgart which was funded by the European Commission. According to that study, hydroelectricity produces the least amount of greenhouse gases and externality of any energy source. Coming in second place was wind, third was nuclear energy, and fourth was solar photovoltaic. The low greenhouse gas impact of hydroelectricity is found especially in temperate climates. Greater greenhouse gas emission impacts are found in the tropical regions, the lower latitude regions of the earth, as it has been noted that the reservoirs of power plants in tropical regions produce a larger amount of the greenhouse gas methane.

Other Uses of the Reservoir

Reservoirs created by hydroelectric schemes often provide facilities for water sports, and become tourist attractions themselves. In some countries, aquaculture in reservoirs is common. Multi-use dams installed for irrigation support agriculture with a relatively constant water supply. Large hydro dams can control floods, which would otherwise affect people living downstream of the project.

Disadvantages

Ecosystem Damage and Loss of Land

Large reservoirs required for the operation of hydroelectric power stations result in submersion of extensive areas upstream of the dams, destroying biologically rich and productive lowland and riverine valley forests, marshland and grasslands. The loss of land is often exacerbated by habitat fragmentation of surrounding areas caused by the reservoir.

Hydroelectric projects can be disruptive to surrounding aquatic ecosystems both upstream and downstream of the plant site. Generation of hydroelectric power changes the downstream river environment. Water exiting a turbine usually contains very little suspended sediment, which can lead to scouring of river beds and loss of riverbanks. Since turbine gates are often opened intermittently, rapid or even daily fluctuations in river flow are observed.

Siltation and Flow Shortage

When water flows it has the ability to transport particles heavier than itself downstream. This has a negative effect on dams and subsequently their power stations, particularly those on rivers or within catchment areas with high siltation. Siltation can fill a reservoir and reduce its capacity to control floods along with causing additional horizontal pressure on the upstream portion of the dam. Eventually, some reservoirs can become full of sediment and useless or over-top during a flood and fail.

Changes in the amount of river flow will correlate with the amount of energy produced by a dam. Lower river flows will reduce the amount of live storage in a reservoir therefore reducing the amount of water that can be used for hydroelectricity. The result of diminished river flow can be power shortages in areas that depend heavily on hydroelectric power. The risk of flow shortage may increase as a result of climate change. One study from the Colorado River in the United States suggest that modest climate changes, such as an increase in temperature in 2 degree Celsius resulting in a 10% decline in precipitation, might reduce river run-off by up to 40%. Brazil in particular is vulnerable due to its heaving reliance on hydroelectricity, as increasing temperatures, lower water flow and alterations in the rainfall regime, could reduce total energy production by 7% annually by the end of the century.

Methane Emissions (from Reservoirs)

Lower positive impacts are found in the tropical regions, as it has been noted that the reservoirs of power plants in tropical regions produce substantial amounts of methane. This is due to plant material in flooded areas decaying in an anaerobic environment, and forming methane, a greenhouse gas. According to the World Commission on Dams report, where the reservoir is large compared to the generating capacity (less than 100 watts per square metre of surface area) and no clearing of the forests in the area was undertaken prior to impoundment of the reservoir, greenhouse gas emissions from the reservoir may be higher than those of a conventional oil-fired thermal generation plant.

In boreal reservoirs of Canada and Northern Europe, however, greenhouse gas emissions are typically only 2% to 8% of any kind of conventional fossil-fuel thermal generation. A new class of underwater logging operation that targets drowned forests can mitigate the effect of forest decay.

Relocation

Another disadvantage of hydroelectric dams is the need to relocate the people living where the reservoirs are planned. In 2000, the World Commission on Dams estimated that dams had physically displaced 40-80 million people worldwide.

Failure Risks

Because large conventional dammed-hydro facilities hold back large volumes of water, a failure due to poor construction, natural disasters or sabotage can be catastrophic to downriver settlements and infrastructure. Dam failures have been some of the largest man-made disasters in history.

The Banqiao Dam failure in Southern China directly resulted in the deaths of 26,000 people, and another 145,000 from epidemics. Millions were left homeless. Also, the creation of a dam in a geologically inappropriate location may cause disasters such as 1963 disaster at Vajont Dam in Italy, where almost 2000 people died.

Smaller dams and micro-hydro facilities create less risk, but can form continuing hazards even after being decommissioned. For example, the small Kelly Barnes Dam failed in 1967, causing 39 deaths with the Toccoa Flood, ten years after its power plant was decommissioned.

Comparison with Other Methods of Power Generation

Hydroelectricity eliminates the flue gas emissions from fossil fuel combustion, including pollutants such as sulfur dioxide, nitric oxide, carbon monoxide, dust, and mercury in the coal. Hydroelectricity also avoids the hazards of coal mining and the indirect health effects of coal emissions. Compared to nuclear power, hydroelectricity generates no nuclear waste, has none of the dangers associated with uranium mining, nor nuclear leaks.

Compared to wind farms, hydroelectricity power plants have a more predictable load factor. If the project has a storage reservoir, it can generate power when needed. Hydroelectric plants can be easily regulated to follow variations in power demand.

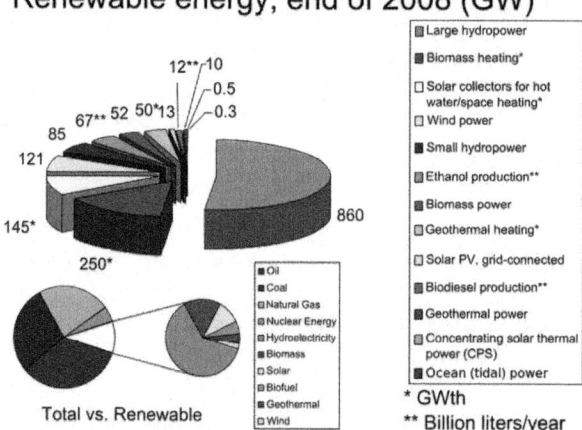

Fig. : World renewable energy share (2008).

World Hydroelectric Capacity

The ranking of hydro-electric capacity is either by actual annual energy production or by installed capacity power rating. Hydro accounted for 16 per cent of global electricity consumption, and 3,427 terawatt-hours of electricity production in 2010, which continues the rapid rate of increase experienced between 2003 and 2009.

Hydropower is produced in 150 countries, with the Asia-Pacific region generated 32 per cent of global hydropower in 2010. China is the largest hydroelectricity producer, with 721 terawatt-hours of production in 2010, representing around 17 per cent of domestic electricity use. Brazil, Canada, New Zealand, Norway, Paraguay, Austria, Switzerland, and Venezuela have a majority of the internal electric energy production from hydroelectric power. Paraguay produces 100% of its electricity from hydroelectric dams, and exports 90% of its production to Brazil and to Argentina. Norway produces 98–99% of its electricity from hydroelectric sources.

There are now three hydroelectric plants larger than 10 GW : the Three Gorges Dam in China, Itaipu Dam across the Brazil/Paraguay border, and Guri Dam in Venezuela.

A hydro-electric plant rarely operates at its full power rating over a full year; the ratio between annual average power and installed capacity rating is the capacity factor. The installed capacity is the sum of all generator nameplate power ratings.

Pumped-storage Hydroelectricity

Pumped-storage hydroelectricity (PSH) is a type of hydroelectric power generation used by some power plants for load balancing. The method stores energy in the form of potential energy of water, pumped from a lower elevation reservoir to a higher elevation. Low-cost off-peak electric power is used to run the pumps.

During periods of high electrical demand, the stored water is released through turbines to produce electric power. Although the losses of the pumping process makes the plant a net consumer of energy overall, the system increases revenue by selling more electricity during periods of *peak demand*, when electricity prices are highest.

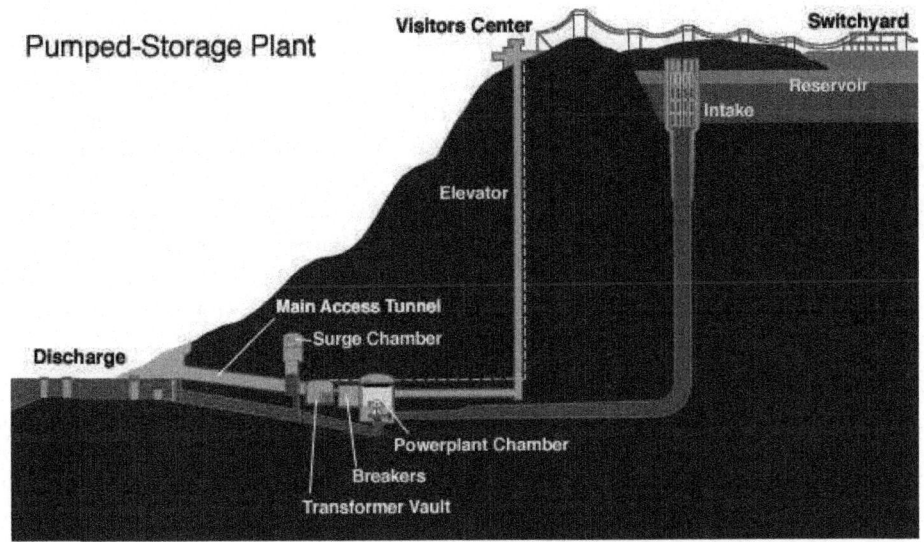

Fig. : Diagram of the TVA pumped storage facility at Raccoon Mountain Pumped-Storage Plant.

Pumped storage is the largest-capacity form of grid energy storage available, and, as of March 2012, the Electric Power Research Institute (EPRI) reports that PSH accounts for more than 99% of bulk storage capacity worldwide, representing around 127,000 MW. PSH reported energy efficiency varies in practice between 70% and 80%, with some claiming up to 87%.

Overview

At times of low electrical demand, excess generation capacity is used to pump water into the higher reservoir. When there is higher demand, water is released back into the lower reservoir through a turbine, generating electricity. Reversible turbine/generator assemblies act as pump and turbine (usually a Francis turbine design). Nearly all facilities use the height difference between two natural bodies of water or artificial reservoirs. Pure pumped-storage plants just shift the water between reservoirs, while the "pump-back" approach is a combination of pumped storage and conventional hydroelectric plants that use natural stream-flow. Plants that do not use pumped-storage are referred to as conventional hydroelectric plants; conventional hydroelectric plants that have significant storage capacity may be able to play a similar role in the electrical grid as pumped storage, by deferring output until needed.

Taking into account evaporation losses from the exposed water surface and conversion losses, approximately 70% to 85% of the electrical energy used to pump the water into the elevated reservoir can be regained. The technique is currently the most cost-effective means of storing large amounts of electrical energy on an operating basis, but capital costs and the presence of appropriate geography are critical decision factors.

The relatively low energy density of pumped storage systems requires either a very large body of water or a large variation in height. For example, 1000 kilograms of water (1 cubic meter) at the top of a 100 meter tower has a potential energy of about 0.272 kW·h (capable of raising the temperature of the same amount of water by only 0.23 Celsius = 0.42 Fahrenheit). The only way to store a significant amount of energy is by having a large body of water located on a hill relatively near, but as high as possible above, a second body of water. In some places this occurs naturally, in others one or both bodies of water have been man-made. Projects in which both reservoirs are artificial and in which no natural waterways are involved are commonly referred to as "closed loop".

This system may be economical because it flattens out load variations on the power grid, permitting thermal power stations such as coal-fired plants and nuclear power plants that provide base-load electricity to continue operating at peak efficiency (Base load power plants), while reducing the need for "peaking" power plants that use the same fuels as many baseload thermal plants, gas and oil, but have been designed for flexibility rather than maximal thermal efficiency. However, capital costs for purpose-built hydrostorage are relatively high.

Along with energy management, pumped storage systems help control electrical network frequency and provide reserve generation. Thermal plants are much less able to respond to sudden changes in electrical demand, potentially causing frequency and voltage instability. Pumped storage plants, like other hydroelectric plants, can respond to load changes within seconds.

The first use of pumped storage was in the 1890s in Italy and Switzerland. In the 1930s reversible hydroelectric turbines became available. These turbines could operate as both turbine-generators and in reverse as electric motor driven pumps. The latest in large-scale engineering technology are variable speed machines for greater efficiency. These machines generate in synchronization with the network frequency, but operate asynchronously (independent of the network frequency) as motor-pumps.

The first use of pumped-storage in the United States was in 1930 by the Connecticut Electric and Power Company, using a large reservoir located near New Milford, Connecticut, pumping water from the Houstatonic River to the storage reservoir 230 feet above.

A new use for pumped storage is to level the fluctuating output of intermittent energy sources. The pumped storage provides a load at times of high electricity output and low electricity demand, enabling additional system peak capacity. In certain jurisdictions, electricity prices may be close to zero or occasionally negative (Ontario in early September, 2006), on occasions that there is more electrical generation than load available to absorb it; although at present this is rarely due to wind alone, increased wind generation may increase the likelihood of such occurrences. It is particularly likely that pumped storage will become especially important as a balance for very large scale photovoltaic generation.

Worldwide Use

In 2009 world pumped storage generating capacity was 104 GW, while other sources claim 127 GW, which comprises the vast majority of all types of utility grade electric storage. The EU had 38.3 GW net capacity (36.8% of world capacity) out of a total of 140 GW of hydropower and representing 5% of total net electrical capacity in the EU. Japan had 25.5 GW net capacity (24.5% of world capacity).

In 2010 the United States had 21.5 GW of pumped storage generating capacity (20.6% of world capacity). PHS generated (net) -5.501 GWh of energy in 2010 in the US because more energy is consumed in pumping than is generated.

The five largest operational pumped-storage plants are listed below :

Station	Country	Location	Capacity (MW)	Ref.
Bath County Pumped Storage Station	United States	38°12'32"N 79°48'00"W	3,003	
Guangdong Pumped Storage Power Station	China	23°45'52"N 113°57'12"E	2,400	
Huizhou Pumped Storage Power Station	China	23°16'07"N 114°18'50"E	2,400	
Okutataragi Pumped Storage Power Station	Japan	35°14'13"N 134°49'55"E	1,932	
Ludington Pumped Storage Power Plant	United States	43°53'37"N 86°26'43"W	1,872	

Note : this table shows Capacity in terms of MW (power) as is usual for power stations. In the case of storage it would also be interesting to know the Capacity in terms of MWh (energy).

Potential Technologies

The use of underground reservoirs has been investigated. Recent examples include the proposed Summit project in Norton, Ohio, the proposed Maysville project in Kentucky (underground limestone mine), and the Mount Hope project in New Jersey, which was to have used a former iron mine as the lower reservoir. Several new underground pumped storage projects have been proposed. Cost-per-kilowatt estimates for these projects can be lower than for surface projects if they use existing underground mine space. There are limited opportunities involving suitable underground space, but the number of underground pumped storage opportunities may expanded if abandoned coal mines prove suitable.

A new concept is to use wind turbines or solar power to drive water pumps directly, in effect an 'Energy Storing Wind or Solar Dam'. This could provide a more efficient process and usefully smooth out the variability of energy captured from the wind or sun.

One can use pumped sea water to store the energy. The 30 MW Yanbaru project in Okinawa was the first demonstration of seawater pumped storage. A 300 MW seawater-based project has recently been proposed on Lanai, Hawaii, and several seawater-based projects have recently been proposed in Ireland. Another potential example of this could be used in a tidal barrage or tidal lagoon. A potential benefit of this arises if seawater is allowed to flow behind the barrage or into the lagoon at high tide when the water level is roughly equal either side of the barrier, when the potential energy difference is close to zero. Then water is released at low tide when a head of water has been built up behind the barrier, when there is a far greater potential energy difference between the two bodies of water. The result being that when the energy used to pump the water is recovered, it will have multiplied to a degree depending on the head of water built up. A further enhancement is to pump more water at high tide further increasing the head with for example intermittent renewables. Two downsides are that the generator must be below sea level, and that marine organisms would tend to grow on the equipment and disrupt operation. This is not a major problem for the EDF La Rance Tidal power station in France.

Small pumped-storage hydropower plants can be built on streams and within infrastructures, such as drinking water networks and artificial snow making infrastructures. Such plants provide distributed energy storage and distributed flexible electricity production and can contribute to the decentralized integration of intermittent renewable energy technologies, such as wind power and photovoltaic. Small pumped-storage hydropower plants have both an upstream and downstream reservoir. Reservoirs that can be used for small pumped-storage hydropower plants could include natural or artificial lakes, reservoirs within other structures such as irrigation, or unused portions of mines or underground military installations. In Switzerland one study suggested that the total installed capacity of small pumped-storage hydropower plants in 2011 could be increased by 3 to 9 times by providing adequate policy instruments.

Instead of pumping water uphill, the pumped storage idea can be inverted, pumping air under water.

Solar Power

Solar power is the conversion of sunlight into electricity, either directly using photovoltaics (PV), or indirectly using concentrated solar power (CSP). Concentrated solar power systems use lenses or mirrors and tracking systems to focus a large area of sunlight into a small beam. Photovoltaics convert light into electric current using the photovoltaic effect.

Photovoltaics were initially, and still are, used to power small and medium-sized applications, from the calculator powered by a single solar cell to off-grid homes powered by a photovoltaic array. They are an important and relatively inexpensive source of electrical energy where grid power is inconvenient, unreasonably expensive to connect, or simply unavailable. However, as the cost of solar electricity is falling, solar power is also increasingly being used even in grid-connected situations as a way to feed low-carbon energy into the grid.

Commercial concentrated solar power plants were first developed in the 1980s. The 392 MW ISEGS CSP installation is the largest solar power plant in the world, located in the Mojave Desert of California. Other large CSP plants include the SEGS (354 MW) in the Mojave Desert of California, the Solnova Solar Power Station (150 MW) and the Andasol solar power station (150 MW), both in Spain. The 250+ MW Agua Caliente Solar Project in the United States, and the 221 MW Charanka Solar Park in India, are the world's largest photovoltaic power stations.

Concentrated Solar Power

Concentrating Solar Power (CSP) systems use lenses or mirrors and tracking systems to focus a large area of sunlight into a small beam. The concentrated heat is then used as a heat source for a conventional power plant. A wide range of concentrating technologies exists : the most developed are the parabolic trough, the concentrating linear fresnel reflector, the Stirling dish and the solar power tower. Various techniques are used to track the sun and focus light. In all of these systems a working fluid is heated by the concentrated sunlight, and is then used for power generation or energy storage. Thermal storage efficiently allows up to 24 hour electricity generation.

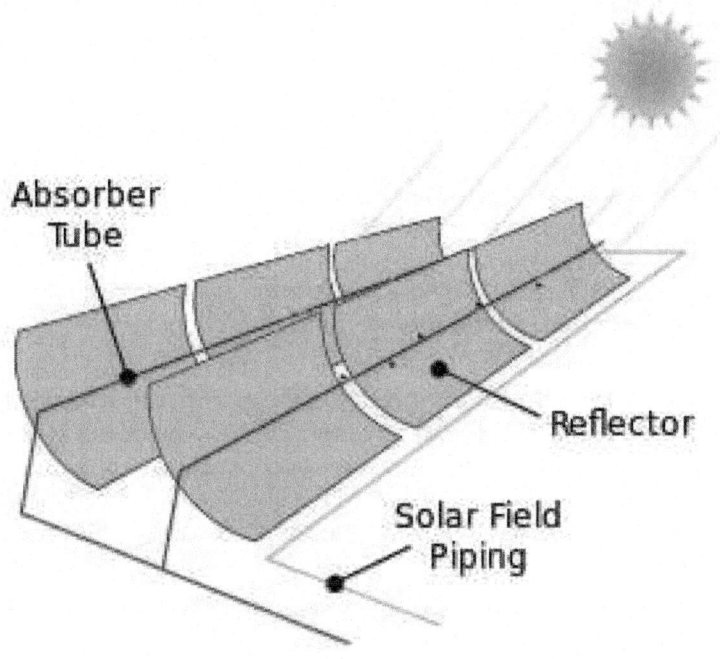

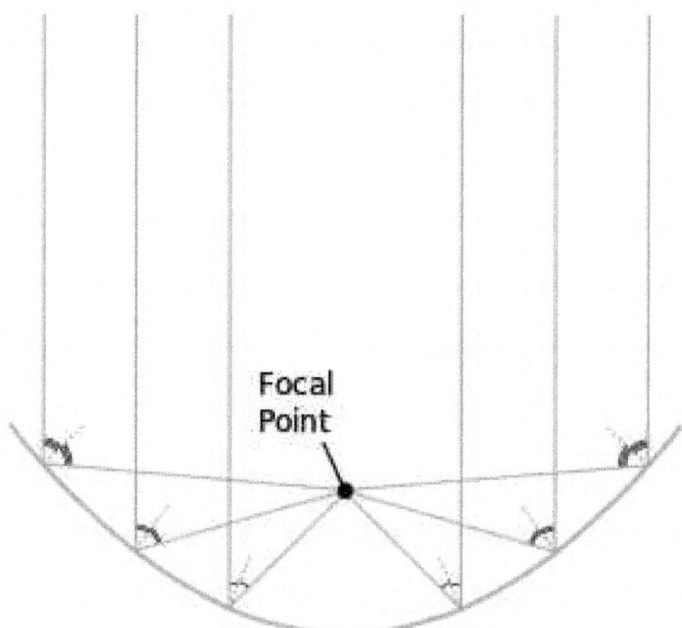

Fig. : A diagram of a parabolic trough solar farm (top), and an end view of how a parabolic collector focuses sunlight onto its focal point.

A parabolic trough consists of a linear parabolic reflector that concentrates light onto a receiver positioned along the reflector's focal line. The receiver is a tube positioned right above the middle of the parabolic mirror and is filled with a working fluid. The reflector is made to follow the sun during daylight hours by tracking along a single axis. Parabolic trough systems provide the best land-use factor of any solar technology. The SEGS plants in California and Acciona's Nevada Solar One near Boulder City, Nevada are representatives of this technology. Compact Linear Fresnel Reflectors are CSP-plants which use many thin mirror strips instead of parabolic mirrors to concentrate sunlight onto two tubes with working fluid. This has the advantage that flat mirrors can be used which are much cheaper than parabolic mirrors, and that more reflectors can be placed in the same amount of space, allowing more of the available sunlight to be used. Concentrating linear fresnel reflectors can be used in either large or more compact plants.

The Stirling solar dish combines a parabolic concentrating dish with a Stirling engine which normally drives an electric generator. The advantages of Stirling solar over photovoltaic cells are higher efficiency of converting sunlight into electricity and longer lifetime. Parabolic dish systems give the highest efficiency among CSP technologies. The 50 kW Big Dish in Canberra, Australia is an example of this technology.

A solar power tower uses an array of tracking reflectors (heliostats) to concentrate light on a central receiver atop a tower. Power towers are more cost effective, offer higher efficiency and better energy storage capability among CSP technologies. The PS10 Solar Power Plant and PS20 solar power plant are examples of this technology.

Photovoltaics

A solar cell, or photovoltaic cell (PV), is a device that converts light into electric current using the photovoltaic effect. The first solar cell was constructed by Charles Fritts in the 1880s. The German industrialist Ernst Werner von Siemens was among those who recognized the importance of this discovery. In 1931, the German engineer Bruno Lange developed a photo cell using silver selenide in place of copper oxide, although the prototype selenium cells converted less than 1% of incident light into electricity. Following the work of Russell Ohl in the 1940s, researchers Gerald Pearson, Calvin Fuller and Daryl Chapin created the silicon solar cell in 1954. These early solar cells cost 286 USD/watt and reached efficiencies of 4.5-6%.

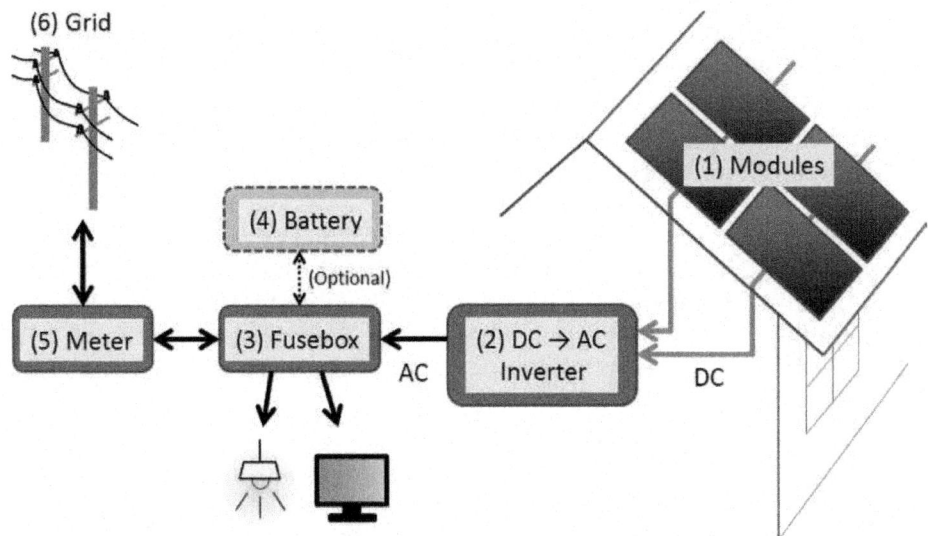

Fig. : Simplified schematics of a grid-connected residential PV power system.

Photovoltaic Power Systems

Solar cells produce direct current (DC) power which fluctuates with the sunlight's intensity. For practical use this usually requires conversion to certain desired voltages or alternating current (AC), through the use of inverters. Multiple solar cells are connected inside modules. Modules are wired together to form arrays, then tied to an inverter, which produces power at the desired voltage, and for AC, the desired frequency/phase.

Many residential systems are connected to the grid wherever available, especially in developed countries with large markets. In these grid-connected PV systems, use of energy storage is optional. In certain applications such as satellites, lighthouses, or in developing countries, batteries or additional power generators are often added as back-ups. Such stand-alone power systems permit operations at night and at other times of limited sunlight.

Development and Deployment

The early development of solar technologies starting in the 1860s was driven by an expectation that coal would soon become scarce. However, development of solar technologies stagnated in the early 20th century in the face of the increasing availability, economy and utility of coal and petroleum. In 1974 it was estimated that only six private homes in all of North America were entirely heated or cooled by functional solar power systems. The 1973 oil embargo and 1979 energy crisis caused a re-organization of energy policies around the world and brought renewed attention to developing solar technologies. Deployment strategies focused on incentive programs such as the Federal Photovoltaic Utilization Program in the US and the Sunshine Program in Japan. Other efforts included the formation of research facilities in the US (SERI, now NREL), Japan (NEDO), and Germany (Fraunhofer Institute for Solar Energy Systems ISE).

Between 1970 and 1983 photovoltaic installations grew rapidly, but falling oil prices in the early 1980s moderated the growth of PV from 1984 to 1996. Since 1997, PV development has accelerated due to supply issues with oil and natural gas, global warming concerns, and the improving economic position of PV relative to other energy technologies. Photovoltaic production growth has averaged 40% per year since 2000 and installed capacity reached 39.8 GW at the end of 2010, of them 17.4 GW in Germany. As of October 2011, the largest photovoltaic (PV) power plants in the world are the Sarnia Photovoltaic Power Plant (Canada, 97 MW), Montalto di Castro Photovoltaic Power Station (Italy, 84.2 MW) and Finsterwalde Solar Park (Germany, 80.7 MW).

There are also many large plants under construction. The Desert Sunlight Solar Farm is a 550 MW solar power plant under construction in Riverside County, California, that will use thin-film solar photovoltaic modules made by First Solar. The Topaz Solar Farm is a 550 MW photovoltaic power plant, being built in San Luis Obispo County, California. The Blythe Solar Power Project is a 500 MW photovoltaic station under construction in Riverside County, California. The Agua Caliente Solar Project is a 290 megawatt photovoltaic solar generating facility being built in Yuma County, Arizona. The California Valley Solar Ranch (CVSR) is a 250 megawatt (MW) solar photovoltaic power plant, which is being built by Sun-Power in the Carrizo Plain, northeast of California Valley. The 230 MW Antelope Valley Solar Ranch is a First Solar photovoltaic project which is under construction in the Antelope Valley area of the Western Mojave Desert, and due to be completed in 2013.

At the end of September 2013, IKEA announced that solar panel packages for houses will be sold at 17 United Kingdom IKEA stores by the end of July 2014. The decision followed a successful pilot project at the Lakeside IKEA store, whereby one photovoltaic (PV) system was sold almost every day. The panels are manufactured by a Chinese company named Hanergy Holding Group Ltd.

Electricity Generation from Solar		
Year	Energy (TWh)	% of Total
2005	3.7	0.02%
2006	5.0	0.03%
2007	6.7	0.03%
2008	11.2	0.06%
2009	19.1	0.09%
2010	30.4	0.14%
2011	58.7	0.27%
2012	93.0	0.41%

Photovoltaic Power Stations

World's largest photovoltaic power stations (50 MW or larger)			
PV power station	Country	DC peak power (MW_p)	Notes
Agua Caliente Solar Project	USA	250 AC	397 MW when complete
Charanka Solar Park	India	221	Completed 2012
Golmud Solar Park	China	200	Completed 2011
Mesquite Solar project	USA	150	up to 700 MW when complete
Neuhardenberg Solar Park	Germany	145	Completed September 2012. A group of 11 co-located plant by the same developer but with different IPPs
Templin Solar Park	Germany	128.48	Completed September 2012
Toul-Rosières Solar Park	France	115	Completed November 2012
Perovo Solar Park	Ukraine	100	Completed 2011
Sarnia Photovoltaic Power Plant	Canada	97	Constructed 2009–2010
Montalto di Castro Photovoltaic Power Station	Italy	84.2	Constructed 2009–2010
Finsterwalde Solar Park	Germany	80.7	Phase I completed 2009, phase II and III 2010
Okhotnykovo Solar Park	Ukraine	80	Completed 2011
Solarpark Senftenberg	Germany	78	Phase II and III completed 2011, another 70 MW phase planned

World's largest photovoltaic power stations (50 MW or larger)			
PV power station	Country	DC peak power (MW_p)	Notes
Lieberose Photovoltaic Park	Germany	71.8	
Rovigo Photovoltaic Power Plant	Italy	70	Completed November 2010
Olmedilla Photovoltaic Park	Spain	60	Completed September 2008
Strasskirchen Solar Park	Germany	54	
Puertollano Photovoltaic Park	Spain	50	opened 2008

Concentrating Solar Thermal Power

Commercial concentrating solar thermal power (CSP) plants were first developed in the 1980s. The 354 MW SEGS CSP installation is the largest solar power plant in the world, located in the Mojave Desert of California. Other large CSP plants include the Solnova Solar Power Station (150 MW), the Andasol solar power station (150 MW), and Extresol Solar Power Station (100 MW), all in Spain. The 370 MW Ivanpah Solar Power Facility, located in California's Mojave Desert, is the world's largest solar thermal power plant project.

Fig.: Ivanpah Solar Electric Generating System with all three towers under load, Feb., 2014. Taken from I-15 in San Bernardino County, California. The Clark Mountain Range can be seen in the distance.

The principal advantage of CSP is the ability to efficiently add thermal storage, allowing the dispatching of electricity over up to a 24-hour period. Since peak electricity demand typically occurs at about 5 pm, many CSP power plants use 3 to 5 hours of thermal storage.

Largest operational solar thermal power stations					
Capacity (MW)	Name	Country	Location	Notes	
354	Solar Energy Generating Systems	USA	Mojave Desert California	Collection of 9 units	
280	Solana Generating Station	USA	Gila Bend, Arizona	Completed in October 2013, with 6h thermal energy storage	
200	Solaben Solar Power Station	Spain	Logrosán	Solaben 3 completed June 2012 Solaben 2 completed October 2012 Solaben 1 and 6 completed September 2013	
150	Solnova Solar Power Station	Spain	Seville	Completed 2010	
150	Andasol solar power station	Spain	Granada	completed 2011, with 7.5h thermal energy storage	
150	Extresol Solar Power Station	Spain	Torre de Miguel Sesmero	Extresol 1 completed February 2010 Extresol 2 completed December 2010 Extresol 3 completed August 2012, with 7.5h thermal energy storage	
100	Palma del Rio Solar Power Station	Spain	Palma del Río	Palma del Rio 2 completed December 2010 Palma del Rio 1 completed July 2011	
100	Manchasol Power Station	Spain	Alcázar de San Juan	Manchasol-1 completed January 2011, with 7.5h heat storage Manchasol-2 completed April 2011, with 7.5h heat storage	
100	Valle Solar Power Station	Spain	San José del Valle	Completed December 2011, with 7.5h heat storage	
100	Helioenergy Solar Power Station	Spain	Écija	Helioenergy 1 completed September 2011 Helioenergy 2 completed January 2012	
100	Aste Solar Power Station	Spain	Alcázar de San Juan	Aste 1A Completed January 2012, with 8h heat storage Aste 1B Completed January 2012, with 8h heat storage	
100	Solacor Solar Power Station	Spain	El Carpio	Solacor 1 completed February 2012 Solacor 2 completed March 2012	
100	Helios Solar Power Station	Spain	Puerto Lápice	Helios 1 completed May 2012 Helios 2 completed August 2012	

Economics

Photovoltaic systems use no fuel and modules typically last 25 to 40 years. The cost of installation is almost the only cost, as there is very little maintenance required. Installation cost is measured in $/watt or €/watt. The electricity generated is sold for ¢/kWh. 1 watt of installed photovoltaics generates roughly 1 to 2 kWh/year, as a result of the local insolation. The product of the local cost of electricity and the insolation determines the break even point for solar power. The International Conference on Solar Photovoltaic Investments, organized by EPIA, has estimated that PV systems will pay back their investors in 8 to 12 years. As a result, since 2006 it has been economical for investors to install photovoltaics for free in return for a long term power purchase agreement. Fifty per cent of commercial systems were installed in this manner in 2007 and over 90% by 2009.

As of 2011, the cost of PV has fallen well below that of nuclear power and is set to fall further. The average retail price of solar cells as monitored by the Solarbuzz group fell from $3.50/watt to $2.43/watt over the course of 2011, and a decline to prices below $2.00/watt seems inevitable :

A U.S. study of the amount of economic installations agrees closely with the actual installations.

For large-scale installations, prices below $1.00/watt are now common. In some locations, PV has reached grid parity, the cost at which it is competitive with coal or gas-fired generation. More generally, it is now evident that, given a carbon price of $50/ton, which would raise the price of coal-fired power by 5c/kWh, solar PV will be cost-competitive in most locations. The declining price of PV has been reflected in rapidly growing installations, totalling about 23 GW in 2011. Although some consolidation is likely in 2012, as firms try to restore profitability, strong growth seems likely to continue for the rest of the decade. Already, by one estimate, total investment in renewables for 2011 exceeded investment in carbon-based electricity generation.

Additionally, governments have created various financial incentives to encourage the use of solar power, such as feed-in tariff programs. Also, Renewable portfolio standards impose a government mandate that utilities generate or acquire a certain percentage of renewable power regardless of increased energy procurement costs. In most states, RPS goals can be achieved by any combination of solar, wind, biomass, landfill gas, ocean, geothermal, municipal solid waste, hydroelectric, hydrogen, or fuel cell technologies.

Shi Zhengrong has said that, as of 2012, unsubsidised solar power is already competitive with fossil fuels in India, Hawaii, Italy and Spain. He said "We are at a tipping point. No longer are renewable power sources like solar and wind a luxury of the rich. They are now starting to compete in the real world without subsidies". "Solar power will be able to compete without subsidies against conventional power sources in half the world by 2015".

Energy Cost

The PV industry is beginning to adopt levelized cost of energy (LCOE) as the unit of cost. For a 10 MW plant in Phoenix, AZ, the LCOE is estimated at $0.15 to 0.22/kWh in 2005.

The table below illustrates the calculated total cost in US cents per kilowatt-hour of electricity generated by a photovoltaic system as function of the investment cost and the efficiency, assuming some accounting parameters such as cost of capital and depreciation period. The row headings on the left show the total cost, per peak kilowatt (kWp), of a photovoltaic installation. The column headings across the top refer to the annual energy output in kilowatt-hours expected from each installed peak kilowatt. This varies by geographic region because the average insolation depends on the average cloudiness and the thickness of atmosphere traversed by the sunlight. It also depends on the path of the sun relative to the panel and the horizon.

Panels can be mounted at an angle based on latitude, or solar tracking can be utilized to access even more perpendicular sunlight, thereby raising the total energy output. The calculated values in the table reflect the total cost in cents per kilowatt-hour produced. They assume a 5%/year total capital cost (for instance 4% interest rate, 1% operating and maintenance cost, and depreciation of the capital outlay over 20 years).

Table showing average cost in cents/kWh over 20 years for solar power panels.

Cost	Insolation								
	2400 kWh/kWp·y	2200 kWh/kWp·y	2000 kWh/kWp·y	1800 kWh/kWp·y	1600 kWh/kWp·y	1400 kWh/kWp·y	1200 kWh/kWp·y	1000 kWh/kWp·y	800 kWh/kWp·y
200 $/kWp	0.8	0.9	1.0	1.1	1.3	1.4	1.7	2.0	2.5
600 $/kWp	2.5	2.7	3.0	3.3	3.8	4.3	5.0	6.0	7.5
1000 $/kWp	4.2	4.5	5.0	5.6	6.3	7.1	8.3	10.0	12.5
1400 $/kWp	5.8	6.4	7.0	7.8	8.8	10.0	11.7	14.0	17.5
1800 $/kWp	7.5	8.2	9.0	10.0	11.3	12.9	15.0	18.0	22.5
2200 $/kWp	9.2	10.0	11.0	12.2	13.8	15.7	18.3	22.0	27.5
2600 $/kWp	10.8	11.8	13.0	14.4	16.3	18.6	21.7	26.0	32.5
3000 $/kWp	12.5	13.6	15.0	16.7	18.8	21.4	25.0	30.0	37.5
3400 $/kWp	14.2	15.5	17.0	18.9	21.3	24.3	28.3	34.0	42.5
3800 $/kWp	15.8	17.3	19.0	21.1	23.8	27.1	31.7	38.0	47.5

	Insolation								
Cost	2400 kWh/ kWp·y	2200 kWh/ kWp·y	2000 kWh/ kWp·y	1800 kWh/ kWp·y	1600 kWh/ kWp·y	1400 kWh/ kWp·y	1200 kWh/ kWp·y	1000 kWh/ kWp·y	800 kWh/ kWp·y
4200 $/ kWp	17.5	19.1	21.0	23.3	26.3	30.0	35.0	42.0	52.5
4600 $/ kWp	19.2	20.9	23.0	25.6	28.8	32.9	38.3	46.0	57.5
5000 $/ kWp	20.8	22.7	25.0	27.8	31.3	35.7	41.7	50.0	62.5

Grid Parity

Grid parity, the point at which the cost of photovoltaic electricity is equal to or cheaper than the price of grid power, is more easily achieved in areas with abundant sun and high costs for electricity such as in California and Japan.

The fully loaded cost (cost not price) of solar electricity in 2008 was $0.25/kWh or less in most of the OECD countries. By late 2011, the fully loaded cost was predicted to fall below $0.15/kWh for most of the OECD and reach $0.10/kWh in sunnier regions. These cost levels are driving three emerging trends :

1. vertical integration of the supply chain;
2. origination of power purchase agreements (PPAs) by solar power companies;
3. unexpected risk for traditional power generation companies, grid operators and wind turbine manufacturers.

Grid parity was first reached in Spain in 2013, Hawaii and other islands that otherwise use fossil fuel (diesel fuel) to produce electricity, and most of the US is expected to reach grid parity by 2015.

General Electric's Chief Engineer predicted grid parity without subsidies in sunny parts of the United States by around 2015. Other companies predict an earlier date : the cost of solar power will be below grid parity for more than half of residential customers and 10% of commercial customers in the OECD, as long as grid electricity prices do not decrease through 2010.

Self Consumption

In cases of self-consumption of the solar energy the payback time is calculated based on how much electricity is not purchased from the grid.

For example, in Germany with electricity prices of 0.25 Euro/KWh and insolation of 900 KWh/KW one KWp will save 225 Euro per year, and with an installation cost of 1700 Euro/KWp the system cost will be returned in less than 7 years.

However, in many cases, the patterns of generation and consumption do not coincide, and some or all of the energy is fed back into the grid; the electricity is sold, and at other times energy is taken from the grid, electricity is bought. The relative costs and prices obtained affect the economics.

Energy Pricing and Incentives

The political purpose of incentive policies for PV is to facilitate an initial small-scale deployment to begin to grow the industry, even where the cost of PV is significantly above grid parity, to allow the industry to achieve the economies of scale necessary to reach grid parity. The policies are implemented to promote national energy independence, high tech job creation and reduction of CO_2 emissions.

Three incentive mechanisms are used (often in combination) :

- investment subsidies : the authorities refund part of the cost of installation of the system,
- Feed-in Tariffs (FIT) : the electricity utility buys PV electricity from the producer under a multi-year contract at a guaranteed rate.
- Solar Renewable Energy Certificates ("SRECs")

Rebates

With investment subsidies, the financial burden falls upon the taxpayer, while with feed-in tariffs the extra-cost is distributed across the utilities' customer bases. While the investment subsidy may be simpler to administer, the main argument in favour of feed-in tariffs is the encouragement of quality. Investment subsidies are paid out as a function of the nameplate capacity of the installed system and are independent of its actual power yield over time, thus rewarding the overstatement of power and tolerating poor durability and maintenance. Some electric companies offer rebates to their customers, such as Austin Energy in Texas, which offers $2.50/watt installed up to $15,000.

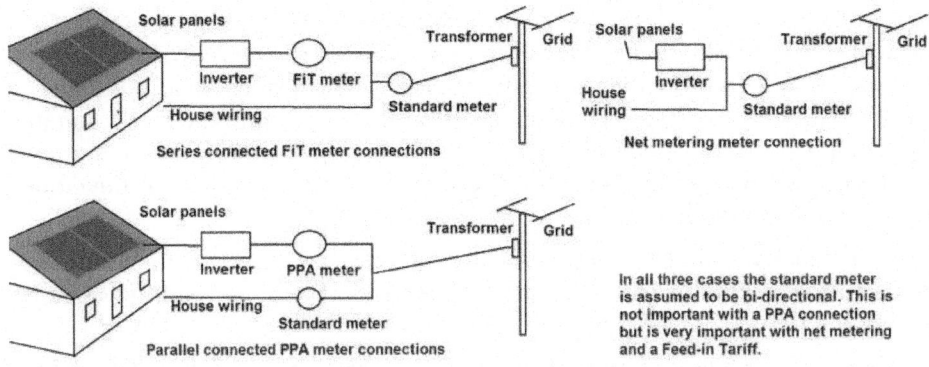

Understanding Feed-in Tariff and Power Purchase Agreement meter connections

Fig. : Net metering, unlike a feed-in tariff, requires only one meter, but it must be bi-directional.

Net Metering

In net metering the price of the electricity produced is the same as the price supplied to the consumer, and the consumer is billed on the difference between production and consumption. Net metering can usually be done with no changes to standard electricity meters, which accurately measure power in both directions and automatically report the difference, and because it allows homeowners and businesses to generate electricity at a different time from consumption, effectively using the grid as a giant storage battery. With net metering, deficits are billed each month while surpluses are rolled over to the following month. Best practices call for perpetual roll over of kWh credits. Excess credits upon termination of service are either lost, or paid for at a rate ranging from wholesale to retail rate or above, as can be excess annual credits. In New Jersey, annual excess credits are paid at the wholesale rate, as are left over credits when a customer terminates service.

Feed-in Tariffs (FiT)

With feed-in tariffs, the financial burden falls upon the consumer. They reward the number of kilowatt-hours produced over a long period of time, but because the rate is set by the authorities, it may result in perceived overpayment. The price paid per kilowatt-hour under a feed-in tariff exceeds the price of grid electricity. Net metering refers to the case where the price paid by the utility is the same as the price charged.

Solar Renewable Energy Credits (SRECs)

Alternatively, SRECs allow for a market mechanism to set the price of the solar generated electricity subsity. In this mechanism, a renewable energy production or consumption target is set, and the utility (more technically the Load Serving Entity) is obliged to purchase renewable energy or face a fine (Alternative Compliance Payment or ACP). The producer is credited for an SREC for every 1,000 kWh of electricity produced. If the utility buys this SREC and retires it, they avoid paying the ACP. In principle this system delivers the cheapest renewable energy, since the all solar facilities are eligible and can be installed in the most economic locations. Uncertainties about the future value of SRECs have led to long-term SREC contract markets to give clarity to their prices and allow solar developers to pre-sell/hedge their SRECs.

Financial incentives for photovoltaics differ across countries, including Australia, China, Germany, Israel, Japan, and the United States and even across states within the US.

The Japanese government through its Ministry of International Trade and Industry ran a successful programme of subsidies from 1994 to 2003. By the end of 2004, Japan led the world in installed PV capacity with over 1.1 GW.

In 2004, the German government introduced the first large-scale feed-in tariff system, under a law known as the 'EEG' (Erneuerbare Energien Gesetz) which resulted in explosive growth of PV installations in Germany. At the outset the

FIT was over 3x the retail price or 8x the industrial price. The principle behind the German system is a 20 year flat rate contract. The value of new contracts is programmed to decrease each year, in order to encourage the industry to pass on lower costs to the end users. The programme has been more successful than expected with over 1GW installed in 2006, and political pressure is mounting to decrease the tariff to lessen the future burden on consumers.

Subsequently, Spain, Italy, Greece (who enjoyed an early success with domestic solar-thermal installations for hot water needs) and France introduced feed-in tariffs. None have replicated the programmed decrease of FIT in new contracts though, making the German incentive relatively less and less attractive compared to other countries. The French and Greek FIT offer a high premium (EUR 0.55/kWh) for building integrated systems. California, Greece, France and Italy have 30-50% more insolation than Germany making them financially more attractive. The Greek domestic "solar roof" programme (adopted in June 2009 for installations up to 10 kW) has internal rates of return of 10-15% at current commercial installation costs, which, furthermore, is tax free.

In 2006 California approved the 'California Solar Initiative', offering a choice of investment subsidies or FIT for small and medium systems and a FIT for large systems. The small-system FIT of $0.39 per kWh (far less than EU countries) expires in just 5 years, and the alternate "EPBB" residential investment incentive is modest, averaging perhaps 20% of cost. All California incentives are scheduled to decrease in the future depending as a function of the amount of PV capacity installed.

At the end of 2006, the Ontario Power Authority (OPA, Canada) began its Standard Offer Program (SOP), the first in North America for small renewable projects (10MW or less). This guarantees a fixed price of $0.42 CDN per kWh over a period of twenty years. Unlike net metering, all the electricity produced is sold to the OPA at the SOP rate. The generator then purchases any needed electricity at the current prevailing rate (*e.g.*, $0.055 per kWh). The difference should cover all the costs of installation and operation over the life of the contract. On 1 October, 2009, OPA issued a Feed-in Tariff (FIT) program, increasing this fixed price to $0.802 per kWh.

The price per kilowatt hour or per peak kilowatt of the FIT or investment subsidies is only one of three factors that stimulate the installation of PV. The other two factors are insolation (the more sunshine, the less capital is needed for a given power output) and administrative ease of obtaining permits and contracts.

Unfortunately the complexity of approvals in California, Spain and Italy has prevented comparable growth to Germany even though the return on investment is better.

In some countries, additional incentives are offered for BIPV compared to stand alone PV.
- France + EUR 0.16 /kWh (compared to semi-integrated) or + EUR 0.27/kWh (compared to stand alone)
- Italy + EUR 0.04-0.09 kWh

- Germany + EUR 0.05/kWh (facades only).

Environmental Impacts

Unlike fossil fuel based technologies, solar power does not lead to any harmful emissions during operation, but the production of the panels leads to some amount of pollution.

Greenhouse Gases

The Life-cycle greenhouse-gas emissions of solar power are in the range of 22 to 46 g/kWh depending on if solar thermal or solar PV is being analyzed, respectively. With this potentially being decreased to 15 g/kWh in the future. For comparison (of weighted averages), a combined cycle gas-fired power plant emits some 400-599 g/kWh, an oil-fired power plant 893 g/kWh, a coal-fired power plant 915-994 g/kWh or with carbon capture and storage some 200 g/kWh, and a geothermal high-temp. power plant 91-122 g/kWh. Similar to all energy sources were their total life cycle emissions primarily lay in the construction and transportation phase, the switch to low carbon power in the manufacturing and transportation of solar devices would further reduce carbon emissions. BP Solar owns two factories built by Solarex (one in Maryland, the other in Virginia) in which all of the energy used to manufacture solar panels is produced by solar panels. A 1-kilowatt system eliminates the burning of approximately 170 pounds of coal, 300 pounds of carbon dioxide from being released into the atmosphere, and saves up to 105 gallons of water consumption monthly.

Energy Payback

The energy payback time of a power generating system is the time required to generate as much energy as was consumed during production of the system. In 2000 the energy payback time of PV systems was estimated as 8 to 11 years and in 2006 this was estimated to be 1.5 to 3.5 years for crystalline silicon PV systems and 1-1.5 years for thin film technologies (S. Europe).

Another economic measure, closely related to the energy payback time, is the energy returned on energy invested (EROEI) or energy return on investment (EROI), which is the ratio of electricity generated divided by the energy required to build *and maintain* the equipment. (This is not the same as the economic return on investment (ROI), which varies according to local energy prices, subsidies available and metering techniques.) With lifetimes of at least 30 years, the EROEI of PV systems are in the range of 10 to 30, thus generating enough energy over their lifetimes to reproduce themselves many times (6-31 reproductions) depending on what type of material, balance of system (BOS), and the geographic location of the system.

Cadmium

One issue that has often raised concerns is the use of cadmium in cadmium telluride solar cells (CdTe is only used in a few types of PV panels). Cadmium in its metallic form is a toxic substance that has the tendency to accumulate in ecological food chains. The amount of cadmium used in thin-film PV modules is relatively small (5-10 g/m²) and with proper emission control techniques in place the cadmium emissions from module production can be almost zero. Current PV technologies lead to cadmium emissions of 0.3-0.9 microgram/kWh over the whole life-cycle. Most of these emissions actually arise through the use of coal power for the manufacturing of the modules, and coal and lignite combustion leads to much higher emissions of cadmium. Life-cycle cadmium emissions from coal is 3.1 microgram/kWh, lignite 6.2, and natural gas 0.2 microgram/kWh.

Note that if electricity produced by photovoltaic panels were used to manufacture the modules instead of electricity from burning coal, cadmium emissions from coal power usage in the manufacturing process could be entirely eliminated.

Energy Storage Methods

Solar energy is not available at night, making energy storage an important issue in order to provide the continuous availability of energy. Both wind power and solar power are intermittent energy sources, meaning that all available output must be taken when it is available and either stored for *when* it can be used, or transported, over transmission lines, to *where* it can be used.

Off-grid PV systems have traditionally used rechargeable batteries to store excess electricity. With grid-tied systems, excess electricity can be sent to the transmission grid. Net metering and feed-in tariff programs give these systems a credit for the electricity they produce. This credit offsets electricity provided from the grid when the system cannot meet demand, effectively using the grid as a storage mechanism. Credits are normally rolled over from month to month and any remaining surplus settled annually. When wind and solar are a small fraction of the grid power, other generation techniques can adjust their output appropriately, but as these forms of variable power grow, this becomes less practical.

Solar energy can be stored at high temperatures using molten salts. Salts are an effective storage medium because they are low-cost, have a high specific heat capacity and can deliver heat at temperatures compatible with conventional power systems. The Solar Two used this method of energy storage, allowing it to store 1.44 TJ in its 68 m³ storage tank, enough to provide full output for close to 39 hours, with an efficiency of about 99%.

Conventional hydroelectricity works very well in conjunction with intermittent electricity sources such as solar and wind, the water can be held back and allowed to flow as required with virtually no energy loss. Where a suitable river is not available, pumped-storage hydroelectricity stores energy in the form of water pumped when surplus electricity is available, from a lower elevation reservoir to a higher elevation one. The energy is recovered when demand is high by releasing the water : the pump becomes a turbine, and the motor a hydroelectric power generator. However, this loses some of the energy to pumpage losses.

Artificial photosynthesis involves the use of nanotechnology to store solar electromagnetic energy in chemical bonds, by splitting water to produce hydrogen fuel or then combining with carbon dioxide to make biopolymers such as methanol. Many large national and regional research projects on artificial photosynthesis are now trying to develop techniques integrating improved light capture, quantum coherence methods of electron transfer and cheap catalytic materials that operate under a variety of atmospheric conditions. Senior researchers in the field have made the public policy case for a Global Project on Artificial Photosynthesis to address critical energy security and environmental sustainability issues.

Wind power and solar power tend to be somewhat complementary, as there tends to be more wind in the winter and more sun in the summer, but on days with no sun and no wind the difference needs to be made up in some manner. Solar power is seasonal, particularly in northern/southern climates, away from the equator, suggesting a need for long term seasonal storage in a medium such as hydrogen. The storage requirements vary and in some cases can be met with biomass. The Institute for Solar Energy Supply Technology of the University of Kassel pilot-tested a combined power plant linking solar, wind, biogas and hydrostorage to provide load-following power around the clock, entirely from renewable sources.

Experimental Solar Power

Concentrated photovoltaics (CPV) systems employ sunlight concentrated onto photovoltaic surfaces for the purpose of electrical power production. Solar concentrators of all varieties may be used, and these are often mounted on a solar tracker in order to keep the focal point upon the cell as the sun moves across the sky. Luminescent solar concentrators (when combined with a PV-solar cell) can also be regarded as a CPV system. Concentrated photovoltaics are useful as they can improve efficiency of PV-solar panels drastically.

Thermoelectric, or "thermovoltaic" devices convert a temperature difference between dissimilar materials into an electric current. First proposed as a method to store solar energy by solar pioneer Mouchout in the 1800s, thermoelectrics re-emerged in the Soviet Union during the 1930s. Under the direction of Soviet scientist Abram Ioffe a concentrating system was used to thermoelectrically generate power for a 1 hp engine. Thermogenerators, but in the following cases powered by the heat source plutonium-238 in radioisotope thermoelectric generators are used in the US space program as an energy conversion technology for powering deep space missions such as the Mars Curiosity rover, Cassini, Galileo and Viking. Research in this area of thermogenerators, which can use any heat source, is focused on raising the efficiency of these devices from 7–8% to 15–20%.

Physicists have claimed that recent technological developments bring the cost of solar energy more in parity with that of fossil fuels. In 2007, David Faiman, the director of the Ben-Gurion National Solar Energy Center of Israel, announced that the Center had entered into a project with Zenith Solar to create a home solar energy system that uses a 10 square meter reflector dish. In testing, the concentrated solar technology proved to be up to five times more cost effective than standard flat photovoltaic silicon panels, which would make it almost the same cost as oil and natural gas. A prototype ready for commercialization achieved a concentration of solar energy that was more than 1,000 times greater than standard flat panels.

Wind Power

Wind power is the conversion of wind energy into a useful form of energy, such as using wind turbines to make electrical power, windmills for mechanical power, windpumps for water pumping or drainage, or sails to propel ships.

Large wind farms consist of hundreds of individual wind turbines which are connected to the electric power transmission network. For new constructions, onshore wind is an inexpensive source of electricity, competitive with or in many places cheaper than fossil fuel plants. Small onshore wind farms provide electricity to isolated locations. Utility companies increasingly buy surplus electricity produced by small domestic wind turbines. Offshore wind is steadier and stronger than on land, and offshore farms have less visual impact, but construction and maintenance costs are considerably higher.

Wind power, as an alternative to fossil fuels, is plentiful, renewable, widely distributed, clean, produces no greenhouse gas emissions during operation and uses little land. The effects on the environment are generally less problematic than those from other power sources. As of 2011, Denmark is generating more than a quarter of its electricity from wind and 83 countries around the world are using wind power to supply the electricity grid. In 2010 wind energy production was over 2.5% of total worldwide electricity usage, and growing rapidly at more than 25% per annum.

Wind power is very consistent from year to year but has significant variation over shorter time scales. As the proportion of windpower in a region increases, a need to upgrade the grid, and a lowered ability to supplant conventional production can occur. Power management techniques such as having excess capacity storage, geographically distributed turbines, dispatchable backing sources, storage such as pumped-storage hydroelectricity, exporting and importing power to neighbouring areas or reducing demand when wind production is low, can greatly mitigate these problems. In addition, weather forecasting permits the electricity network to be readied for the predictable variations in production that occur.

History

Wind power has been used as long as humans have put sails into the wind. For more than two millennia wind-powered machines have ground grain and pumped water. Wind power was widely available and not confined to the banks of fast-flowing streams, or later, requiring sources of fuel. Wind-powered pumps drained the polders of the Netherlands, and in arid regions such as the American mid-west or the Australian outback, wind pumps provided water for live stock and steam engines.

With the development of electric power, wind power found new applications in lighting buildings remote from centrally-generated power. Throughout the 20th century parallel paths developed small wind plants suitable for farms or residences, and larger utility-scale wind generators that could be connected to electricity grids for remote use of power. Today wind powered generators operate in every size range between tiny plants for battery charging at isolated residences, up to near-gigawatt sized offshore wind farms that provide electricity to national electrical networks.

Wind Energy

Wind energy is the kinetic energy of air in motion, also called wind. Total wind energy flowing through an imaginary area A d $E = \frac{1}{2}mv^2 = \frac{1}{2}(Avt\rho)v^2 = \frac{1}{2}A\rho v^3$ uring the time t is :

where ρ is the density of air; v is the wind speed; Avt is the volume of air passing through A (which is considered perpendicular to the direction of the wind); $Avt\rho$ is therefore the mass m passing per unit time. Note that ½ ρv^2 is the kinetic energy of the moving air per unit volume.

Power is energy per unit time, so the wind power incident on A (e.g. equal to the rotor area of a wind turbine) is :

$$P = \frac{E}{t} = \frac{1}{2}A\rho v^3.$$

Wind power in an open air stream is thus *proportional* to the *third power* of the wind speed; the available power increases eightfold when the wind speed doubles. Wind turbines for grid electricity therefore need to be especially efficient at greater wind speeds.

Wind is the movement of air across the surface of the Earth, affected by areas of high pressure and of low pressure. The surface of the Earth is heated unevenly by the Sun, depending on factors such as the angle of incidence of the sun's rays at the surface (which differs with latitude and time of day) and whether the land is open or covered with vegetation. Also, large bodies of water, such as the oceans, heat up and cool down slower than the land. The heat energy absorbed at the Earth's surface is transferred to the air directly above it and, as warmer air is less dense than cooler air, it rises above the cool air to form areas of high pressure and thus pressure differentials. The rotation of the Earth drags the atmosphere around with it causing turbulence. These effects combine to cause a constantly varying pattern of winds across the surface of the Earth.

The total amount of economically extractable power available from the wind is considerably more than present human power use from all sources. Axel Kleidon of the Max Planck Institute in Germany, carried out a "top down" calculation on how much wind energy there is, starting with the incoming solar radiation that drives the winds by creating temperature differences in the atmosphere. He concluded that somewhere between 18 TW and 68 TW could be extracted. Cristina Archer and Mark Z. Jacobson presented a "bottom-up" estimate, which unlike Kleidon's are based on actual measurements of wind speeds, and found that there is 1700 TW of wind power at an altitude of 100 metres over land and sea. Of this, "between 72 and 170 TW could be extracted in a practical and cost-competitive manner". They later estimated 80 TW. However research at Harvard University estimates 1 Watt/m^2 on average and 2-10 MW/km^2 capacity for large scale wind farms, suggesting that these estimates of total global wind resources are too high by a factor of about 4.

Distribution of Wind Speed

The strength of wind varies, and an average value for a given location does not alone indicate the amount of energy a wind turbine could produce there. To assess the frequency of wind speeds at a particular location, a probability distribution function is often fit to the observed data. Different locations will have different wind speed distributions. The Weibull model closely mirrors the actual distribution of hourly/ten-minute wind speeds at many locations. The Weibull factor is often close to 2 and therefore a Rayleigh distribution can be used as a less accurate, but simpler model.

High Altitude Winds

Power generation from winds usually comes from winds very close to the surface of the earth. Winds at higher altitudes are stronger and more consistent, and may have a global capacity of 380 TW. Recent years have seen significant advances in technologies meant to generate electricity from high altitude winds.

Wind Farms

A wind farm is a group of wind turbines in the same location used for production of electricity. A large wind farm may consist of several hundred individual wind turbines distributed over an extended area, but the land between the turbines may be used for agricultural or other purposes. A wind farm may also be located offshore.

Almost all large wind turbines have the same design — a horizontal axis wind turbine having an upwind rotor with three blades, attached to a nacelle on top of a tall tubular tower. In a wind farm, individual turbines are interconnected with a medium voltage (often 34.5 kV), power collection system and communications network. At a substation, this medium-voltage electric current is increased in voltage with a transformer for connection to the high voltage electric power transmission system.

Many of the largest operational onshore wind farms are located in the US. As of 2012, the Alta Wind Energy Center is the largest onshore wind farm in the world at 1020 MW, followed by the Shepherds Flat Wind Farm (845 MW), and the Roscoe Wind Farm (781.5 MW). As of September 2012, the Sheringham Shoal

Offshore Wind Farm and the Thanet Wind Farm in the UK are the largest offshore wind farms in the world at 317 MW and 300 MW, followed by Horns Rev II (209 MW) in Denmark.

There are many large wind farms under construction including; The London Array (offshore) (630 MW), BARD Offshore 1 (400 MW), Sheringham Shoal Offshore Wind Farm (317 MW), Lincs Wind Farm (offshore), Clyde Wind Farm (548 MW), Greater Gabbard wind farm (500 MW), Macarthur Wind Farm (420 MW), Lower Snake River Wind Project (343 MW) and Walney Wind Farm (367 MW).

Feeding into Grid

Induction generators, often used for wind power, require reactive power for excitation so substations used in wind-power collection systems include substantial capacitor banks for power factor correction. Different types of wind turbine generators behave differently during transmission grid disturbances, so extensive modelling of the dynamic electromechanical characteristics of a new wind farm is required by transmission system operators to ensure predictable stable behaviour during system faults. In particular, induction generators cannot support the system voltage during faults, unlike steam or hydro turbine-driven synchronous generators. Doubly fed machines generally have more desirable properties for grid interconnection. Transmission systems operators will supply a wind farm developer with a *grid code* to specify the requirements for interconnection to the transmission grid. This will include power factor, constancy of frequency and dynamic behaviour of the wind farm turbines during a system fault.

Offshore Wind Power

Offshore wind power refers to the construction of wind farms in large bodies of water to generate electricity. These installations can utilise the more frequent and powerful winds that are available in these locations and have less aesthetic impact on the landscape than land based projects. However, the construction and the maintenance costs are considerably higher. As of 2011, offshore wind farms were at least 3 times more expensive than onshore wind farms of the same nominal power but these costs are expected to fall as the industry matures.

Siemens and Vestas are the leading turbine suppliers for offshore wind power. DONG Energy, Vattenfall and E.ON are the leading offshore operators. As of October 2010, 3.16 GW of offshore wind power capacity was operational, mainly in Northern Europe. According to BTM Consult, more than 16 GW of additional capacity will be installed before the end of 2014 and the UK and Germany will become the two leading markets. Offshore wind power capacity is expected to reach a total of 75 GW worldwide by 2020, with significant contributions from China and the US.

At the end of 2012, 1,662 turbines at 55 offshore wind farms in European are generating 18 TWh, which can power almost five million households.

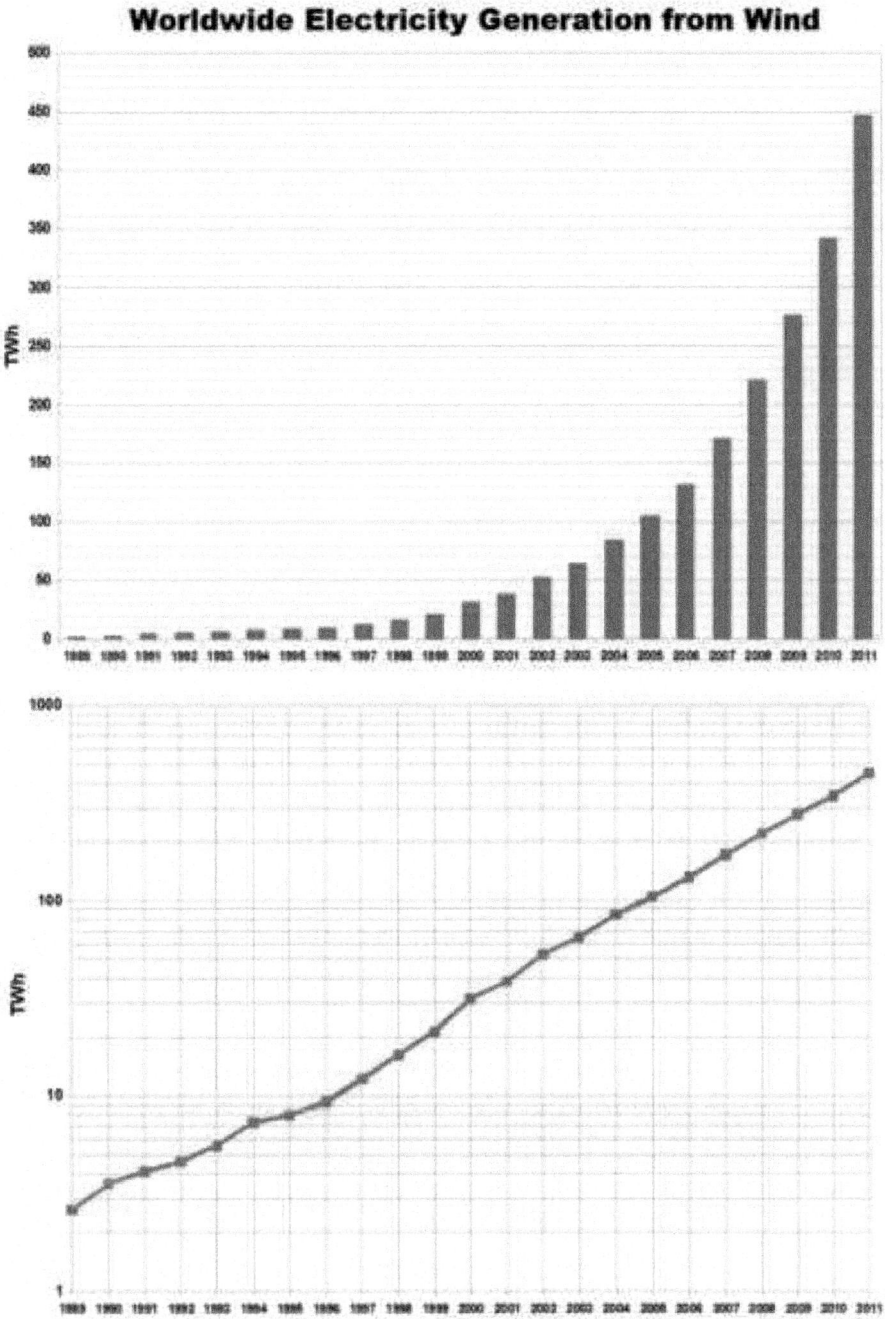

Fig. : Worldwide wind generation up to 2010.

Wind Power Capacity and Production

Worldwide there are now over two hundred thousand wind turbines operating, with a total nameplate capacity of 282,482 MW as of end 2012. The European Union alone passed some 100,000 MW nameplate capacity in September 2012, while the United States surpassed 50,000 MW in August 2012 and China's grid connected capacity passed 50,000 MW the same month.

World wind generation capacity more than quadrupled between 2000 and 2006, doubling about every three years. The United States pioneered wind farms and led the world in installed capacity in the 1980s and into the 1990s. In 1997 German installed capacity surpassed the U.S. and led until once again overtaken by the U.S. in 2008. China has been rapidly expanding its wind installations in the late 2000s and passed the U.S. in 2010 to become the world leader.

At the end of 2012, worldwide nameplate capacity of wind-powered generators was 282 gigawatts (GW), growing by 44 GW over the preceding year. According to the World Wind Energy Association, an industry organization, in 2010 wind power generated 430 TWh or about 2.5% of worldwide electricity usage, up from 1.5% in 2008 and 0.1% in 1997. Between 2005 and 2010 the average annual growth in new installations was 27.6%. Wind power market penetration is expected to reach 3.35% by 2013 and 8% by 2018.

Several countries have already achieved relatively high levels of penetration, such as 28% of stationary (grid) electricity production in Denmark (2011), 19% in Portugal (2011), 16% in Spain (2011), 16% in Ireland (2012) and 8% in Germany (2011). As of 2011, 83 countries around the world were using wind power on a commercial basis.

Europe accounted for 48% of the world total wind power generation capacity in 2009. In 2010, Spain became Europe's leading producer of wind energy, achieving 42,976 GWh. Germany held the top spot in Europe in terms of installed capacity, with a total of 27,215 MW as of 31 December 2010.

Top 10 countries by nameplate wind-power capacity (2012 year-end)			
Country	New 2012 capacity (MW)	Wind-power total capacity (MW)	% world total
China	12,960	75,324	26.7
United States	13,124	60,007	21.2
Germany	2,145	31,308	11.1
Spain	1,122	22,796	8.1
India	2,336	18,421	6.5
UK	1,897	8,845	3.0
Italy	1,273	8,144	2.9

Top 10 countries by nameplate wind-power capacity (2012 year-end)			
Country	New 2012 capacity (MW)	Wind-power total capacity (MW)	% world total
France	757	7,564	2.7
Canada	935	6,200	2.2
Portugal	145	4,525	1.6
(rest of world)	6,737	39,853	14.1
World total	44,799 MW	282,587 MW	100%

Top 10 countries by wind-power electricity production (2011 totals)		
Country	Windpower production (TWh)	% world total
United States	120.5	26.2
China	88.6	19.3
Germany	48.9	10.6
Spain	42.4	9.2
India	24.9	5.4
Canada	19.7	4.3
UK	15.5	3.4
France	12.2	2.7
Italy	9.9	2.1
Denmark	9.8	2.1
(rest of world)	67.7	14.7
World total	459.9 TWh	100%

Growth Trends

In 2010, more than half of all new wind power was added outside of the traditional markets in Europe and North America. This was largely from new construction in China, which accounted for nearly half the new wind installations (16.5 GW).

Global Wind Energy Council (GWEC) figures show that 2007 recorded an increase of installed capacity of 20 GW, taking the total installed wind energy capacity to 94 GW, up from 74 GW in 2006. Despite constraints facing supply chains for wind turbines, the annual market for wind continued to increase at an estimated rate of 37%, following 32% growth in 2006. In terms of economic value, the wind energy sector has become one of the important players in the energy markets, with the total value of new generating equipment installed in 2007 reaching €25 billion, or US$36 billion.

Although the wind power industry was affected by the global financial crisis in 2009 and 2010, a BTM Consult five-year forecast up to 2013 projects substantial growth. Over the past five years the average growth in new installations has been 27.6% each year. In the forecast to 2013 the expected average annual growth rate is 15.7%. More than 200 GW of new wind power capacity could come on line before the end of 2013. Wind power market penetration is expected to reach 3.35% by 2013 and 8% by 2018.

Capacity Factor

Since wind speed is not constant, a wind farm's annual energy production is never as much as the sum of the generator nameplate ratings multiplied by the total hours in a year. The ratio of actual productivity in a year to this theoretical maximum is called the capacity factor. Typical capacity factors are 15–50%; values at the upper end of the range are achieved in favourable sites and are due to wind turbine design improvements.

Online data is available for some locations, and the capacity factor can be calculated from the yearly output. For example, the German nation-wide average wind power capacity factor over all of 2012 was just under 17.5% (45867 GW·h/yr / (29.9 GW × 24 × 366) = 0.1746), and the capacity factor for Scottish wind farms averaged 24% between 2008 and 2010.

Unlike fueled generating plants, the capacity factor is affected by several parameters, including the variability of the wind at the site and the size of the generator relative to the turbine's swept area. A small generator would be cheaper and achieve a higher capacity factor but would produce less electricity (and thus less profit) in high winds. Conversely, a large generator would cost more but generate little extra power and, depending on the type, may stall out at low wind speed. Thus an optimum capacity factor of around 40–50% would be aimed for.

In a 2008 study released by the U.S. Department of Energy's Office of Energy Efficiency and Renewable Energy, the capacity factor achieved by the U.S. wind turbine fleet is shown to be increasing as the technology improves. The capacity factor achieved by new wind turbines in 2010 reached almost 40%.

Penetration

Wind energy penetration refers to the fraction of energy produced by wind compared with the total available generation capacity. There is no generally accepted maximum level of wind penetration. The limit for a particular grid will depend on the existing generating plants, pricing mechanisms, capacity for energy storage, demand management and other factors. An interconnected electricity grid will already include reserve generating and transmission capacity to allow for equipment failures. This reserve capacity can also serve to compensate for the varying power generation produced by wind plants. Studies have indicated that 20% of the total annual electrical energy consumption may be incorporated with minimal difficulty. These studies have been for locations with geographically dispersed wind farms, some degree of dispatchable energy or hydropower with storage capacity, demand management, and interconnected to a large grid area enabling the export of electricity when needed. Beyond the 20% level, there are few technical limits, but the economic implications become more significant. Electrical utilities continue to study the effects of large scale penetration of wind generation on system stability and economics.

A wind energy penetration figure can be specified for different durations of time. On an annual basis, as of 2011, few grid systems have penetration levels above 5% : Denmark – 29%, Portugal – 19%, Spain – 19%, Ireland – 18%, and Germany – 11%. For the U.S. in 2011, the penetration level was estimated at 3.3%. To obtain 100% from wind annually requires substantial long-term storage. On a monthly, weekly, daily, or hourly basis – or less – wind can supply as much as or more than 100% of current use, with the rest stored or exported. Seasonal industry can take advantage of high wind and low usage times such as at night when wind output can exceed normal demand. Such industry can include production of silicon, aluminum, steel, or of natural gas, and hydrogen, which allow long term storage, facilitating 100% energy from variable renewable energy. Homes can also be programmed to accept extra-electricity on demand, for example by remotely turning up water heater thermostats.

Variability

Electricity generated from wind power can be highly variable at several different timescales : hourly, daily, or seasonally. Annual variation also exists, but is not as significant. Because instantaneous electrical generation and consumption must remain in balance to maintain grid stability, this variability can present substantial challenges to incorporating large amounts of wind power into a grid system. Intermittency and the non-dispatchable nature of wind energy production can raise costs for regulation, incremental operating reserve, and (at high penetration levels) could require an increase in the already existing energy demand management, load shedding, storage solutions or system interconnection with HVDC cables.

Fluctuations in load and allowance for failure of large fossil-fuel generating units require reserve capacity that can also compensate for variability of wind generation.

Increase in system operation costs, Euros per MWh, for 10% & 20% wind share		
Country	10%	20%
Germany	2.5	3.2
Denmark	0.4	0.8
Finland	0.3	1.5
Norway	0.1	0.3
Sweden	0.3	0.7

Wind power is however, variable, but during low wind periods it can be replaced by other power sources. Transmission networks presently cope with outages of other generation plants and daily changes in electrical demand, but the variability of intermittent power sources such as wind power, are unlike those of conventional power generation plants which, when scheduled to be operating, may be able to deliver their nameplate capacity around 95% of the time.

Presently, grid systems with large wind penetration require a small increase in the frequency of usage of natural gas spinning reserve power plants to prevent a loss of electricity in the event that conditions are not favourable for power production from the wind. At lower wind power grid penetration, this is less of an issue.

GE has installed a prototype wind turbine with onboard battery similar to that of an electric car, equivalent of 1 minute of production. Despite the small capacity, it is enough to guarantee that power output complies with forecast for 15 minutes, as the battery is used to eliminate the difference rather than provide full output. The increased predictability can be used to take wind power penetration from 20 to 30 or 40 per cent. The battery cost can be retrieved by selling burst power on demand and reducing backup needs from gas plants.

A report on Denmark's wind power noted that their wind power network provided less than 1% of average demand on 54 days during the year 2002. Wind power advocates argue that these periods of low wind can be dealt with by simply restarting existing power stations that have been held in readiness, or interlinking with HVDC. Electrical grids with slow-responding thermal power plants and without ties to networks with hydroelectric generation may have to limit the use of wind power. According to a 2007 Stanford University study published in the Journal of Applied Meteorology and Climatology, interconnecting ten or more wind farms can allow an average of 33% of the total energy produced (*i.e.* about 8% of total nameplate capacity) to be used as reliable, baseload electric power which can be relied on to handle peak loads, as long as minimum criteria are met for wind speed and turbine height.

Conversely, on particularly windy days, even with penetration levels of 16%, wind power generation can surpass all other electricity sources in a country. In Spain, on 16 April, 2012 wind power production reached the highest percentage of electricity production till then, with wind farms covering 60.46% of the total demand. In Denmark, which had power market penetration of 30% in 2013, over 90 hours, wind power generated 100% of the countries power, peaking at 122% of the countries demand at 2 am on the 28th October.

Three reports on the wind variability in the UK issued in 2009, generally agree that variability of wind needs to be taken into account, but it does not make the grid unmanageable. The additional costs, which are modest, can be quantified.

The combination of diversifying variable renewables by type and location, forecasting their variation, and integrating them with dispatchable renewables, flexible fueled generators, and demand response can create a power system that has the potential to meet power supply needs reliably. Integrating ever-higher levels of renewables is being successfully demonstrated in the real world :

In 2009, eight American and three European authorities, writing in the leading electrical engineers' professional journal, didn't find "a credible and firm technical limit to the amount of wind energy that can be accommodated by electricity grids". In fact, not one of more than 200 international studies, nor official studies for the eastern and western U.S. regions, nor the International Energy Agency, has found major costs or technical barriers to reliably integrating up to 30% variable renewable supplies into the grid, and in some studies much more. – *Reinventing Fire*

Solar power tends to be complementary to wind. On daily to weekly timescales, high pressure areas tend to bring clear skies and low surface winds, whereas low pressure areas tend to be windier and cloudier. On seasonal timescales, solar energy peaks in summer, whereas in many areas wind energy is lower in summer and higher in winter. Thus the intermittencies of wind and solar power tend to cancel each other somewhat. In 2007 the Institute for Solar Energy Supply Technology of the University of Kassel pilot-tested a combined power plant linking solar, wind, biogas and hydrostorage to provide load-following power around the clock and throughout the year, entirely from renewable sources.

Predictability

Wind power forecasting methods are used, but predictability of any particular wind farm is low for short-term operation. For any particular generator there is an 80% chance that wind output will change less than 10% in an hour and a 40% chance that it will change 10% or more in 5 hours.

However, studies by Graham Sinden (2009) suggest that, in practice, the variations in thousands of wind turbines, spread out over several different sites and wind regimes, are smoothed. As the distance between sites increases, the correlation between wind speeds measured at those sites, decreases.

Thus, while the output from a single turbine can vary greatly and rapidly as local wind speeds vary, as more turbines are connected over larger and larger areas the average power output becomes less variable and more predictable.

Wind speeds can be accurately forecast over large areas, and hence wind is a predictable source of power for feeding into an electrical grid. However, due to the variability, although predictable, wind energy availability must be scheduled.

Energy Storage

Typically, conventional hydroelectricity complements wind power very well. When the wind is blowing strongly, nearby hydroelectric plants can temporarily hold back their water. When the wind drops they can, provided they have the generation capacity, rapidly increase production to compensate. This gives a very even overall power supply and virtually no loss of energy and uses no more water.

Alternatively, where a suitable head of water is not available, pumped-storage hydroelectricity or other forms of grid energy storage can store energy developed by high-wind periods and release it when needed. The type of storage needed depends on the wind penetration level – low penetration requires daily storage, and high penetration requires both short and long term storage – as long as a month or more. Stored energy increases the economic value of wind energy since it can be shifted to displace higher cost generation during peak demand periods. The potential revenue from this arbitrage can offset the cost and losses of storage; the cost of storage may add 25% to the cost of any wind energy stored but it is not envisaged that this would apply to a large proportion of wind energy generated. For example, in the UK, the 1.7 GW Dinorwig pumped storage plant evens out electrical demand peaks, and allows base-load suppliers to run their plants more efficiently. Although pumped storage power systems are only about 75% efficient, and have high installation costs, their low running costs and ability to reduce the required electrical base-load can save both fuel and total electrical generation costs.

In particular geographic regions, peak wind speeds may not coincide with peak demand for electrical power. In the US states of California and Texas, for example, hot days in summer may have low wind speed and high electrical demand due to the use of air conditioning. Some utilities subsidize the purchase of geothermal heat pumps by their customers, to reduce electricity demand during the summer months by making air conditioning up to 70% more efficient; widespread adoption of this technology would better match electricity demand to wind availability in areas with hot summers and low summer winds. Another option is to interconnect widely dispersed geographic areas with an HVDC "Super grid". In the U.S. it is estimated that to upgrade the transmission system to take in planned or potential renewables would cost at least $60 billion.

Germany has an installed capacity of wind and solar that exceeds daily demand, and has been exporting peak power to neighbouring countries. A more practical solution is the installation of thirty days storage capacity able to supply 80% of demand, which will become necessary when most of Europe's energy is obtained from wind power and solar power. Just as the EU requires member countries to maintain 90 days strategic reserves of oil it can be expected that countries will provide electricity storage, instead of expecting to use their neighbors for net metering.

Wind power hardly ever suffers major technical failures, since failures of individual wind turbines have hardly any effect on overall power, so that the distributed wind power is highly reliable and predictable, whereas conventional generators, while far less variable, can suffer major unpredictable outages.

Capacity Credit and Fuel Savings

The capacity credit of wind is estimated by determining the capacity of conventional plants displaced by wind power, whilst maintaining the same degree of system security. However, the precise value is irrelevant since the main value of wind is its fuel and CO_2 savings, and wind is not expected to be constantly available.

Economics

Wind turbines reached grid parity (the point at which the cost of wind power matches traditional sources) in some areas of Europe in the mid-2000s, and in the US around the same time. Falling prices continue to drive the levelized cost down and it has been suggested that it has reached general grid parity in Europe in 2010, and will reach the same point in the US around 2016 due to an expected reduction in capital costs of about 12%. Nevertheless, a significant amount of the wind power resource in North America remains above grid parity due to the long transmission distances involved.

Cost Trends

Wind power is capital intensive, but has no fuel costs. The price of wind power is therefore much more stable than the volatile prices of fossil fuel sources. The marginal cost of wind energy once a plant is constructed is usually less than 1-cent per kW·h. This cost has additionally reduced as wind turbine technology has improved. There are now longer and lighter wind turbine blades, improvements in turbine performance and increased power generation efficiency. Also, wind project capital and maintenance costs have continued to decline.

The estimated average cost per unit incorporates the cost of construction of the turbine and transmission facilities, borrowed funds, return to investors (including cost of risk), estimated annual production, and other components, averaged over the projected useful life of the equipment, which may be in excess of twenty years. Energy cost estimates are highly dependent on these assumptions so published cost figures can differ substantially. In 2004, wind energy cost a fifth of what it did in the 1980s, and some expected that downward trend to continue as larger multi-megawatt turbines were mass-produced. As of 2012 capital costs for wind turbines are substantially lower than 2008–2010 but are still above 2002 levels. A 2011 report from the American Wind Energy Association stated, "Wind's costs have dropped over the past two years, in the range of 5 to 6 cents per kilowatt-hour recently... about 2 cents cheaper than coal-fired electricity, and more projects were financed through debt arrangements than tax equity structures last year... winning more mainstream acceptance from Wall Street's banks.... Equipment makers can also deliver products in the same year that they are ordered instead of waiting up to three years as was the case in previous cycles... 5,600 MW of new installed capacity is under construction in the United States, more than double the number at this point in 2010. Thirty-five per cent of all new power generation built in the United States since 2005 has come from wind, more than new gas and coal plants combined, as power providers are increasingly enticed to wind as a convenient hedge against unpredictable commodity price moves."

A British Wind Energy Association report gives an average generation cost of onshore wind power of around 3.2 pence (between US 5 and 6 cents) per kW·h

(2005). Cost per unit of energy produced was estimated in 2006 to be comparable to the cost of new generating capacity in the US for coal and natural gas : wind cost was estimated at $55.80 per MW·h, coal at $53.10/MW·h and natural gas at $52.50. Similar comparative results with natural gas were obtained in a governmental study in the UK in 2011. The presence of wind energy, even when subsidised, can reduce costs for consumers (€5 billion/yr in Germany) by reducing the marginal price, by minimising the use of expensive peaking power plants.

In February 2013 Bloomberg New Energy Finance reported that the cost of generating electricity from new wind farms is cheaper than new coal or new baseload gas plants. When including the current Australian federal government carbon pricing scheme their modelling gives costs (in Australian dollars) of $80/MWh for new wind farms, $143/MWh for new coal plants and $116/MWh for new baseload gas plants. The modeling also shows that "even without a carbon price (the most efficient way to reduce economy-wide emissions) wind energy is 14% cheaper than new coal and 18% cheaper than new gas." Part of the higher costs for new coal plants is due to high financial lending costs because of "the reputational damage of emissions-intensive investments". The expense of gas fired plants is partly due to "export market" effects on local prices. Costs of production from coal fired plants built in "the 1970s and 1980s" are cheaper than renewable energy sources because of depreciation.

The wind industry in the USA is now able to produce more power at lower cost by using taller wind turbines with longer blades, capturing the faster winds at higher elevations. This has opened up new opportunities and in Indiana, Michigan, and Ohio, the price of power from wind turbines built 300 feet to 400 feet above the ground can now compete with conventional fossil fuels like coal. Prices have fallen to about 4 cents per kilowatt-hour in some cases and utilities have been increasing the amount of wind energy in their portfolio, saying it is their cheapest option.

Incentives and Community Benefits

The U.S. wind industry generates tens of thousands of jobs and billions of dollars of economic activity. Wind projects provide local taxes, or payments in lieu of taxes and strengthen the economy of rural communities by providing income to farmers with wind turbines on their land. Wind energy in many jurisdictions receives financial or other support to encourage its development. Wind energy benefits from subsidies in many jurisdictions, either to increase its attractiveness, or to compensate for subsidies received by other forms of production which have significant negative externalities.

In the US, wind power receives a production tax credit (PTC) of 1.5¢/kWh in 1993 dollars for each kW·h produced, for the first ten years; at 2.2 cents per kW·h in 2012, the credit was renewed on 2 January 2012, to include construction begun in 2013. A 30% tax credit can be applied instead of receiving the PTC. Another tax benefit is accelerated depreciation. Many American states also provide incentives, such as exemption from property tax, mandated purchases, and additional markets for "green credits". The Energy Improvement and Extension Act of 2008 contains extensions of credits for wind, including micro-turbines. Countries such as Canada and Germany also provide incentives for wind turbine construction, such as tax credits or minimum purchase prices for wind generation, with assured grid access (sometimes referred to as feed-in tariffs). These feed-in tariffs are typically set well above average electricity prices. As of December 2013, US Senator Lamar Alexander and other senators are arguing that the "wind energy production tax credit should be allowed to expire at the end of 2013."

Secondary market forces also provide incentives for businesses to use wind-generated power, even if there is a premium price for the electricity. For example, socially responsible manufacturers pay utility companies a premium that goes to subsidize and build new wind power infrastructure. Companies use wind-generated power, and in return they can claim that they are undertaking strong "green" efforts. In the US the organization Green-e monitors business compliance with these renewable energy credits.

Environmental Effects

Compared to the environmental impact of traditional energy sources, the environmental impact of wind power is relatively minor in terms of pollution. Wind power consumes no fuel, and emits no air pollution, unlike fossil fuel power sources. The energy consumed to manufacture and transport the materials used to build a wind power plant is equal to the new energy produced by the plant within a few months. While a wind farm may cover a large area of land, many land uses such as agriculture are compatible, with only small areas of turbine foundations and infrastructure made unavailable for use.

There are reports of bird and bat mortality at wind turbines as there are around other artificial structures. The scale of the ecological impact may or may not be significant, depending on specific circumstances. Although many artificial structures can kill birds, wind power has a disproportionate effect on certain endangered bird species. An especially vulnerable group are raptors, which are slow to reproduce and favour the high wind speed corridors that wind turbine companies build turbines in, to maximize energy production. Although they have a negligible effect on most birds, in some locations there is a disproportionate effects on some birds of conservation concern, such as the golden eagle and raptor species.

However, a large meta-analysis of 616 individual studies on electricity production and its effects on avian mortality concluded that the most visible impacts of wind technology are not necessarily the most flagrant ones, as :

"Wind turbines seem to present a significant threat as all their negative externalities are concentrated in one place, while those from conventional and nuclear fuel cycles are spread out across space and time. Avian mortality and wind energy has consequently received far more attention and research than the avian deaths associated with coal, oil, natural gas and nuclear power generators [although] study suggests that wind energy may be the least harmful to birds."

Prevention and mitigation of wildlife fatalities, and protection of peat bogs, affect the siting and operation of wind turbines.

There are anecdotal reports of negative effects from noise on people who live very close to wind turbines. Peer-reviewed research has generally not supported these statements.

Politics

Central Government

Fossil fuels are subsidized by many governments, and wind power and other forms of renewable energy are also often subsidized. For example a 2009 study by the Environmental Law Institute assessed the size and structure of U.S. energy subsidies over the 2002–2008 period. The study estimated that subsidies to fossil-fuel based sources amounted to approximately $72 billion over this period and subsidies to renewable fuel sources totalled $29 billion. In the United States, the federal government has paid US$74 billion for energy subsidies to support R & D for nuclear power ($50 billion) and fossil fuels ($24 billion) from 1973 to 2003. During this same time frame, renewable energy technologies and energy efficiency received a total of US$26 billion. It has been suggested that a subsidy shift would help to level the playing field and support growing energy sectors, namely solar power, wind power, and biofuels. History shows that no energy sector was developed without subsidies.

According to the International Energy Agency (IEA) (2011), energy subsidies artificially lower the price of energy paid by consumers, raise the price received by producers or lower the cost of production. "Fossil fuels subsidies costs generally outweigh the benefits. Subsidies to renewables and low-carbon energy technologies can bring long-term economic and environmental benefits". In November 2011, an IEA report entitled *Deploying Renewables 2011* said "subsidies in green energy technologies that were not yet competitive are justified in order to give an incentive to investing into technologies with clear environmental and energy security benefits". The IEA's report disagreed with claims that renewable energy technologies are only viable through costly subsidies and not able to produce energy reliably to meet demand.

In the US, the wind power industry has recently increased its lobbying efforts considerably, spending about $5 million in 2009 after years of relative obscurity in Washington. By comparison, the US nuclear industry alone spent over $650 million on its lobbying efforts and campaign contributions during a single ten-year period ending in 2008.

Following the 2011 Japanese nuclear accidents, Germany's federal government is working on a new plan for increasing energy efficiency and renewable energy commercialization, with a particular focus on offshore wind farms. Under the plan, large wind turbines will be erected far away from the coastlines, where the wind blows more consistently than it does on land, and where the enormous turbines won't bother the inhabitants. The plan aims to decrease Germany's dependence on energy derived from coal and nuclear power plants.

Commenting on the EU's 2020 renewable energy target, economist Professor Dieter Helm is critical of how the costs of wind power are cited by lobbyists. Helm also says that the problem of intermittent supply will probably lead to another dash for gas or dash for coal in Europe, possibly with a negative impact on energy security. A House of Lords Select Committee report (2008) on renewable energy in the UK reported a "concern over the prospective role of wind generated and other intermittent sources of electricity in the UK, in the absence of a breakthrough in electricity storage technology or the integration of the UK grid with that of continental Europe".

Public Opinion

Surveys of public attitudes across Europe and in many other countries show strong public support for wind power. About 80% of EU citizens support wind power. In Germany, where wind power has gained very high social acceptance, hundreds of thousands of people have invested in citizens' wind farms across the country and thousands of small and medium sized enterprises are running successful businesses in a new sector that in 2008 employed 90,000 people and generated 8% of Germany's electricity. Although wind power is a popular form of energy generation, the construction of wind farms is not universally welcomed, often for aesthetic reasons.

In Spain, with some exceptions, there has been little opposition to the installation of inland wind parks. However, the projects to build offshore parks have been more controversial. In particular, the proposal of building the biggest offshore wind power production facility in the world in southwestern Spain in the coast of Cádiz, on the spot of the 1805 Battle of Trafalgar. has been met with strong opposition who fear for tourism and fisheries in the area, and because the area is a war grave.

In a survey conducted by Angus Reid Strategies in October 2007, 89 per cent of respondents said that using renewable energy sources like wind or solar power was positive for Canada, because these sources were better for the environment. Only 4 per cent considered using renewable sources as negative since they can be unreliable and expensive. According to a Saint Consulting survey in April 2007, wind power was the alternative energy source most likely to gain public support for future development in Canada, with only 16% opposed to this type of energy. By contrast, 3 out of 4 Canadians opposed nuclear power developments.

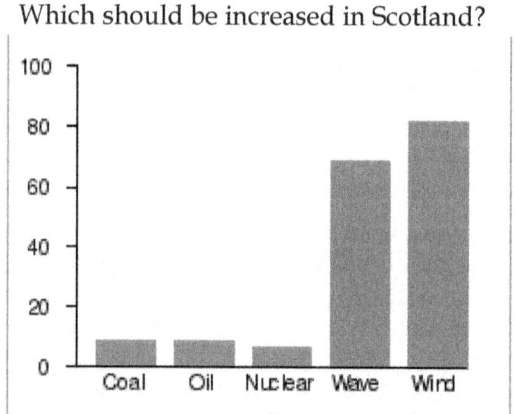

Which should be increased in Scotland?

A 2003 survey of residents living around Scotland's 10 existing wind farms found high levels of community acceptance and strong support for wind power, with much support from those who lived closest to the wind farms. The results of this survey support those of an earlier Scottish Executive survey 'Public attitudes to the Environment in Scotland 2002', which found that the Scottish public would prefer the majority of their electricity to come from renewables, and which rated wind power as the cleanest source of renewable energy. A survey conducted in 2005 showed that 74% of people in Scotland agree that wind farms are necessary to meet current and future energy needs. When people were asked the same question in a Scottish renewables study conducted in 2010, 78% agreed. The increase is significant as there were twice as many wind farms in 2010 as there were in 2005. The 2010 survey also showed that 52% disagreed with the statement that wind farms are "ugly and a blot on the landscape". 59% agreed that wind farms were necessary and that how they looked was unimportant. Scotland is planning to obtain 100% of electricity from renewable sources by 2020.

In other cases there is direct community ownership of wind farm projects. In Germany, hundreds of thousands of people have invested in citizens' wind farms across the country and thousands of small and medium sized enterprises are running successful businesses in a new sector that in 2008 employed 90,000 people and generated 8 per cent of Germany's electricity. Wind power has gained very high social acceptance in Germany. Surveys of public attitudes across Europe and in many other countries show strong public support for wind power.

Community

Many wind power companies work with local communities to reduce environmental and other concerns associated with particular wind farms. In other cases there is direct community ownership of wind farm projects. Appropriate government consultation, planning and approval procedures also help to minimize environmental risks. Some may still object to wind farms but, according to The Australia Institute, their concerns should be weighed against the need to address the threats posed by climate change and the opinions of the broader community.

Opinion on increase in number of wind farms, 2010 Harris Poll						
	U.S.	Great Britain	France	Italy	Spain	Germany
	%	%	%	%	%	%
Strongly oppose	3	6	6	2	2	4
Oppose more than favour	9	12	16	11	9	14
Favour more than oppose	37	44	44	38	37	42
Strongly favour	50	38	33	49	53	40

In America, wind projects are reported to boost local tax bases, helping to pay for schools, roads and hospitals. Wind projects also revitalize the economy of rural communities by providing steady income to farmers and other landowners.

In the UK, both the National Trust and the Campaign to Protect Rural England have expressed concerns about the effects on the rural landscape caused by inappropriately sited wind turbines and wind farms.

Some wind farms have become tourist attractions. The Whitelee Wind Farm Visitor Centre has an exhibition room, a learning hub, a café with a viewing deck and also a shop. It is run by the Glasgow Science Centre.

In Denmark, a loss-of-value scheme gives people the right to claim compensation for loss of value of their property if it is caused by proximity to a wind turbine. The loss must be at least 1% of the property's value.

Despite this general support for the concept of wind power in the public at large, local opposition often exists and has delayed or aborted a number of projects.

While aesthetic issues are subjective and some find wind farms pleasant and optimistic, or symbols of energy independence and local prosperity, protest groups are often formed to attempt to block new wind power sites for various reasons.

This type of opposition is often described as NIMBYism, but research carried out in 2009 found that there is little evidence to support the belief that residents only object to renewable power facilities such as wind turbines as a result of a "Not in my Back Yard" attitude.

Small-scale Wind Power

Small-scale wind power is the name given to wind generation systems with the capacity to produce up to 50 kW of electrical power. Isolated communities, that may otherwise rely on diesel generators, may use wind turbines as an alternative. Individuals may purchase these systems to reduce or eliminate their dependence on grid electricity for economic reasons, or to reduce their carbon footprint. Wind turbines have been used for household electricity generation in conjunction with battery storage over many decades in remote areas.

Grid-connected domestic wind turbines may use grid energy storage, thus replacing purchased electricity with locally produced power when available. The surplus power produced by domestic micro-generators can, in some jurisdictions, be fed into the network and sold to the utility company, producing a retail credit for the micro-generators' owners to offset their energy costs.

Off-grid system users can either adapt to intermittent power or use batteries, photovoltaic or diesel systems to supplement the wind turbine. Equipment such as parking meters, traffic warning signs, street lighting, or wireless Internet gateways may be powered by a small wind turbine, possibly combined with a photovoltaic system, that charges a small battery replacing the need for a connection to the power grid.

A Carbon Trust study into the potential of small-scale wind energy in the UK, published in 2010, found that small wind turbines could provide up to 1.5 terawatt hours (TW·h) per year of electricity (0.4% of total UK electricity consumption), saving 0.6 million tonnes of carbon dioxide (Mt CO_2) emission savings. This is based on the assumption that 10% of households would install turbines at costs competitive with grid electricity, around 12 pence (US 19 cents) a kW·h. A report prepared for the UK's government-sponsored Energy Saving Trust in 2006, found that home power generators of various kinds could provide 30 to 40% of the country's electricity needs by 2050.

Distributed generation from renewable resources is increasing as a consequence of the increased awareness of climate change. The electronic interfaces required to connect renewable generation units with the utility system can include additional functions, such as the active filtering to enhance the power quality.

Marine Energy

Marine energy or **marine power** (also sometimes referred to as **ocean energy** or **ocean power**) refers to the energy carried by ocean waves, tides, salinity, and ocean temperature differences. The movement of water in the world's oceans creates a vast store of kinetic energy, or energy in motion. This energy can be harnessed to generate electricity to power homes, transport and industries.

The term marine energy encompasses both wave power — power from surface waves, and tidal power — obtained from the kinetic energy of large bodies of moving water. Offshore wind power is not a form of marine energy, as wind power is derived from the wind, even if the wind turbines are placed over water.

The oceans have a tremendous amount of energy and are close to many if not most concentrated populations. Ocean energy has the potential of providing a substantial amount of new renewable energy around the world.

Potential of Ocean Energy

The theoretical potential is equivalent to 4-18 million ToE.

Theoretical global ocean energy resource		
Capacity (GW)	Annual gen. (TW h)	Form
5,000	50,000	Marine current power
20	2,000	Osmotic power
1,000	10,000	Ocean thermal energy
90	800	Tidal energy
1,000 – 9,000	8,000 – 80,000	Wave energy

Indonesia as archipelagic country with three quarter of the area is ocean, has 49 GW recognized potential ocean energy and has 727 GW theoretical potential ocean energy.

Forms of Ocean Energy

Renewable

The oceans represent a vast and largely untapped source of energy in the form of surface waves, fluid flow, salinity gradients, and thermal.

Marine Current Power

The energy obtained from ocean currents

Tidal power, also called tidal energy, is a form of hydropower that converts the energy of tides into useful forms of power - mainly electricity.

The operating principle behind tidal energy converters is that the energy contained within the moving current is harnessed by a device that extracts kinetic energy from the flow and imparts this into a mechanical motion of a rotor or foil. The device then converts the mechanical motion of the structure into electrical energy by means of a power take-off system. Before connection to the electricity grid, the electrical power output from the device will need to be conditioned in order to make it compliant with grid code regulations. In essence, tidal device operation is synonymous to that of a wind turbine, albeit operating within a different fluid medium.

Osmotic Power

At the mouth of rivers where fresh water mixes with salt water, energy associated with the salinity gradient can be harnessed using pressure-retarded reverse osmosis process and associated conversion technologies. Another system is based on using freshwater upwelling through a turbine immersed in seawater, and one involving electrochemical reactions is also in development.

Significant research took place from 1975 to 1985 and gave various results regarding the economy of PRO and RED plants. It is important to note that small-scale investigations into salinity power production take place in other countries like Japan, Israel, and the United States. In Europe the research is concentrated in Norway and the Netherlands, in both places small pilots are tested. Salinity gradient energy is the energy available from the difference in salt concentration between freshwater with saltwater. This energy source is not easy to understand, as it is not directly occurring in nature in the form of heat, waterfalls, wind, waves, or radiation.

Ocean Thermal Energy

The power from temperature differences at varying depths.

Tidal Power

The energy from moving masses of water — a popular form of hydroelectric power generation. Tidal power generation comprises three main forms, namely : tidal stream power, tidal barrage power, and dynamic tidal power.

Wave Power

Wave energy forms as kinetic energy from the wind is transmitted to the upper surface of the ocean. The height and period of resulting waves will vary depending on the energy flux between the wind and the ocean surface. Much work has been carried out in the field of research and development of technology capable of harnessing energy from the waves. At present there is limited design consensus surrounding the design of wave energy technology, and there are several areas in which a wave energy converter can be placed in order to harness the energy most efficiently.

The wave energy sector is reaching a significant milestone in the development of the industry, with positive steps towards commercial viability being taken. The more advanced device developers are now progressing beyond single unit demonstration devices and are proceeding to array development and multi-megawatt projects. The backing of major utility companies is now manifesting itself through partnerships within the development process, unlocking further investment and, in some cases, international co-operation.

At a simplified level, wave energy technology can be located near-shore and offshore. Wave energy converters can also be designed for operation in specific water depth conditions : deep water, intermediate water or shallow water. The fundamental device design will be dependent on the location of the device and the intended resource characteristics.

Non-renewable

Petroleum and natural gas beneath the ocean floor are also sometimes considered a form of ocean energy. An ocean engineer directs all phases of discovering, extracting, and delivering offshore petroleum (via oil tankers and pipelines,) a complex and demanding task. Also centrally important is the development of new methods to protect marine wildlife and coastal regions against the undesirable side effects of offshore oil extraction.

Osmotic Power

Osmotic power or **salinity gradient power** is the energy available from the difference in the salt concentration between seawater and river water. Two practical methods for this are reverse electrodialysis (RED) and pressure retarded osmosis (PRO). Both processes rely on osmosis with ion specific membranes. The key waste product is brackish water. This by-product is the result of natural forces that are being harnessed : the flow of fresh water into seas that are made up of salt water.

The method of generating power by pressure retarded osmosis was invented by Prof. Sidney Loeb in 1973 at the Ben-Gurion University of the Negev, Beersheba, Israel. The idea came to Prof. Loeb, in part, as he observed the Jordan River flowing into the Dead Sea. He wanted to harvest the energy of mixing of the two aqueous solutions (the Jordan River being one and the Dead Sea being the other) that was going to waste in this natural mixing process. In 1977 Prof. Loeb invented a method of producing power by a reverse electrodialysis heat engine.

The technologies have been confirmed in laboratory conditions. They are being developed into commercial use in the Netherlands (RED) and Norway (PRO). The cost of the membrane has been an obstacle. A new, lower cost membrane, based on an electrically modified polyethylene plastic, made it fit for potential commercial use. Other methods have been proposed and are currently under development. Among them, a method based on electric double-layer capacitor technology. and a method based on vapour pressure difference.

The world's first osmotic power plant with capacity of 4 kW was opened by Statkraft on 24 November, 2009 in Tofte, Norway. The plant utilized the original schematic proposed by Loeb. This plant uses polyimide as a membrane, and is able to produce $1W/m^2$ of membrane. This amount of power is obtained with water flow through the membrane of 10 L/s, at a pressure of 1 MPa. Both the increasing of the pressure as well as the flow rate of the water would make it possible to increase the power output. Hypothetically, the output of the SGP-plant could easily be doubled. I

Basics of Salinity Gradient Power

Salinity gradient power is a specific renewable energy alternative that creates renewable and sustainable power by using naturally occurring processes. This practice does not contaminate or release carbon dioxide (CO_2) emissions (vapour pressure methods will release dissolved air containing CO_2 at low pressures – these non-condensable gases can be re-dissolved of course, but with an energy penalty). Also as stated by Jones and Finley within their article "Recent Development in Salinity Gradient Power", there is basically no fuel cost.

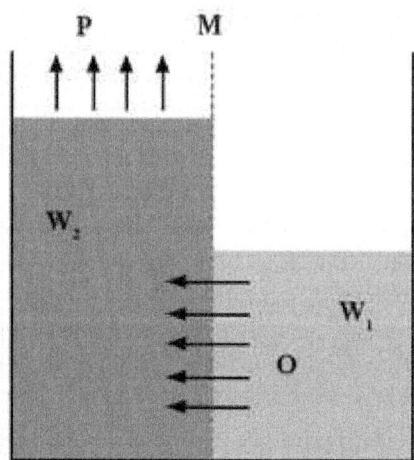

Fig. : Pressure-retarded osmosis.

Salinity gradient energy is based on using the resources of "osmotic pressure difference between fresh water and sea water." All energy that is proposed to use salinity gradient technology relies on the evaporation to separate water from salt. Osmotic pressure is the "chemical potential of concentrated and dilute solutions of salt". When looking at relations between high osmotic pressure and low, solutions with higher concentrations of salt have higher pressure.

Differing salinity gradient power generations exist but one of the most commonly discussed is pressure-retarded osmosis (PRO). Within PRO seawater is pumped into a pressure chamber where the pressure is lower than the difference between fresh and salt water pressure. Fresh water moves in a semi-permeable membrane and increases its volume in the chamber. As the pressure in the chamber is compensated a turbine spins to generate electricity. In Braun's article he states that this process is easy to understand in a more broken down manner. Two solutions, A being salt water and B being fresh water are separated by a membrane. He states "only water molecules can pass the semi-permeable membrane. As a result of the osmotic pressure difference between both solutions, the water from solution B thus will diffuse through the membrane in order to dilute the solution". The pressure drives the turbines and power the generator that produces the electrical energy. Osmosis might be used directly to "pump" fresh water out of The Netherlands into the sea. This is currently done using electric pumps.

Efficiency

A recent study on efficiency from Yale university concluded that the highest extractable work in constant-pressure PRO with a seawater draw solution and river water feed solution is 0.75 kWh/m^3 while the free energy of mixing is 0.81 kWh/m^3 – a thermodynamic extraction efficiency of 91.1%.

Methods

While the mechanics and concepts of salinity gradient power are still being studied, the power source has been implemented in several different locations. Most of these are experimental, but thus far they have been predominantly successful. The various companies that have utilized this power have also done so in many different ways as there are several concepts and processes that harness the power from salinity gradient.

Pressure-retarded Osmosis

One method to utilize salinity gradient energy is called pressure-retarded osmosis. In this method, seawater is pumped into a pressure chamber that is at a pressure lower than the difference between the pressures of saline water and fresh water. Freshwater is also pumped into the pressure chamber through a membrane, which increase both the volume and pressure of the chamber. As the pressure differences are compensated, a turbine is spun, providing kinetic energy. This method is being specifically studied by the Norwegian utility Statkraft, which has calculated that up to 25 TWh/yr would be available from this process in Norway. Statkraft has built the world's first prototype osmotic power plant on the Oslo fjord which was opened by Princess Mette-Marit of Norway on November 24, 2009. It aims to produce enough electricity to light and heat a small town within five years by osmosis. At first it will produce a minuscule 4 kilowatts – enough to heat a large electric kettle, but by 2015 the target is 25 megawatts – the same as a small wind farm.

Reversed Electrodialysis

A second method being developed and studied is reversed electrodialysis or reverse dialysis, which is essentially the creation of a salt battery. This method was described by Weinstein and Leitz as "an array of alternating anion and cation exchange membranes can be used to generate electric power from the free energy of river and sea water."

The technology related to this type of power is still in its infant stages, even though the principle was discovered in the 1950s. Standards and a complete understanding of all the ways salinity gradients can be utilized are important goals to strive for in order make this clean energy source more viable in the future.

Capacitive Method

A third method is Doriano Brogioli's capacitive method, which is relatively new and has so far only been tested on lab scale. With this method energy can be extracted out of the mixing of saline water and freshwater by cyclically charging up electrodes in contact with saline water, followed by a discharge in freshwater. Since the amount of electrical energy which is needed during the charging step is less than one gets out during the discharge step, each completed cycle effectively produces energy. An intuitive explanation of this effect is that the great number of ions in the saline water efficiently neutralizes the charge on each electrode by forming a thin layer of opposite charge very close to the electrode surface, known as an electric double layer. Therefore, the voltage over the electrodes remains low during the charge step and charging is relatively easy. In between the charge and discharge step, the electrodes are brought in contact with freshwater. After this, there are less ions available to neutralize the charge on each electrode such that the voltage over the electrodes increases. The discharge step which follows is therefore able to deliver a relatively high amount of energy. A physical explanation is that on an electrically charged capacitor, there is a mutually attractive electric force between the electric charge on the electrode, and the ionic charge in the liquid. In order to pull ions away from the charged electrode, osmotic pressure must do work. This work done increases the electrical potential energy in the capacitor. An electronic explanation is that capacitance is a function of ion density. By introducing a salinity gradient and allowing some of the ions to diffuse out of the capacitor, this reduces the capacitance, and so the voltage must increase, since the voltage equals the ratio of charge to capacitance.

Vapour Pressure Differences : Open Cycle and Absorption Refrigeration Cycle (Closed Cycle)

Both of these methods do not rely on membranes, so filtration requirements are not as important as they are in the PRO & RED schemes.

Open Cycle

Similar to the open cycle in ocean thermal energy conversion (OTEC). The disadvantage of this cycle is the cumbersome problem of a large diameter turbine (75 meters +) operating at below atmospheric pressure to extract the power between the water with less salinity & the water with greater salinity.

Absorption Refrigeration Cycle (Closed Cycle)

For the purpose of dehumidifying air, in a water-spray absorption refrigeration system, water vapour is dissolved into a deliquescent salt water mixture using osmotic power as an intermediary. The primary power source originates from a thermal difference, as part of a thermodynamic heat engine cycle.

Solar Pond

At the Eddy Potash Mine in New Mexico, a technology called "salinity gradient solar pond" (SGSP) is being utilized to provide the energy needed by the mine. **This method does not harness osmotic power**, only solar power. Sunlight reaching the bottom of the saltwater pond is absorbed as heat. The effect of natural convection, wherein "heat rises", is blocked using density differences between the three layers that make up the pond, in order to trap heat. The upper convection zone is the uppermost zone, followed by the stable gradient zone, then the bottom thermal zone. The stable gradient zone is the most important. The saltwater in this layer can not rise to the higher zone because the saltwater above has lower salinity and is therefore less-dense and more buoyant; and it can not sink to the lower level because that saltwater is denser. This middle zone, the stable gradient zone, effectively becomes an "insulator" for the bottom layer (although the main purpose is to block natural convection, since water is a poor insulator). This water from the lower layer, the storage zone, is pumped out and the heat is used to produce energy, usually by turbine in an organic Rankine cycle.

In theory a solar pond *could* be used to generate osmotic power if evaporation from solar heat is used to create a salinity gradient, *and* the potential energy in this salinity gradient is *harnessed directly* using one of the first three methods above, such as the capacitive method.

Boron Nitride Nanotubes

A research team built an experimental system using boron nitride that produced much greater power than the Statoil prototype. It used an impermeable and electrically insulating membrane that was pierced by a single boron nitride nanotube with an external diameter of a few dozen nanometers. With this membrane separating a salt water reservoir and a fresh water reservoir, the team measured the electric current passing through the membrane using two electrodes immersed in the fluid either side of the nanotube.

The results showed the device was able to generate an electric current on the order of a nanoampere. The researchers claim this is 1,000 times the yield of other known techniques for harvesting osmotic energy and makes boron nitride nanotubes an extremely efficient solution for harvesting the energy of salinity gradients for usable electrical power.

The team claimed that a 1 square metre (11 sq ft) membrane could generate around 4 kW and be capable of generating up to 30 MWh per year.

Possible Negative Environmental Impact

Marine and river environments have obvious differences in water quality, namely salinity. Each species of aquatic plant and animal is adapted to survive in either marine, brackish, or freshwater environments. There are species that can tolerate both, but these species usually thrive best in a specific water environment. The main waste product of salinity gradient technology is brackish water. The discharge of brackish water into the surrounding waters, if done in large quantities and with any regularity, will cause salinity fluctuations. While some variation in salinity is usual, particularly where fresh water (rivers) empties into an ocean or sea anyway, these variations become less important for both bodies of water with the addition of brackish waste waters. Extreme salinity changes in an aquatic environment may result in findings of low densities of both animals and plants due to intolerance of sudden severe salinity drops or spikes. According to the prevailing environmentalist opinions, the possibility of these negative effects should be considered by the operators of future large blue energy establishments.

The impact of brackish water on ecosystems can be minimized by puping it out to sea and releasing it into the mid-layer, away from the surface and bottom ecosystems.

Impingement and entrainment at intake structures are a concern due to large volumes of both river and sea water utilized in both PRO and RED schemes. Intake construction permits must meet strict environmental regulations and desalination plants and power plants that utilize surface water are sometimes involved with various local, state and federal agencies to obtain permission that can take upwards to 18 months.

Finally, some scientists have predicted that if China does not check their irrigation withdrawals from rivers, ALL Chinese rivers will fail to reach the ocean at least during part of the year by 2025. This has already happened with the mother of Chinese rivers, the Yellow River. An investment in osmotic power must consider future upstream use in the long-run

Fossil-fuel Power Station

Fossil fuel power stations have rotating machinery to convert the heat energy of combustion into mechanical energy, which then operates an electrical generator. The prime mover may be a steam turbine, a gas turbine or, in small plants, a reciprocating internal combustion engine. All plants use the energy extracted from expanding gas-steam or combustion gases. Very few MHD generators have been built which directly convert the energy of moving hot gas into electricity.

By-products of thermal power plant operation must be considered in their design and operation. Waste heat energy, which remains due to the finite efficiency of the Carnot, Rankine, or Diesel power cycle, is released directly to the atmosphere, directly to river or lake water, or indirectly to the atmosphere using a cooling tower with river or lake water used as a cooling medium. The flue gas from combustion of the fossil fuels is discharged to the air. This gas contains carbon dioxide and water vapour, as well as other substances such as nitrogen oxides (NO_x), sulfur oxides (SO_x), mercury, traces of other metals, and, for coal-fired plants, fly ash. Solid waste ash from coal-fired boilers must also be removed. Some coal ash can be recycled for building materials.

Fossil fueled power stations are major emitters of CO_2, a greenhouse gas (GHG) which according to a consensus opinion of scientific organisations is a contributor to global warming as it has been observed over the last 100 years. Per unit of electric energy, brown coal emits about 3 times as much CO_2 as natural gas, and black coal emits about twice as much CO_2. Carbon capture and storage of emissions is not expected to be available until governmental regulations force big polluters to reduce or eliminate their CO_2 emissions.

Basic Concepts

In a fossil fuel power plant the chemical energy stored in fossil fuels such as coal, fuel oil, natural gas or oil shale and oxygen of the air is converted successively into thermal energy, mechanical energy and, finally, electrical energy. Each fossil fuel power plant is a complex, custom-designed system. Construction costs, as of 2004, run to US$1,300 per kilowatt, or $650 million for a 500 MWe unit. Multiple generating units may be built at a single site for more efficient use of land, natural resources and labour. Most thermal power stations in the world use fossil fuel, out-numbering nuclear, geothermal, biomass, or solar thermal plants.

Heat into Mechanical Energy

The second law of thermodynamics states that any closed-loop cycle can only convert a fraction of the heat produced during combustion into mechanical work. The rest of the heat, called waste heat, must be released into a cooler environment during the return portion of the cycle. The fraction of heat released into a cooler medium must be equal or larger than the ratio of absolute temperatures of the cooling system (environment) and the heat source (combustion furnace). Raising the furnace temperature improves the efficiency but complicates the design, primarily by the selection of alloys used for construction, making the furnace more expensive. The waste heat cannot be converted into mechanical energy without an even cooler cooling system. However, it may be used in cogeneration plants to heat buildings, produce hot water, or to heat materials on an industrial scale, such as in some oil refineries, plants, and chemical synthesis plants.

Typical thermal efficiency for utility-scale electrical generators is around 33% for coal and oil-fired plants, and 56–60% (LHV) for combined-cycle gas-fired plants. Plants designed to achieve peak efficiency while operating at capacity will be less efficient when operating off-design (*i.e.* temperatures too low.)

Practical fossil-fuel stations operating as heat engines cannot exceed the Carnot cycle limit for conversion of heat energy into useful work. Fuel cells do not have the same thermodynamic limits as they are not heat engines.

Coal

Coal is the most abundant fossil fuel on the planet. It is a relatively cheap fuel, with some of the largest deposits in regions that are relatively stable politically, such as China, India and the United States. This contrasts with natural gas and petroleum, the largest deposits of which are located in the politically volatile Persian Gulf. Solid coal cannot directly replace natural gas or petroleum in most applications, petroleum is mostly used for transportation and the natural gas not used for electricity generation is used for space, water and industrial heating. Coal can be converted to gas or liquid fuel, but the efficiencies and economics of such processes can make them unfeasible. Vehicles or heaters may require modification to use coal-derived fuels. Coal can produce more pollution than petroleum or natural gas.

As of 2009 the largest coal-fired power station is Taichung Power Plant in Taiwan. The world's most energy-efficient coal-fired power plant is the Avedøre Power Station in Denmark.

Fuel Transport and Delivery

Coal is delivered by highway truck, rail, barge, collier ship or coal slurry pipeline. Some plants are even built near coal mines and coal is delivered by conveyors. A large coal train called a "unit train" may be two kilometers (over a mile) long, containing 130-140 cars with 100 short tons of coal in each one, for a total load of over 15,000 tons. A large plant under full load requires at least one coal delivery this size every day. Plants may get as many as three to five trains a day, especially in "peak season" during the hottest summer or coldest winter months (depending on local climate) when power consumption is high. A large thermal power plant such as the one in Nanticoke, Ontario stores several million metric tons of coal for winter use when the lakes are frozen.

Modern unloaders use rotary dump devices, which eliminate problems with coal freezing in bottom dump cars. The unloader includes a train positioner arm that pulls the entire train to position each car over a coal hopper. The dumper clamps an individual car against a platform that swivels the car upside down to dump the coal. Swiveling couplers enable the entire operation to occur while the cars are still coupled together. Unloading a unit train takes about three hours.

Shorter trains may use rail cars with an "air-dump", which relies on air pressure from the engine plus a "hot shoe" on each car. This "hot shoe" when it comes into contact with a "hot rail" at the unloading trestle, shoots an electric charge through the air dump apparatus and causes the doors on the bottom of the car to open, dumping the coal through the opening in the trestle. Unloading one of these trains takes anywhere from an hour to an hour and a half. Older unloaders may still use manually operated bottom-dump rail cars and a "shaker" attached to dump the coal. Generating stations adjacent to a mine may receive coal by conveyor belt or massive diesel-electric-drive trucks.

A collier (cargo ship carrying coal) may hold 40,000 long tons of coal and takes several days to unload. Some colliers carry their own conveying equipment to unload their own bunkers; others depend on equipment at the plant. Colliers are large, seaworthy, self-powered ships. For transporting coal in calmer waters, such as rivers and lakes, flat-bottomed vessels called barges are often used. Barges are usually unpowered and must be moved by tugboats or towboats.

For start up or auxiliary purposes, the plant may use fuel oil as well. Fuel oil can be delivered to plants by pipeline, tanker, tank car or truck. Oil is stored in vertical cylindrical steel tanks with capacities as high as 90,000 barrels (14,000 m^3)' worth. The heavier No. 5 "bunker" and No. 6 fuels are typically steam-heated before pumping in cold climates.

Fuel Processing

Coal is prepared for use by crushing the rough coal to pieces less than 2 inches (5 cm) in size. The coal is then transported from the storage yard to in-plant storage silos by rubberized conveyor belts at rates up to 4,000 short tons per hour.

In plants that burn pulverized coal, silos feed coal pulverizers (coal mills) that take the larger 2-inch (51 mm) pieces, grind them to the consistency of talcum powder, sort them, and mix them with primary combustion air which transports the coal to the boiler furnace and preheats the coal in order to drive off excess moisture content. A 500 MWe plant may have six such pulverizers, five of which can supply coal to the furnace at 250 tons per hour under full load.

In plants that do not burn pulverized coal, the larger 2-inch (51 mm) pieces may be directly fed into the silos which then feed either mechanical distributors that drop the coal on a travelling grate or the cyclone burners, a specific kind of combustor that can efficiently burn larger pieces of fuel.

Steam-electric

Most electric power made from fossil fuel is produced by thermal power stations. Reciprocating steam engines fell out of use rapidly after the first steam turbines were introduced around 1906.

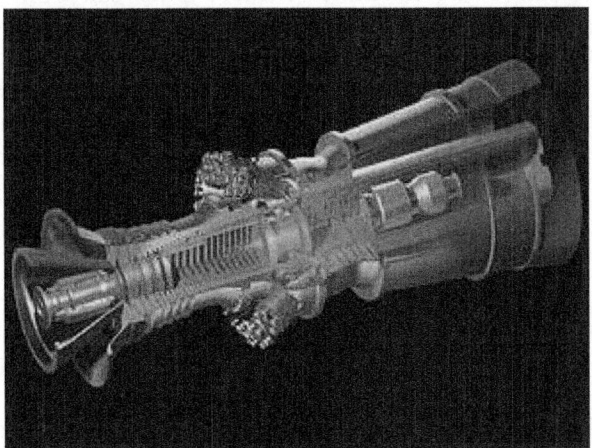

Fig. : 480 megawatt GE H series power generation gas turbine.

Gas Turbine Plants

One type of fossil fuel power plant uses a gas turbine in conjunction with a heat recovery steam generator (HRSG). It is referred to as a combined cycle power plant because it combines the Brayton cycle of the gas turbine with the Rankine cycle of the HRSG. The thermal efficiency of these plants has reached a record heat rate of 5690 Btu/(kW·h), or just under 60%, at a facility in Baglan Bay, Wales.

The turbines are fueled either with natural gas, syngas or fuel oil. While more efficient and faster to construct (a 1,000 MW plant may be completed in as little as 18 months from start of construction), the economics of such plants is heavily influenced by the volatile cost of fuel, normally natural gas. The combined cycle plants are designed in a variety of configurations composed of the number of gas turbines followed by the steam turbine. For example, a 3-1 combined cycle facility has three gas turbines tied to one steam turbine. The configurations range from (1-1), (2-1), (3-1), (4-1), (5-1), to (6-1)

Simple-cycle or open cycle gas turbine plants, without a steam cycle, are sometimes installed as emergency or peaking capacity; their thermal efficiency is much lower. The high running cost per hour is offset by the low capital cost and the intention to run such units only a few hundred hours per year. Other gas turbine plants are installed in stages, with an open cycle gas turbine the first stage and additional turbines or conversion to a closed cycle part of future project plans.

Reciprocating Engines

Diesel engine generator sets are often used for prime power in communities not connected to a widespread power grid. Emergency (standby) power systems may use reciprocating internal combustion engines operated by fuel oil or natural gas. Standby generators may serve as emergency power for a factory or data center, or may also be operated in parallel with the local utility system to reduce peak power demand charge from the utility. Diesel engines can produce strong torque at relatively low rotational speeds, which is generally desirable when driving an alternator, but diesel fuel in long-term storage can be subject to problems resulting from water accumulation and chemical decomposition. Rarely used generator sets may correspondingly be installed as natural gas or LPG to minimize the fuel system maintenance requirements.

Spark-ignition internal combustion engines operating on gasoline (petrol), propane, or LPG are commonly used as portable temporary power sources for construction work, emergency power, or recreational uses.

Reciprocating external combustion engines such as the Stirling engine can be run on a variety of fossil fuels, as well as renewable fuels or industrial waste heat. Installations of Stirling engines for power production are relatively uncommon.

Environmental Impacts

The world's power demands are expected to rise 60% by 2030. In 2007 there were over 50,000 active coal plants worldwide and this number is expected to grow. In 2004, the International Energy Agency (IEA) estimated that fossil fuels will account for 85% of the energy market by 2030.

World organizations and international agencies, like the IEA, are concerned about the environmental impact of burning fossil fuels, and coal in particular. The combustion of coal contributes the most to acid rain and air pollution, and has been connected with global warming. Due to the chemical composition of coal there are difficulties in removing impurities from the solid fuel prior to its combustion. Modern day coal power plants pollute less than older designs due to new "scrubber" technologies that filter the exhaust air in smoke stacks; however emission levels of various pollutants are still on average several times greater than natural gas power plants. In these modern designs, pollution from coal-fired power plants comes from the emission of gases such as carbon dioxide, nitrogen oxides, and sulfur dioxide into the air.

Acid rain is caused by the emission of nitrogen oxides and sulfur dioxide. These gases may be only mildly acidic themselves, yet when they react with the atmosphere, they create acidic compounds such as sulfurous acid, nitric acid and sulfuric acid which fall as rain, hence the term acid rain. In Europe and the U.S.A., stricter emission laws and decline in heavy industries have reduced the environmental hazards associated with this problem, leading to lower emissions after their peak in 1960s.

In 2008, the European Environment Agency (EEA) documented fuel-dependent emission factors based on actual emissions from power plants in the European Union.

Pollutant	Hard coal	Brown coal	Fuel oil	Other oil	Gas
CO_2 (g/GJ)	94,600	101,000	77,400	74,100	56,100
SO_2 (g/GJ)	765	1,361	1,350	228	0.68
NO_x (g/GJ)	292	183	195	129	93.3
CO (g/GJ)	89.1	89.1	15.7	15.7	14.5
Non-methane organic compounds (g/GJ)	4.92	7.78	3.70	3.24	1.58
Particulate matter (g/GJ)	1,203	3,254	16	1.91	0.1
Flue gas volume total (m^3/GJ)	360	444	279	276	272

Carbon Dioxide

Electricity generation using carbon based fuels is responsible for a large fraction of carbon dioxide (CO_2) emissions worldwide and for 34% of U.S. man-made carbon dioxide emissions in 2010. In the U.S., 70% of electricity generation is produced from combustion of fossil fuels.

Of the fossil fuels, coal is much more carbon intensive than oil or natural gas, resulting in greater volumes of carbon dioxide emissions per unit of electricity generated. In 2010, coal contributed about 81% of CO_2 emissions from generation and contributed about 45% of the electricity generated in the United States. In 2000, the carbon intensity of U.S. coal thermal combustion was 2249 lbs/MWh (1,029 kg/MWh).) while the carbon intensity of U.S. oil thermal generation was 1672 lb/MWh (758 kg/MWh or 211 kg/GJ) and the carbon intensity of U.S. natural gas thermal production was 1135 lb/MWh (515 kg/MWh or 143 kg/GJ).)

The Intergovernmental Panel on Climate Change states that carbon dioxide is a greenhouse gas and that increased quantities within the atmosphere will "very likely" lead to higher average temperatures on a global scale (global warming); concerns regarding the potential for such warming to change the global climate prompted IPCC recommendations calling for large cuts to CO_2 emissions worldwide.

Emissions may be reduced through more efficient and higher combustion temperature and through more efficient production of electricity within the cycle. Carbon capture and storage (CCS) of emissions from coal-fired power stations is another alternative but the technology is still being developed and will increase the cost of fossil fuel-based production of electricity. CCS may not be economically viable, unless the price of emitting CO_2 to the atmosphere rises.

Particulate Matter

Another problem related to coal combustion is the emission of particulates that have a serious impact on public health. Power plants remove particulate from the flue gas with the use of a bag house or electrostatic precipitator. Several newer plants that burn coal use a different process, Integrated Gasification Combined Cycle in which synthesis gas is made out of a reaction between coal and water. The synthesis gas is processed to remove most pollutants and then used initially to power gas turbines. Then the hot exhaust gases from the gas turbines are used to generate steam to power a steam turbine. The pollution levels of such plants are drastically lower than those of "classic" coal power plants.

Particulate matter from coal-fired plants can be harmful and have negative health impacts. Studies have shown that exposure to particulate matter is related to an increase of respiratory and cardiac mortality. Particulate matter can irritate small airways in the lungs, which can lead to increased problems with asthma, chronic bronchitis, airway obstruction, and gas exchange.

There are different types of particulate matter, depending on the chemical composition and size. The dominant form of particulate matter from coal-fired plants is coal fly ash, but secondary sulfate and nitrate also comprise a major portion of the particulate matter from coal-fired plants. Coal fly ash is what remains after the coal has been combusted, so it consists of the incombustible materials that are found in the coal.

The size and chemical composition of these particles affects the impacts on human health. Currently coarse (diameter greater than 2.5 µm) and fine (diameter between 0.1 µm and 2.5 µm) particles are regulated, but ultra-fine particles (diameter less than 0.1 µm) are currently unregulated, yet they pose many dangers. Unfortunately much is still unknown as to which kinds of particulate matter pose the most harm, which makes it difficult to come up with adequate legislation for regulating particulate matter.

There are several methods of helping to reduce the particulate matter emissions from coal-fired plants. Roughly 80% of the ash falls into an ash hopper, but the rest of the ash then gets carried into the atmosphere to become coal-fly ash. Methods of reducing these emissions of particulate matter include :

1. a baghouse
2. an electrostatic precipitator (ESP)
3. cyclone collector.

The baghouse has a fine filter that collects the ash particles, electrostatic precipitators use an electric field to trap ash particles on high-voltage plates, and cyclone collectors use centrifugal force to trap particles to the walls. A recent study indicates that sulfur emissions from fossil fueled power stations in China may have caused a 10-year lull in global warming (1998-2008)

Radioactive Trace Elements

Coal is a sedimentary rock formed primarily from accumulated plant matter, and it includes many inorganic minerals and elements which were deposited along with organic material during its formation. As the rest of the Earth's crust, coal also contains low levels of uranium, thorium, and other naturally occurring radioactive isotopes whose release into the environment leads to radioactive contamination. While these substances are present as very small trace impurities, enough coal is burned that significant amounts of these substances are released. A 1,000 MW coal-burning power plant could have an uncontrolled release of as much as 5.2 metric tons per year of uranium (containing 74 pounds (34 kg) of uranium-235) and 12.8 metric tons per year of thorium. In comparison, a 1,000 MW nuclear plant will generate about 30 short tons of high-level radioactive solid packed waste per year. It is estimated that during 1982, US coal burning released 155 times as much uncontrolled radioactivity into the atmosphere as the Three Mile Island incident. The collective radioactivity resulting from all coal burning worldwide between 1937 and 2040 is estimated to be 2,700,000 curies or 0.101 EBq. It should also be noted that during normal operation, the effective dose equivalent from coal plants is 100 times that from nuclear plants. But it is also worth noting that normal operation is a deceiving baseline for comparison : just the Chernobyl nuclear disaster released, in iodine-131 alone, an estimated 1.76 EBq . of radioactivity, a value one order of magnitude above this value for total emissions from all coal burned within a century. But at the same time, it shall also be understood that the iodine-131, the major radioactive substance which comes out in accident situations, has a half life of just 8 days.

Water and Air Contamination by Coal Ash

A study released in August 2010 that examined state pollution data in the United States by the organizations Environmental Integrity Project, the Sierra Club and Earth justice found that coal ash produced by coal-fired power plants dumped at sites across 21 U.S. states has contaminated ground water with toxic elements. The contaminants including the poisons arsenic and lead.

Arsenic has been shown to cause skin cancer, bladder cancer and lung cancer, and lead damages the nervous system. Coal ash contaminants are also linked to respiratory diseases and other health and developmental problems, and have disrupted local aquatic life. Coal ash also releases a variety of toxic contaminants into nearby air, posing a health threat to those who breath in fugitive coal dust.

Currently, the EPA does not regulate the disposal of coal ash; regulation is up to the states and the electric power industry has been lobbying to maintain this *status quo*. Most states require no monitoring of drinking water near coal ash dump sites. The study found an additional 39 contaminated U.S. sites and concluded that the problem of coal ash-caused water contamination is even more extensive in the United States than has been estimated. The study brought to 137 the number of ground water sites across the United States that are contaminated by power plant-produced coal ash.

Mercury Contamination

U.S. government scientists tested fish in 291 streams around the country for mercury contamination. They found mercury in every fish tested, according to the study by the U.S. Department of the Interior. They found mercury even in fish of isolated rural waterways. Twenty five per cent of the fish tested had mercury levels above the safety levels determined by the U.S. Environmental Protection Agency for people who eat the fish regularly. The largest source of mercury contamination in the United States is coal-fueled power plant emissions.

Greening of Fossil Fuel Power Plants

Several methods exist to improve the efficiency of fossil fuel power plants. A frequently used and cost-efficient method is to convert a plant to run on a different fuel. This includes conversions for biomass and waste. Conversions to waste-fired power plants have the benefit of reducing landfilling. In addition, waste-fired power plants can be equipped with material recovery, which is also beneficial to the environment.

Regardless of the conversion, a truly green fossil fuel power plant implements carbon capture and storage (CCS). CCS means that the exhaust CO_2 is not released into the environment and the fossil fuel power plant becomes an emissionless power plant. A 2006 example of a CCS fossil fuel power plant is the pilot Elsam power station near Esbjerg, Denmark.

Low NOx Burners

A common retrofit in fossil fueled power stations is the replacement of original burners with Low NOx burners. Careful consideration of fluid dynamics and flame thermodynamics has enabled substantial reduction in flame temperature, leading to reduced formation of Nitrous Oxides.

Clean Coal

"Clean coal" is the name attributed to a process whereby coal is chemically washed of minerals and impurities, sometimes gasified, burned and the resulting flue gases treated with steam, with the purpose of removing sulfur dioxide, and reburned so as to make the carbon dioxide in the flue gas economically recoverable. The coal industry uses the term "clean coal" to describe technologies designed to enhance both the efficiency and the environmental acceptability of coal extraction, preparation and use, but has provided no specific quantitative limits on any emissions, particularly carbon dioxide. Whereas contaminants like sulfur or mercury can be removed from coal, carbon cannot be effectively removed while still leaving a usable fuel, and clean coal plants without carbon sequestration and storage do not significantly reduce carbon dioxide emissions. James Hansen in an open letter to U.S. President Barack Obama has advocated a "moratorium and phase-out of coal plants that do not capture and store CO_2". In his book *Storms of My Grandchildren*, similarly, Hansen discusses his *Declaration of Stewardship* the first principle of which requires "a moratorium on coal-fired power plants that do not capture and sequester carbon dioxide".

Combined Heat and Power

Combined heat and power (CHP), also known as cogeneration, is the use of a power station to provide both electric power and process heat or district heating. While rejecting heat at a higher than normal temperature to enable building heating lowers overall plant electric power efficiency, the extra fuel burnt is more than offset by the reduction in fossil fuel that would otherwise be used for heating buildings. This technology is widely practiced in for example Denmark, other Scandinavian countries and parts of Germany. Calculations show that CHPDH is the cheapest method of carbon emissions reductions.

Alternatives to Fossil Fuel Power Plants

Alternatives to fossil fuel power plants include nuclear power, solar power, geothermal power, wind power, tidal power, hydroelectric power (hydroelectricity) and other renewable energies. Some of these are proven technologies on an industrial scale (*i.e.* nuclear, wind, tidal and hydroelectric power) others are still in prototype form.

Nuclear power, and geothermal power may be classed as heat pollutants as they add heat energy to the biosphere that would not otherwise be released. The net quantity of energy conversion within the biosphere due to the utilisation of wind power, solar power, tidal power, hydroelectric power (hydroelectricity) is static and is derived from the effects of sunlight and the movement of the moon and planets.

Generally, the cost of electrical energy produced by non-fossil fuel burning power plants is greater than that produced by burning fossil fuels. This statement however only includes the cost to produce the electrical energy and does not take into account indirect costs associated with the many pollutants created by burning fossil fuels (*e.g.* increased hospital admissions due respiratory diseases caused by fine smoke particles).

Relative Cost by Generation Source

When comparing power plant costs, it is customary to start by calculating the cost of power at the generator terminals by considering several main factors. External costs such as connections costs, the effect of each plant on the distribution grid are considered separately as an additional cost to the calculated power cost at the terminals.

Initial factors considered are :

- Capital costs (including waste disposal and decommissioning costs for nuclear energy)
- Operating and maintenance costs
- Fuel costs (for fossil fuel and biomass sources, and which may be negative for wastes)
- Likely annual hours per year run or load factor (may be 30% for wind energy, but 90% for nuclear energy)
- Offset sales of heat (for example in combined heat and power district heating (CHP/DH)).

These costs occur over the 30-50 year life of the fossil fuel power plants, using discounted cash flows. In general large fossil plants are attractive due to their low initial capital costs — typically around £750-£1000 per kilowatt electrical compared to perhaps £1500 per kilowatt for onshore wind.

Nuclear Power Plant

A **nuclear power plant** is a thermal power station in which the heat source is a nuclear reactor. As is typical in all conventional thermal power stations the heat is used to generate steam which drives a steam turbine connected to a generator which produces electricity. As of 16 January, 2013, the IAEA report there are 439 nuclear power reactors in operation operating in 31 countries.

Nuclear power plants are usually considered to be base load stations, since fuel is a small part of the cost of production.

History

Electricity was generated by a nuclear reactor for the first time ever on September 3, 1948 at the X-10 Graphite Reactor in Oak Ridge, Tennessee in the United States, and was the first nuclear power plant to power a light bulb. The second, larger experiment occurred on December 20, 1951 at the EBR-I experimental station near Arco, Idaho in the United States. On June 27, 1954, the world's first nuclear power plant to generate electricity for a power grid started operations at the Soviet city of Obninsk. The world's first full scale power station, Calder Hall in England opened on October 17, 1956.

Systems

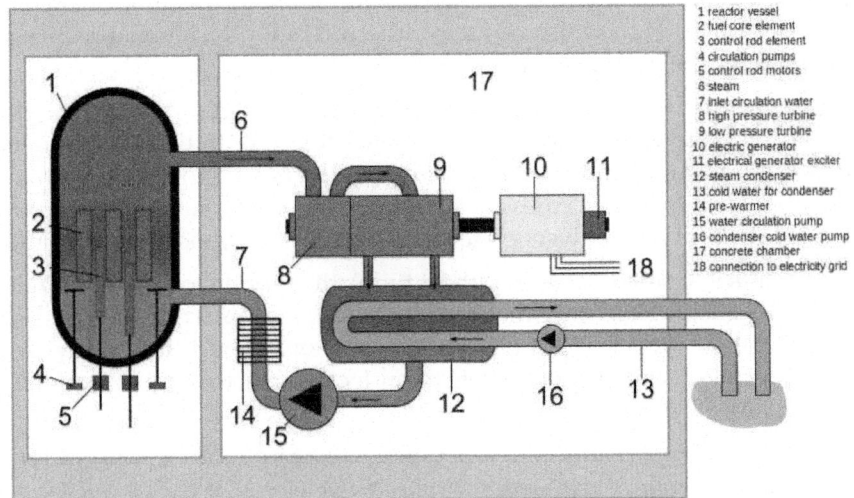

Fig. : BWR schematic.

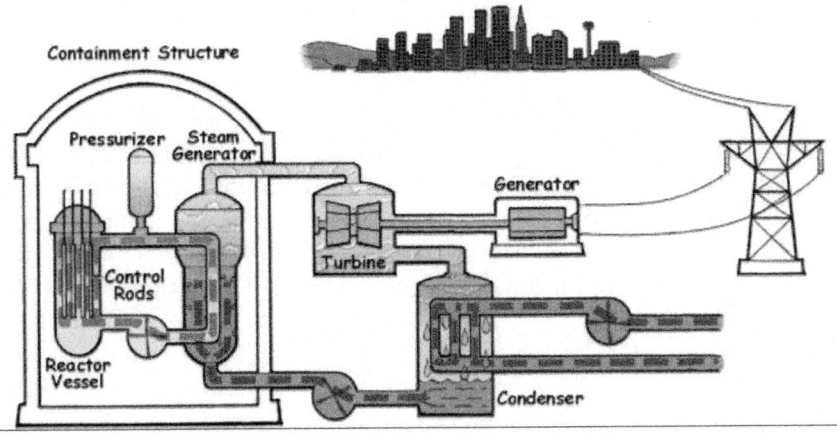

Fig. : Pressurized water reactor.

The conversion to electrical energy takes place indirectly, as in conventional thermal power plants. The heat is produced by fission in a nuclear reactor (a light water reactor). Directly or indirectly, water vapour (steam) is produced. The pressurized steam is then usually fed to a multi-stage steam turbine. Steam turbines in Western nuclear power plants are among the largest steam turbines ever. After the steam turbine has expanded and partially condensed the steam, the remaining vapour is condensed in a condenser. The condenser is a heat exchanger which is connected to a secondary side such as a river or a cooling tower. The water is then pumped back into the nuclear reactor and the cycle begins again. The water-steam cycle corresponds to the Rankine cycle.

Nuclear Reactors

A **nuclear reactor** is a device to initiate and control a sustained nuclear chain reaction. The most common use of nuclear reactors is for the generation of electric energy and for the propulsion of ships.

The nuclear reactor is the heart of the plant. In its central part, the reactor core's heat is generated by controlled nuclear fission. With this heat, a coolant is heated as it is pumped through the reactor and thereby removes the energy from the reactor. Heat from nuclear fission is used to raise steam, which runs through turbines, which in turn powers either ship's propellers or electrical generators.

Since nuclear fission creates radioactivity, the reactor core is surrounded by a protective shield. This containment absorbs radiation and prevents radioactive material from being released into the environment. In addition, many reactors are equipped with a dome of concrete to protect the reactor against both internal casualties and external impacts.

In nuclear power plants, different types of reactors, nuclear fuels, and cooling circuits and moderators are used.

Steam Turbine

The purpose of the steam turbine is to convert the heat contained in steam into mechanical energy. The engine house with the steam turbine is usually structurally separated from the main reactor building. It is so aligned to prevent debris from the destruction of a turbine in operation from flying towards the reactor.

In the case of a pressurized water reactor, the steam turbine is separated from the nuclear system. To detect a leak in the steam generator and thus the passage of radioactive water at an early stage, an activity meter is mounted to track the outlet steam of the steam generator. In contrast, boiling water reactors pass radioactive water through the steam turbine, so the turbine is kept as part of the control area of the nuclear power plant.

Generator

The generator converts kinetic energy supplied by the turbine into electrical energy. Low-pole AC synchronous generators of high rated power are used.

Cooling System

A cooling system removes heat from the reactor core and transports it to another area of the plant, where the thermal energy can be harnessed to produce electricity or to do other useful work. Typically the hot coolant is used as a heat source for a boiler, and the pressurized steam from that boiler powers one or more steam turbine driven electrical generators.

Safety Valves

In the event of an emergency, two independent safety valves can be used to prevent pipes from bursting or the reactor from exploding. The valves are designed so that they can derive all of the supplied flow rates with little increase in pressure. In the case of the BWR, the steam is directed into the condensate chamber and condenses there. The chambers on a heat exchanger are connected to the intermediate cooling circuit.

Feedwater Pump

The water level in the steam generator and nuclear reactor is controlled using the feedwater system. The feedwater pump has the task of taking the water from the condensate system, increasing the pressure and forcing it into either the steam generators (in the case of a pressurized water reactor) or directly into the reactor vessel (for boiling water reactors).

Emergency Power Supply

Most nuclear plants require two distinct sources of offsite power feeding station service transformers that are sufficiently separated in the plant's switchyard and can receive power from multiple transmission lines. In addition in some nuclear plants the turbine generator can power the plant's house loads while the plant is online via station service transformers which tap power from the generator output bus bars before they reach the step-up transformer (these plants also have station service transformers that receive offsite power directly from the switchyard.) Even with the redundancy of two power sources total loss of offsite power is still possible. Nuclear power plants are equipped with emergency power systems to maintain safety in the event of unit shutdown and loss of offsite power. Batteries provide uninterruptible power to instrumentation, control systems, and valves. Emergency diesel generators provide direct AC power to charge the batteries and to provide power to systems requiring AC power such as motor driven pumps. The emergency diesel generators do not power all plant systems, only those required to shut the reactor down safely, remove decay heat from the reactor, provide emergency core cooling, and, in some plants, spent fuel pool cooling. The large power generation pumps such as the main feedwater, condensate, circulating water, and (in pressurized water reactors) reactor coolant pumps are not backed up by the diesels. Some plants are also equipped with gas turbines to provide an additional layer of redundancy and can also provide power to non-safety-related loads during long outages.

People in a Nuclear Power Plant

In the United States and Canada, workers except for management, professional (such as engineers) and security personnel are likely to be members of either the International Brotherhood of Electrical Workers (IBEW) or the Utility Workers Union of America (UWUA), or one of the various trades and labour unions representing Machinist, labourers, boilermakers, millwrights, iron workers etc.

Economics

The economics of new nuclear power plants is a controversial subject, and multi-billion dollar investments ride on the choice of an energy source. Nuclear power plants typically have high capital costs, but low direct fuel costs, with the costs of fuel extraction, processing, use and spent fuel storage internalized costs. Therefore, comparison with other power generation methods is strongly dependent on assumptions about construction timescales and capital financing for nuclear plants. Cost estimates take into account plant decommissioning and nuclear waste storage or recycling costs in the United States due to the Price Anderson Act. With the prospect that all spent nuclear fuel/"nuclear waste" could potentially be recycled by using future reactors, generation IV reactors, that are being designed to completely close the nuclear fuel cycle.

On the other hand, construction, or capital cost aside, measures to mitigate global warming such as a carbon tax or carbon emissions trading, increasingly favour the economics of nuclear power. Further efficiencies are hoped to be achieved through more advanced reactor designs, Generation III reactors promise to be at least 17% more fuel efficient, and have lower capital costs, while futuristic Generation IV reactors promise 10000-30000% greater fuel efficiency and the elimination of nuclear waste.

In Eastern Europe, a number of long-established projects are struggling to find finance, notably Belene in Bulgaria and the additional reactors at Cernavoda in Romania, and some potential backers have pulled out. Where cheap gas is available and its future supply relatively secure, this also poses a major problem for nuclear projects.

Analysis of the economics of nuclear power must take into account who bears the risks of future uncertainties. To date all operating nuclear power plants were developed by state-owned or regulated utility monopolies where many of the risks associated with construction costs, operating performance, fuel price, and other factors were borne by consumers rather than suppliers. Many countries have now liberalized the electricity market where these risks, and the risk of cheaper competitors emerging before capital costs are recovered, are borne by plant suppliers and operators rather than consumers, which leads to a significantly different evaluation of the economics of new nuclear power plants.

Following the 2011 Fukushima I nuclear accidents, costs are likely to go up for currently operating and new nuclear power plants, due to increased requirements for on-site spent fuel management and elevated design basis threats. However many designs, such as the currently under construction AP1000, use passive nuclear safety cooling systems, unlike those of Fukushima I which required active cooling systems, this largely eliminates the necessity to spend more on redundant back up safety equipment.

Safety

There are trades to be made between safety, economic and technical properties of different reactor designs for particular applications. Historically these decisions were often made in private by scientists, regulators and engineers, but this may be considered problematic, and since Chernobyl and Three Mile Island, many involved now consider informed consent and morality to be primary considerations.

Complexity

Nuclear power plants are some of the most sophisticated and complex energy systems ever designed. Any complex system, no matter how well it is designed and engineered, cannot be deemed failure-proof. Veteran anti-nuclear activist and author Stephanie Cooke has argued :

The reactors themselves were enormously complex machines with an incalculable number of things that could go wrong. When that happened at Three Mile Island in 1979, another fault line in the nuclear world was exposed. One malfunction led to another, and then to a series of others, until the core of the reactor itself began to melt, and even the world's most highly trained nuclear engineers did not know how to respond. The accident revealed serious deficiencies in a system that was meant to protect public health and safety.

The 1979 Three Mile Island accident inspired Perrow's book *Normal Accidents*, where a nuclear accident occurs, resulting from an unanticipated interaction of multiple failures in a complex system. TMI was an example of a normal accident because it was "unexpected, incomprehensible, uncontrollable and unavoidable".

Perrow concluded that the failure at Three Mile Island was a consequence of the system's immense complexity. Such modern high-risk systems, he realized, were prone to failures however well they were managed. It was inevitable that they would eventually suffer what he termed a 'normal accident'. Therefore, he suggested, we might do better to contemplate a radical redesign, or if that was not possible, to abandon such technology entirely.

A fundamental issue contributing to a nuclear power system's complexity is its extremely long lifetime. The timeframe from the start of construction of a commercial nuclear power station through the safe disposal of its last radioactive waste, may be 100 to 150 years.

Failure Modes of Nuclear Power Plants

There are concerns that a combination of human and mechanical error at a nuclear facility could result in significant harm to people and the environment :

Operating nuclear reactors contain large amounts of radioactive fission products which, if dispersed, can pose a direct radiation hazard, contaminate soil and vegetation, and be ingested by humans and animals. Human exposure at high enough levels can cause both short-term illness and death and longer-term death by cancer and other diseases.

It is impossible for a commercial nuclear reactor to explode like a nuclear bomb since the fuel is never sufficiently enriched for this to occur.

Nuclear reactors can fail in a variety of ways. Should the instability of the nuclear material generate unexpected behaviour, it may result in an uncontrolled power excursion. Normally, the cooling system in a reactor is designed to be able to handle the excess heat this causes; however, should the reactor also experience a loss-of-coolant accident, then the fuel may melt or cause the vessel in which it is contained to overheat and melt. This event is called a nuclear meltdown.

After shutting down, for some time the reactor still needs external energy to power its cooling systems. Normally this energy is provided by the power grid to which that plant is connected, or by emergency diesel generators. Failure to provide power for the cooling systems, as happened in Fukushima I, can cause serious accidents.

Nuclear safety rules in the United States "do not adequately weigh the risk of a single event that would knock out electricity from the grid and from emergency generators, as a quake and tsunami recently did in Japan", Nuclear Regulatory Commission officials said in June 2011.

Vulnerability of Nuclear Plants to Attack

Nuclear reactors become preferred targets during military conflict and, over the past three decades, have been repeatedly attacked during military air strikes, occupations, invasions and campaigns :

- In September 1980, Iran bombed the Al Tuwaitha nuclear complex in Iraq, in Operation Scorch Sword.
- In June 1981, an Israeli air strike completely destroyed Iraq's Osirak nuclear research facility.
- Between 1984 and 1987, Iraq bombed Iran's Bushehr nuclear plant six times.
- In 1991, the U.S. bombed three nuclear reactors and an enrichment pilot facility in Iraq.
- In 1991, Iraq launched Scud missiles at Israel's Dimona nuclear power plant.
- In September 2007, Israel bombed a Syrian reactor under construction.

In the U.S., plants are surrounded by a double row of tall fences which are electronically monitored. The plant grounds are patrolled by a sizeable force of armed guards. The NRC's "Design Basis Threat" criteria for plants is a secret, and so what size of attacking force the plants are able to protect against is unknown. However, to scram (make an emergency shutdown) a plant takes fewer than 5 seconds while unimpeded restart takes hours, severely hampering a terrorist force in a goal to release radioactivity.

Attack from the air is an issue that has been highlighted since the September 11 attacks in the U.S. However, it was in 1972 when three hijackers took control of a domestic passenger flight along the east coast of the U.S. and threatened to crash the plane into a U.S. nuclear weapons plant in Oak Ridge, Tennessee. The plane got as close as 8,000 feet above the site before the hijackers' demands were met.

The most important barrier against the release of radioactivity in the event of an aircraft strike on a nuclear power plant is the containment building and its missile shield. Current NRC Chairman Dale Klein has said "Nuclear power plants are inherently robust structures that our studies show provide adequate protection in a hypothetical attack by an airplane. The NRC has also taken actions that require nuclear power plant operators to be able to manage large fires or explosions—no matter what has caused them."

In addition, supporters point to large studies carried out by the U.S. Electric Power Research Institute that tested the robustness of both reactor and waste fuel storage and found that they should be able to sustain a terrorist attack comparable to the September 11 terrorist attacks in the U.S. Spent fuel is usually housed inside the plant's "protected zone" or a spent nuclear fuel shipping cask; stealing it for use in a "dirty bomb" would be extremely difficult. Exposure to the intense radiation would almost certainly quickly incapacitate or kill anyone who attempts to do so.

Plant Location

In many countries, plants are often located on the coast, in order to provide a ready source of cooling water for the essential service water system. As a consequence the design needs to take the risk of flooding and tsunamis into account. The World Energy Council (WEC) argues disaster risks are changing and increasing the likelihood of disasters such as earthquakes, cyclones, hurricanes, typhoons, flooding. High temperatures, low precipitation levels and severe droughts may lead to fresh water shortages. Seawater is corrosive and so nuclear energy supply is likely to be negatively affected by the fresh water shortage. This generic problem may become increasingly significant over time. Failure to calculate the risk of flooding correctly lead to a Level 2 event on the International Nuclear Event Scale during the 1999 Blayais Nuclear Power Plant flood, while flooding caused by the 2011 Tōhoku earthquake and tsunami lead to the Fukushima I nuclear accidents.

The design of plants located in seismically active zones also requires the risk of earthquakes and tsunamis to be taken into account. Japan, India, China and the USA are among the countries to have plants in earthquake-prone regions. Damage caused to Japan's Kashiwazaki-Kariwa Nuclear Power Plant during the 2007 Chūetsu offshore earthquake underlined concerns expressed by experts in Japan prior to the Fukushima accidents, who have warned of a *genpatsu-shinsai* (domino-effect nuclear power plant earthquake disaster).

Multiple Reactors

The Fukushima nuclear disaster illustrated the dangers of building multiple nuclear reactor units close to one another. This proximity triggered the parallel, chain-reaction accidents that led to hydrogen explosions damaging reactor buildings and water draining from open-air spent fuel pools -- a situation that was potentially more dangerous than the loss of reactor cooling itself. Because of the closeness of the reactors, Plant Director Masao Yoshida "was put in the position of trying to cope simultaneously with core meltdowns at three reactors and exposed fuel pools at three units".

Nuclear Safety Systems

The three primary objectives of nuclear safety systems as defined by the Nuclear Regulatory Commission are to shut down the reactor, maintain it in a shutdown condition, and prevent the release of radioactive material during events and accidents. These objectives are accomplished using a variety of equipment, which is part of different systems, of which each performs specific functions.

Routine Emissions of Radioactive Materials

During everyday routine operations, emissions of radioactive materials from nuclear plants are released to the outside of the plants although they are quite slight amounts. The daily emissions go into the air, water and soil.

NRC says, "nuclear power plants sometimes release radioactive gases and liquids into the environment under controlled, monitored conditions to ensure that they pose no danger to the public or the environment", and "routine emissions during normal operation of a nuclear power plant are never lethal".

According to the United Nations (UNSCEAR), regular nuclear power plant operation including the nuclear fuel cycle amounts to 0.0002 mSv (milli-Sievert) annually in average public radiation exposure; the legacy of the Chernobyl disaster is 0.002 mSv/yr as a global average as of a 2008 report; and natural radiation exposure averages 2.4 mSv annually although frequently varying depending on an individual's location from 1 to 13 mSv.

The Japanese Myth f Absolute Safety

In Japan, many government agencies and nuclear companies have promoted a public myth of "absolute safety" that nuclear power proponents had nurtured over decades. The tsunami that began the Fukushima nuclear disaster could have been anticipated and in March 2012, Prime Minister Yoshihiko Noda acknowledged that the Japanese government shared the blame for the Fukushima disaster, saying that officials had been blinded to the country's "technological infallibility", and were all too steeped in a "safety myth".

In Japan, a national program to develop robots for use in nuclear emergencies was terminated in midstream because it "smacked too much of underlying danger". Japan, supposedly a major power in robotics, had none to send in to Fukushima during the disaster. Similarly, Japan's Nuclear Safety Commission stipulated in its safety guidelines for light-water nuclear facilities that "the potential for extended loss of power need not be considered." However, it was exactly such an extended loss of power to the cooling pumps that caused the meltdown at the Fukushima nuclear facilities.

Controversy

The nuclear power debate is about the controversy which has surrounded the deployment and use of nuclear fission reactors to generate electricity from nuclear fuel for civilian purposes. The debate about nuclear power peaked during the 1970s and 1980s, when it "reached an intensity unprecedented in the history of technology controversies", in some countries.

Proponents argue that nuclear power is a sustainable energy source which reduces carbon emissions and can increase energy security if its use supplants a dependence on imported fuels. Proponents advance the notion that nuclear power produces virtually no air pollution, in contrast to the chief viable alternative of fossil fuel. Proponents also believe that nuclear power is the only viable course to achieve energy independence for most Western countries. They emphasize that the risks of storing waste are small and can be further reduced by using the latest technology in newer reactors, and the operational safety record in the Western world is excellent when compared to the other major kinds of power plants.

Opponents say that nuclear power poses many threats to people and the environment. These threats include health risks and environmental damage from uranium mining, processing and transport, the risk of nuclear weapons proliferation or sabotage, and the unsolved problem of radioactive nuclear waste. They also contend that reactors themselves are enormously complex machines where many things can and do go wrong, and there have been many serious nuclear accidents. Critics do not believe that these risks can be reduced through new technology. They argue that when all the energy-intensive stages of the nuclear fuel chain are considered, from uranium mining to nuclear decommissioning, nuclear power is not a low-carbon electricity source.

Reprocessing

Nuclear reprocessing technology was developed to chemically separate and recover fissionable plutonium from irradiated nuclear fuel. Reprocessing serves multiple purposes, whose relative importance has changed over time. Originally reprocessing was used solely to extract plutonium for producing nuclear weapons. With the commercialization of nuclear power, the reprocessed plutonium was recycled back into MOX nuclear fuel for thermal reactors. The reprocessed uranium, which constitutes the bulk of the spent fuel material, can in principle also be re-used as fuel, but that is only economic when uranium prices are high or disposal is expensive. Finally, the breeder reactor can employ not only the recycled plutonium and uranium in spent fuel, but all the actinides, closing the nuclear fuel cycle and potentially multiplying the energy extracted from natural uranium by more than 60 times.

Nuclear reprocessing reduces the volume of high-level waste, but by itself does not reduce radioactivity or heat generation and therefore does not eliminate the need for a geological waste repository. Reprocessing has been politically controversial because of the potential to contribute to nuclear proliferation, the potential vulnerability to nuclear terrorism, the political challenges of repository siting (a problem that applies equally to direct disposal of spent fuel), and because of its high cost compared to the once-through fuel cycle. In the United States, the Obama administration stepped back from President Bush's plans for commercial-scale reprocessing and reverted to a program focused on reprocessing-related scientific research.

Accident Indemnification

The Vienna Convention on Civil Liability for Nuclear Damage puts in place an international framework for nuclear liability. However states with a majority of the world's nuclear power plants, including the U.S., Russia, China and Japan, are not party to international nuclear liability conventions.

In the U.S., insurance for nuclear or radiological incidents is covered (for facilities licensed through 2025) by the Price-Anderson Nuclear Industries Indemnity Act.

Under the Energy policy of the United Kingdom through its Nuclear Installations Act of 1965, liability is governed for nuclear damage for which a UK nuclear licensee is responsible. The Act requires compensation to be paid for damage up to a limit of £150 million by the liable operator for ten years after the incident. Between ten and thirty years afterwards, the Government meets this obligation. The Government is also liable for additional limited cross-border liability (about £300 million) under international conventions (Paris Convention on Third Party Liability in the Field of Nuclear Energy and Brussels Convention supplementary to the Paris Convention).

Decommissioning

Nuclear decommissioning is the dismantling of a nuclear power plant and decontamination of the site to a state no longer requiring protection from radiation for the general public. The main difference from the dismantling of other power plants is the presence of radioactive material that requires special precautions.

Warranty period of operation of nuclear power plants is 30 years. One from factors wear is the destruction of the reactors shell under the action of ionizing radiation.

Generally speaking, nuclear plants were designed for a life of about 30 years. Newer plants are designed for a 40 to 60-year operating life.

Decommissioning involves many administrative and technical actions. It includes all clean-up of radioactivity and progressive demolition of the plant. Once a facility is decommissioned, there should no longer be any danger of a radioactive accident or to any persons visiting it. After a facility has been completely decommissioned it is released from regulatory control, and the licensee of the plant no longer has responsibility for its nuclear safety.

Historic Accidents

The nuclear industry says that new technology and oversight have made nuclear plants much safer, but 57 small accidents have occurred since the Chernobyl disaster in 1986 until 2008. Two thirds of these mishaps occurred in the US. The French Atomic Energy Agency (CEA) has concluded that technical innovation cannot eliminate the risk of human errors in nuclear plant operation.

According to Benjamin Sovacool, an interdisciplinary team from MIT in 2003 estimated that given the expected growth of nuclear power from 2005 – 2055, at least four serious nuclear accidents would be expected in that period. However the MIT study Benjamin Sovacool references does not state this, instead the authors of the MIT study acknowledge that this estimated accident number does not take into account the existing, as of the time of publishing in 2003, and future, improvements in safety, with the authors basing this assumed high accident rate, cited by Sovacool, only if none of the improvements in safety technology from 1970 to 2003 were implemented. Therefore the figure the MIT team suggest, by their own account, would only be possible if the world nuclear fleet were to continue to operate and build old Generation II reactors and not learn from past mistakes, and that the world fleet would not include any of the presently (as of 2013) built, Generation III or in design phase Generation IV nuclear power plants between 2005 and 2055. The interdisciplinary team from MIT also went on to endorse that it was possible to make nuclear power safe, going on to state that the substantial safety features of then under construction Generation III reactors appear "plausible" at reducing the serious accident rate to near zero with, amongst other features, the use of passive nuclear safety features, which are now part of the state of the art in nuclear reactor safety.

Flexibility of Nuclear Power Plants

It is often claimed that nuclear stations are inflexible in their output, implying that other forms of energy would be required to meet peak demand. While that is true for the vast majority of reactors, this may no longer true of at least some modern designs.

Nuclear plants are routinely used in load following mode on a large scale in France, although "it is generally accepted that this is not an ideal economic situation for nuclear plants." Unit A at the German Biblis Nuclear Power Plant is designed to in- and decrease its output 15% per minute between 40 and 100% of its nominal power. Boiling water reactors normally have load-following capability, implemented by varying the recirculation water flow.

Future Power Plants

A number of new designs for nuclear power generation, collectively known as the Generation IV reactors, are the subject of active research and may be used for practical power generation in the future. Many of these new designs specifically attempt to make fission reactors cleaner, safer and/or less of a risk to the proliferation of nuclear weapons. Passively safe plants (such as the ESBWR) are available to be built and other designs that are believed to be nearly fool-proof are being pursued. Fusion reactors, which may be viable in the future, diminish or eliminate many of the risks associated with nuclear fission.

The 1600 MWe European Pressurized Reactor reactor is being built in Olkiluoto, Finland. A joint effort of French AREVA and German Siemens AG, it will be the largest reactor in the world. In December 2006 construction was about 18 months behind schedule so completion was expected 2010-2011.

As of March 2007, there are seven nuclear power plants under construction in India, and five in China.

In November 2011 Gulf Power stated that by the end of 2012 it hopes to finish buying off 4000 acres of land north of Pensacola, Florida in order to build a possible nuclear power plant.

Russia has begun building the world's first floating nuclear power plant. The £100 million vessel, the *Lomonosov*, is the first of seven plants that Moscow says will bring vital energy resources to remote Russian regions.

By 2025, Southeast Asia nations would have a total of 29 nuclear power plants, Indonesia will have 4 nuclear power plants, Malaysia 4, Thailand 5 and Vietnam 16 from nothing at all in 2011.

Expansion at two Nuclear Power Plants in the United States, Plant Vogtle and V. C. Summer Nuclear Power Plant, located in Georgia and South Carolina, respectively, are scheduled to be completed between 2016 and 2019. The two new Plant Vogtle reactors, and the two new reactors at Virgil C. Summer Nuclear Plant, represent the first nuclear power construction projects in the United States since the Three Mile Island nuclear accident in 1979.

Chapter 6

ELECTRONIC POWER TRANSFORMER CONTROL STRATEGY IN WIND ENERGY CONVERSION SYSTEMS FOR LOW VOLTAGE RIDE-THROUGH CAPABILITY ENHANCEMENT OF DIRECTLY DRIVEN WIND TURBINES WITH PERMANENT MAGNET SYNCHRONOUS GENERATORS (D-PMSGs)

Hui Huang, Chengxiong Mao, Jiming Lu and Dan Wang*

School of Electrical & Electronic Engineering, Huazhong University of Science and Technology, Wuhan 430074, Hubei, China; E-Mails: dh-1126@163.com (H.H.); cxmao @mail.hust.edu.cn (C.M.); lujiming@mail.hust.edu.cn (J.L.)

* Author to whom correspondence should be addressed; E-Mail: wangdan@mail.hust.edu.cn; Tel.: +86-27-8779-3169; Fax: +86-27-8754-2669.

External Editor: Frede Blaabjerg

ABSTRACT

This paper investigates the use of an Electronic Power Transformer (EPT) incorporated with an energy storage system to smooth the wind power fluctuations and enhance the low voltage ride-through (LVRT) capability of directly driven wind turbines with permanent magnet synchronous generators (D-PMSGs). The decoupled control schemes of the system, including the grid side converter control scheme, generator side converter control scheme and the control scheme of the energy storage system, are presented in detail. Under normal operating conditions, the energy storage system absorbs the high frequency component of the D-PMSG output power to smooth the wind power fluctuations. Under grid fault conditions, the energy storage system absorbs the redundant power, which could not be transferred to the grid by the EPT, to help the D-PMSG to ride through low voltage conditions. This coordinated control strategy is validated by simulation studies using MATLAB/Simulink. With the proposed control strategy, the output wind power quality is improved and the D-PMSG can ride through severe grid fault conditions.

Keywords

Electronic power transformer; permanent magnet synchronous generator; power smoothing; low voltage ride through; super-capacitor energy storage system.

1. INTRODUCTION

With the increasing urgency of both energy crisis and environmental pollution, there is a pressing urgent need to find alternative fuel sources which are clean, environmental-friendly and reproducible. Wind power is assumed to offer the most favorable technical and economic prospects in this respect, and is the most rapidly growing one among the various renewable energy sources [1,2].

There are mainly two kinds of wind power generators in wind farms, directly driven wind turbines with permanent magnet synchronous generators (D-PMSGs) and doubly fed induction generators (DFIGs). The D-PMSG type has gained much attention in wind power generation recently due to its advantages such as lower mechanical consumption, higher efficiency and energy yield, higher reliability, higher power to weight ratio and easier maintenance, compared with DFIGs [3]. However, the intermittent and fluctuant active power output of wind farms will have different impacts on various aspects of the power system, such as power quality, frequency control, voltage support, system reserve capacity, etc. especially under grid fault conditions, and if the wind power generators trip offline for self-protection, this will deteriorate the grid conditions and make the grid more difficult to recover [4]. To address these issues, on the one hand, wind farms generally need to install reactive power compensation devices, such as Static VAR (volt ampere reactive) compensators (SVCs), static compensators (STATCOMs) *etc.*, for wind generators to provide quick reactive power compensation and grid voltage support [5–7]; on the other hand, some control strategies based on the wind turbine pitch angle control or energy storage system have been proposed to smooth the wind power fluctuations or enhance the low voltage ride-through (LVRT) capability of wind generators [8].

As a multiple-functional FACTS (flexible alternating current transmission system) devices with the basic functions of the conventional power transformer [9], electronic power transformer (EPT), also called power electronic transformer (PET) [10,11] or solid-state transformer (SST) [12], have been attracting much attention from both academia and industry. On the one hand, a number of recent investigations have been done on the circuit topology design, and establishing the mathematical model and control strategy design of EPTs [13–17]. On the other hand, efforts are focusing on the applications of EPTs in areas where conventional power transformers are dominating, such as solar farms, wind farms, charge stations and smart grids [18–23]. In [20], a family of wind energy systems with integrated functions of active power transfer, reactive power compensation, and voltage conversion were proposed. In [21] researchers put forward a new D-PMSG grid-connected system based on SST, and a crowbar circuit was add to the DC (direct current) bus to enhance the LVRT capability of the D-PMSG.

In all the previous applications of EPT in wind farms, only the benefits of voltage conversion, reduced volume and weight, decoupling control of active and reactive power, and reactive power compensation were considered. Although it is known that EPTs incorporated with energy storage systems could smooth wind power fluctuations and enhance the LVRT capability of wind generators, to the best of the authors' knowledge, no literature has explored EPTs in these applications.

This paper investigates the use of EPTs incorporated with energy storage systems to smooth the wind power fluctuations and enhance the LVRT capability of D-PMSGs. The EPT control schemes for wind power fluctuation smoothing and LVRT capability enhancement are presented in detail. Case studies during wind speed increase, normal conditions, unbalanced grid voltage conditions and three phase ground fault conditions are conducted to verify the effectiveness of the proposed control strategy using MATLAB/Simulink.

2. SYSTEM DESCRIPTION AND SYSTEM MODEL

The configuration for the EPT considered in this work is presented in Figure 1, where the D-PMSG is connected to the grid network via an EPT, which is equipped with a super-capacitor energy storage system. The EPT is a three-stage design that includes a generator side stage, an isolation stage and a high voltage side stage (grid side stage). The generator side voltage source converter (VSC) controls the speed of the generator according to a maximum power point tracking (MPPT) algorithm, to extract the maximum amount of power with the actual wind force, the grid side VSC operates as an inverter to keep the generator side DC bus voltage constant. Considering the high grid AC (alternating current) voltage, a cascaded H-bridge topology is used for the high voltage stage. The isolation stage consists of nine DC/DC converters connected in parallel on the secondary side. The two H-bridge converters and a medium frequency transformer (MFT) in each

DC/DC converter constitute a medium-frequency modulating-demodulating block for voltage transformation and isolation, the DC voltage is modulated to a medium-frequency square wave by one of the H-bridge, then coupled to the secondary by the MFT, and reconverted into DC voltage via the other H-bridge. The energy storage system is connected to the EPT generator side DC bus via a DC/DC converter.

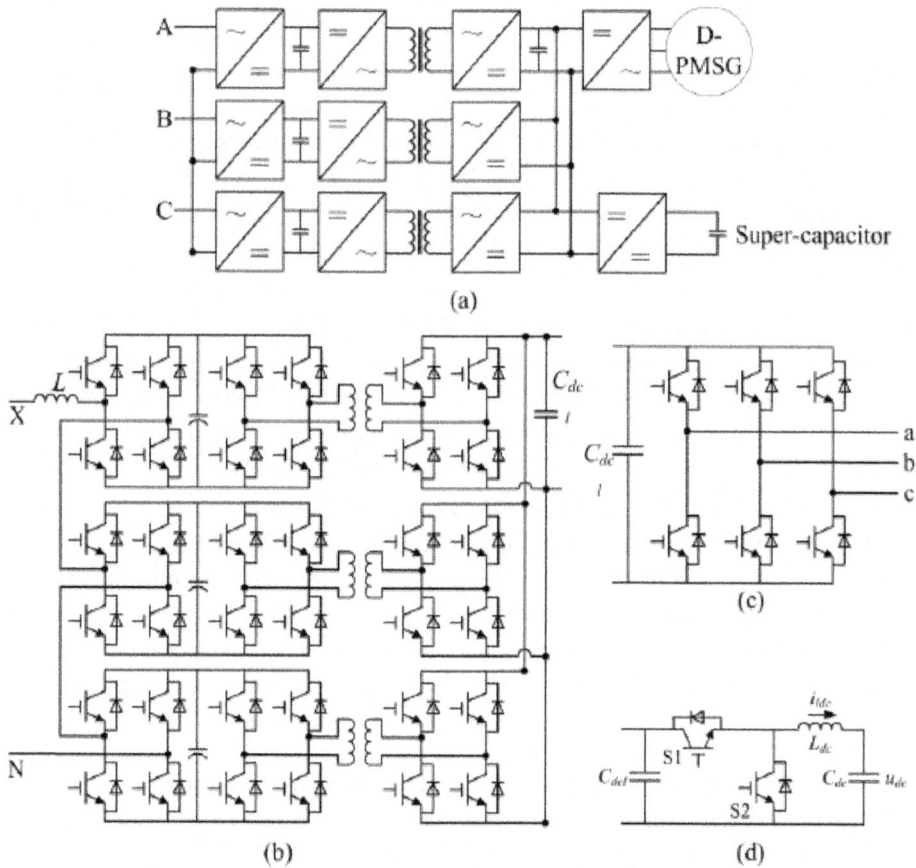

Figure 1. Topology of the system. (a) Basic diagram of the system; (b) Single phase topology of the high voltage side and isolation stage converters; (c) Topology of the generator side 3 phase converter; and (d) Topology of the DC/DC converter.

2.1. Low Voltage Ride through Requirement

Nowadays most grid codes require LVRT capability from large wind power plants. This means that they have to remain connected to the network during faults. Figure 2 gives the latest LVRT requirements in China. Wind turbines must remain connected to the network for voltages above the curve and may be disconnected otherwise.

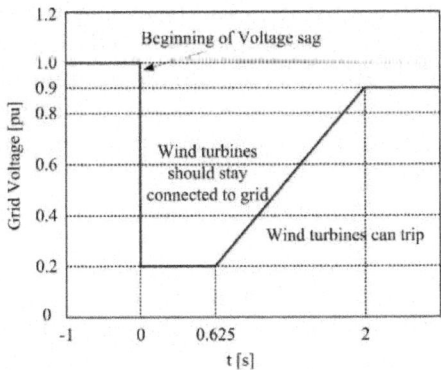

Figure 2. The LVRT (low voltage ride-through) requirements in China.

2.2. Mechanical Drive Train Model

The output mechanical power of the wind turbine is given by the following algebraic Equation (1):

$$\begin{cases} P_m = \frac{1}{2}\pi\rho R^2 v_w^3 C_p(\lambda, \beta) \\ \lambda = \omega_r R / v_w \end{cases} \quad (1)$$

where P_m is the extract mechanical power; ρ is the air density; R is the turbine radius; v_w is the wind speed; ω_r is the mechanical angular velocity of the generator; C_p is the power coefficient, which is a function of tip speed ratio λ and blade angle β. The tip speed ratio λ is given by Equation (1) and the blade angle β is controlled by the pitch angle controller. A two-mass shaft model [24] is used to represent torsional oscillations in the drive train.

2.3. PMSG Model

The PMSG model in the d-q frame can be described by the following equations ($L_d = L_q$ in this paper which corresponds to a non-salient machine):

$$\begin{cases} v_d = R_s i_d + L_d \dfrac{di_d}{dt} - \omega_e L_q i_q \\ v_q = R_s i_q + L_q \dfrac{di_q}{dt} + \omega_e L_d i_d + \omega_e \psi_m \\ T_e = \dfrac{3}{2} n_p \psi_m i_q \\ J \dfrac{d\omega_m}{dt} = T_m - T_e - f_B \omega_r \end{cases} \quad (2)$$

where v_d, v_q and i_d, i_q are stator voltages and currents in the d-q frame respectively; Rs is the stator resistance; L_d, L_q are inductances in the d-q frame; ω_e, ω_r are machine electrical and mechanical speed; T_e, T_m are the machine electro-magnetic torque and the turbine mechanical torque respectively; n_p is the machine pole pair number; ψ_m is the flux linkage created by the rotor permanent magnets; J is the total system inertia and f_B is the friction coefficient associated to the mechanical drive train.

2.4. The Isolation Stage Model

For the isolation stage, the amount and direction of the active power flow, presented in Equation (3) is determined by the phase-shift angle between the two AC square-wave signals [25]:

$$P_O = \frac{\frac{V_{dch}}{m} V_{dcl}}{2\pi f_T L_T} \varphi (1 - \frac{|\varphi|}{\pi}) \qquad (3)$$

where PO is the average power through the transformer; V_{dch} and V_{dcl} are the capacitor DC voltages; φ is the phase-shift angle between the two AC square-wave signals; m is the high frequency transformer ratio; f_T is the switching frequency; and L_T is the transformer leakage inductance referred to the secondary side.

2.5. Grid Side VSC Model

During grid fault conditions, the unbalanced three-phase grid voltage can be represented as the orthogonal sum of positive and negative sequences. The dynamics of the grid side VSC can be written in positive and negative d-q frame as follows [26]:

$$\begin{cases} u_d^P = R i_d^P + L \frac{di_d^P}{dt} - \omega L i_q^P + v_d^P \\ u_q^P = R i_q^P + L \frac{di_q^P}{dt} + \omega L i_d^P + v_q^P \\ u_d^N = R i_d^N + L \frac{di_d^N}{dt} - \omega L i_q^N + v_d^N \\ u_q^N = R i_q^N + L \frac{di_q^N}{dt} + \omega L i_d^N + v_q^N \end{cases} \qquad (4)$$

where u_d^P, u_q^P, u_d^N, u_q^N and i_d^P, i_q^P, i_d^N, i_q^N are positive and negative sequences of the grid voltages and currents in the d-q frame respectively; v_d^P, v_q^P, v_d^N, v_q^N are positive and negative sequences of the VSC pole voltage, which are generated by converter switches; R and L are the resistance and inductance of the VSC linked inductor; ω is the grid voltage frequency.

With the unbalanced input voltage, the power transferred to the grid is presented as follows:

$$\begin{bmatrix} P_{T0} \\ P_{Ts2} \\ P_{Tc2} \\ Q_{T0} \\ Q_{Ts2} \\ Q_{Tc2} \end{bmatrix} = \begin{bmatrix} u_d^P & u_q^P & u_d^N & u_q^N \\ u_q^N & -u_d^N & -u_q^P & u_d^P \\ u_d^N & u_q^N & u_d^P & u_q^P \\ u_q^P & -u_d^P & u_q^N & -u_d^N \\ -u_d^N & -u_q^N & u_d^P & u_q^P \\ u_q^N & -u_d^N & u_q^P & -u_d^P \end{bmatrix} \begin{bmatrix} i_d^P \\ i_q^P \\ i_d^N \\ i_q^N \end{bmatrix} \quad (5)$$

where P_{T0} and Q_{T0} are the mean active and reactive power, P_{Ts2} and Q_{Ts2} are the sine components of the second harmonic active and reactive power, P_{Tc2} and Q_{Tc2} are the cosine components of the second harmonic active and reactive power.

2.6. DC/DC Converter Model

The DC/DC converter presented in Figure 1d has two operation modes: Buck and Boost. By triggering insulated-gate bipolar transistors (IGBT) S1, the converter works in Buck mode and energy is transferred from the EPT low voltage side capacitor to the energy storage system. By triggering IGBT S2, the converter works in Boost mode and energy is transferred from the energy storage system to the EPT low voltage side capacitor. The dynamic model of the converter is presented as follows:

$$\begin{cases} L_{dc} \dfrac{di_{ldc}}{dt} = u_{pwm} - u_{sc} \\ C_{sc} \dfrac{du_{sc}}{dt} = i_{ldc} \end{cases} \quad (6)$$

where L_{dc} is the inductance of the inductor, C_{sc} is the capacitance of the supercapacitor stack, i_{ldc} is the current of the inductor, u_{sc} is the voltage of the supercapacitor stack, u_{pwm} is the voltage of IGBT S2.

3. CONTROL STRATEGY OF THE SYSTEM

3.1. Control Strategy of the Generator Side Converter

The main purpose of the generator side converter is to extract the maximum amount of power with the actual wind force, and the control strategy of the generator side converter is presented in Figure 3.

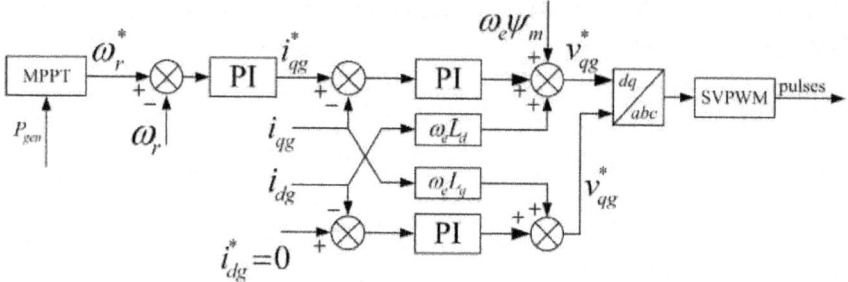

Figure 3. Control strategy of the generator side converter.

The outer control loop is the speed loop while the inner control loop is the current control loop [27]. In the outer speed loop, the speed reference is given by a MPPT algorithm presented in Equation (7) [28], where P_{gen} is output power of D-PMSG. The q axis reference current i_{qg}^* is obtained when the speed control error is adjusted by a PI (proportional integral) controller. The d axis reference current i_{dg}^* is set to zero:

$$\omega_r^* = \begin{cases} 1.2 & P_{gen} \geq 0.46 \\ -0.67 P_{gen}^2 + 1.42 P_{gen} + 0.51 & P_{gen} < 0.46 \end{cases} \tag{7}$$

3.2. Wind Turbine Pitch Angle Controller

The control strategy of the turbine pitch angle is depicted in Figure 4. The main purpose of the pitch angle controller is that when the available wind power is above the equipment rating (1.0 pu.), the blade pitch angle controller increases the pitch angle to limit the mechanical power delivered to the shaft to the equipment rating, and when the available wind power is less than equipment rating, the blades are set at minimum pitch to maximize the mechanical power [29].

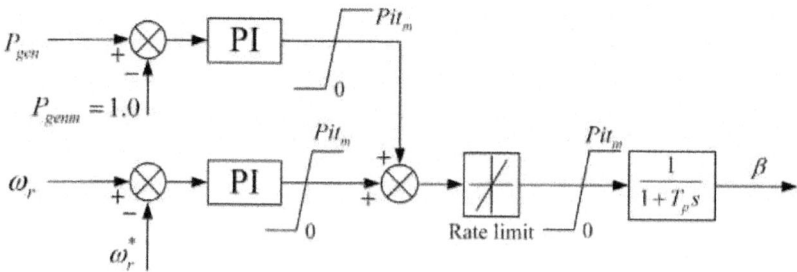

Figure 4. Control strategy of the turbine pitch angle.

3.3. Control Strategy of the Isolation Stage

The two H-bridges of the isolation stage operate at fixed 50% duty ratio and the frequency of the H-bridges drive signal is 2 kHz. The control block diagram is shown in Figure 5.

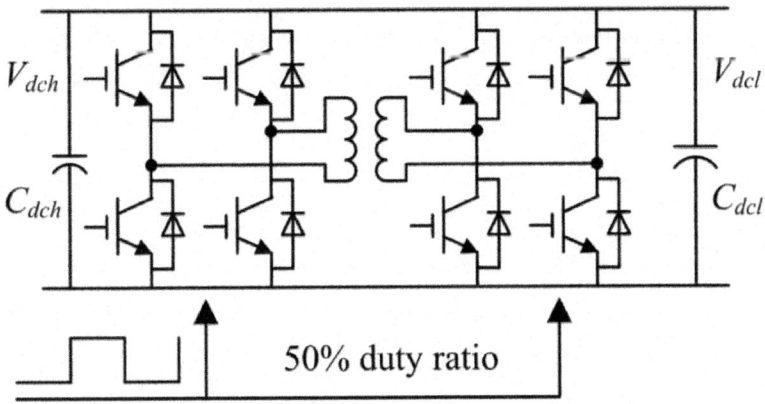

Figure 5. Control strategy of the isolation stage converter.

3.4. Control Strategy of the Grid Side Converter

During grid fault conditions, voltage unbalance causes performance deterioration of a pulse width modulation (PWM) converter by producing 100-Hz voltage ripples in the dc link and by increasing the reactive power [30]. To eliminate the negative sequence currents and the dc component of the reactive power, a dual current control scheme is introduced to control both positive and negative sequence currents [26], as shown in Figure 6, where the positive sequence voltage and current are measured in the positive synchronous reference frame (SRF) by eliminating the negative sequence with a 100-Hz notch filter, and the negative sequence voltage and current are measured in the negative SRF. The DC current reference i^*_{DC} is determined by the DC-link voltage controller, and the output active power command P^*_{To} is obtained by multiplying the DC-link voltage and the output of the DC voltage controller. As the grid side controller has to control the negative sequence currents and the DC component of the reactive power, in this paper the command values of these components are set to zero. Hence with Equation (5), the positive sequence current command $i^{P\,*}_d$ and $i^{P\,*}_q$ are obtained as follows:

$$\begin{cases} i^{P*}_d = \dfrac{\left(u^P_d\right)^2}{\left(u^P_d\right)^2 + \left(u^P_q\right)^2} P^*_{T0} \\ \\ i^{P*}_q = \dfrac{\left(u^P_q\right)^2}{\left(u^P_d\right)^2 + \left(u^P_q\right)^2} P^*_{T0} \end{cases} \qquad (8)$$

where $(u^P_d)^2 + (u^P_q)^2 \neq 0$ is assumed.

PI regulators are introduced to independently control the positive and negative sequence currents in the positive SRF and negative SRF. In positive SRF, the positive sequence converter pole voltages is calculated in Equation (9), and the negative sequence converter pole voltages are calculated in Equation (10) in negative SRF:

$$\begin{cases} V_d^{P*} = -\left(K_P + \dfrac{K_i}{S}\right)\left(i_d^{P*} - i_d^{P}\right) + u_d^{P} + \omega L i_q^{P} \\ V_q^{P*} = -\left(K_P + \dfrac{K_i}{S}\right)\left(i_q^{P*} - i_q^{P}\right) + u_q^{P} - \omega L i_d^{P} \end{cases} \quad (9)$$

$$\begin{cases} V_d^{N*} = -\left(K_P + \dfrac{K_i}{S}\right)\left(i_d^{N*} - i_d^{N}\right) + u_d^{N} - \omega L i_q^{N} \\ V_q^{N*} = -\left(K_P + \dfrac{K_i}{S}\right)\left(i_q^{N*} - i_q^{N}\right) + u_q^{N} + \omega L i_d^{N} \end{cases} \quad (10)$$

3.5. Control Strategy of the Energy Sorage System

The control strategy of the energy storage system is presented in Figure 7. The outer control loop is the power loop while the inner control loop is the current control loop. Under normal conditions, the energy storage system smoothes the power injected to the grid by absorbing the relatively high frequency components of the D-PMSG output power, and its reference power is calculated by submitting the output power of D-PMSG to a filter.

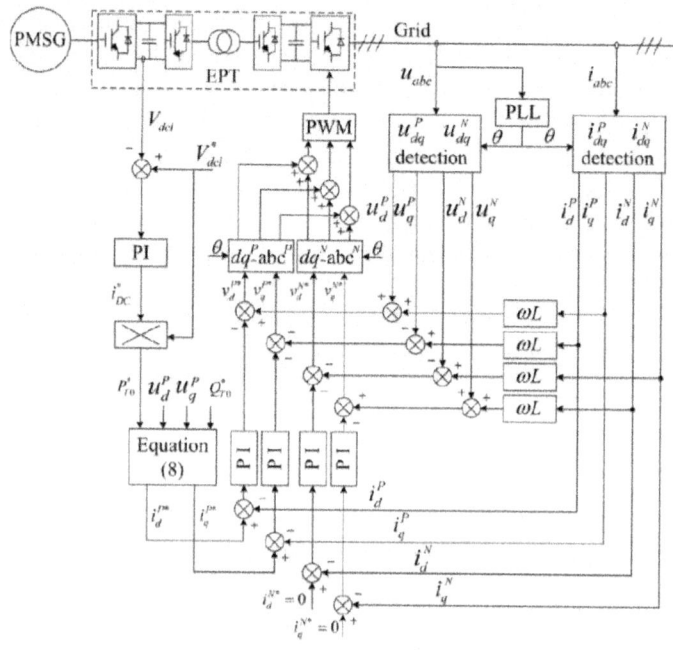

Figure 6. Control strategy of the grid side converter.

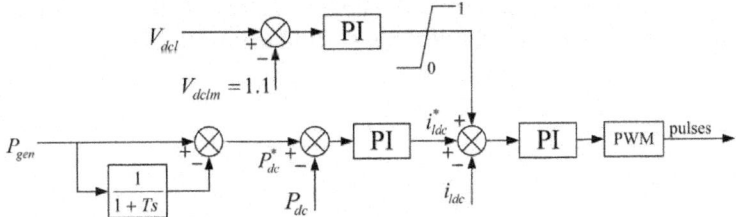

Figure 7. Control strategy of the DC/DC convert

During grid fault conditions, the generator side converter works to extract maximum amount of power from the wind while the power transferred to the grid is limited as the grid side converter current is limited by its current controller, this may cause the DC-link voltage increase and may damage the DC capacitance. Using the proposed control strategy shown in Figure 7, the DC-link voltage is limited to 1.1 pu as a supplementary controller is add to the current control loop to prevent the DC capacitance from overvoltage and the redundant power is absorbed by the energy storage system.

4. SIMULATION RESULTS

A simulation model is built in Matlab/Simulink to verify the effectiveness of the proposed control strategy. The D-PMSG parameters are: J = 33,000 kgm²; ψ_m = 1.245 Wb; p = 48; R_s = 0.01 Ω; $L_d = L_q$ = 8.35 mH; The rated power is 1.5 MVA; the rated output wind speed is 12 m/s. The 1.65 MVA 690 V/10 kV EPT parameters are: R = 0.0001 Ω; L = 20 mH; C_{dch} = 6400 µF; C_{dc2} = 27.2 mF; The generator side converter switching frequency: $(f_s)_{ac/dc}$ = 4 kHz; The isolation stage converter frequency: $(f_s)_{dc/dc}$ = 2 kHz; The grid side converter switching frequency: $(f_s)_{dc/ac}$ = 5 kHz. The medium frequency transformer ratio is 1500 V/3300 V. The rated capacity of the 600 V super-capacitor stack is 1.5 MVA/1 MWh. The smoothing period is seconds and the time constant in Figure 7 for calculating the reference super-capacitor stack power is 50 s.

4.1. Response to Wind Speed Step up

A step change of wind speed is applied to the system to test the dynamic performance of the whole system. Figure 8 shows the response of the system to a step change of wind speed V_w, which changes from 10 to 14 m/s at 20 s. As the rated output wind speed is 12 m/s, before the wind speed changes, the generator output power is less than 1.0 pu and the pitch angle is nearly 0°. After the wind speed changes to 14 m/s, the pitch controller adjusts the pitch angle to 5.07° to keep the generator speed at 1.2 pu and the output power at 1.0 pu, as shown in Figure 8b-d. As shown in Figure 8d, the D-PMSG output power P_{gen} increase from 0.523 to 1.0 pu in less than 2 s after the wind speed steps, while the power transferred to the grid P_{grid} increases more slowly and the redundant power is absorbed by the super-capacitor stack, the reactive power transferred to the grid by EPT is kept at 0 pu during all the simulation time. The low voltage side DC voltage presented in Figure 8e is kept constant during the simulation time.

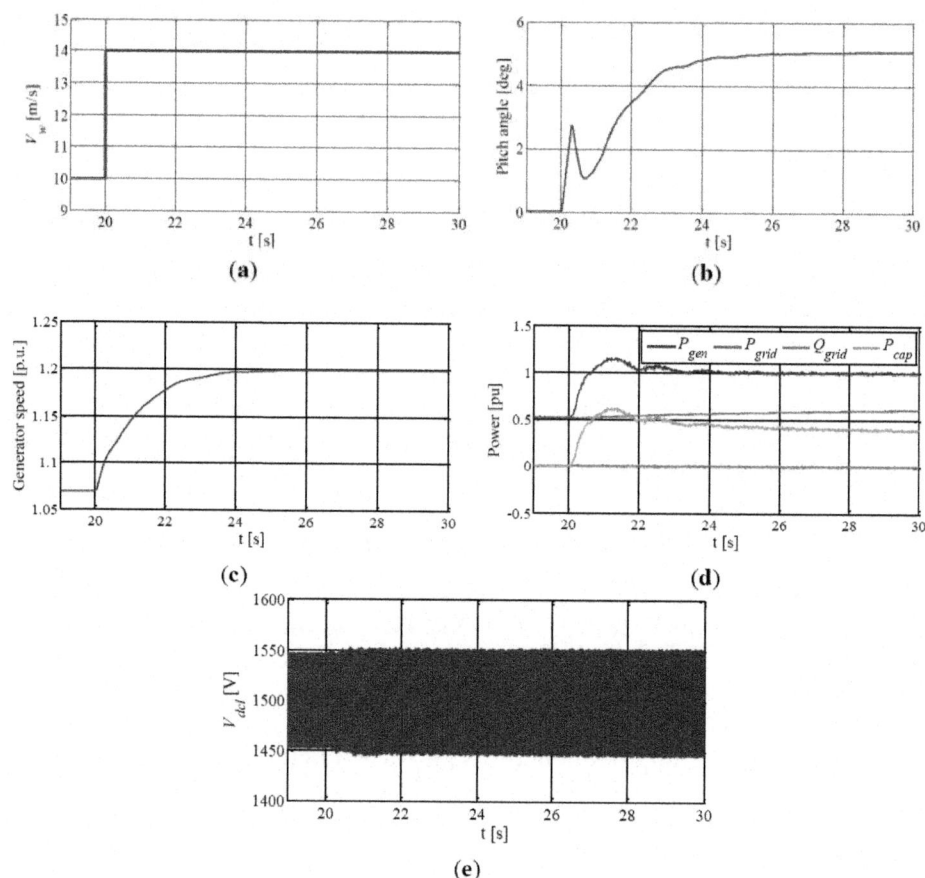

Figure 8. Simulation results of wind speed step up. (a) Wind speed; (b) Pitch angle; (c) Generator speed; (d) Power of D-PMSG EPT and super-capacitor; and (e) EPT low voltage side DC voltage.

4.2. Simulation Results under Normal Conditions

Under normal conditions, the energy storage system smoothes the power injected to the grid by absorbing the relatively high frequency components of the D-PMSG output power. Wind power with turbulence is used to test the power smoothing performance of the energy storage system under normal conditions, where the active power data of the D-PMSG is obtained from a wind farm located in central China. The D-PMSG output power P_{gen}, the power transferred to the grid P_{grid} and the power absorbed by the super-capacitor stack P_{cap} are presented in Figure 9. Figure 9a is a comparison of real D-PMSG output power and smoothed power profiles. It can be seen from Figure 9 that the D-PMSG output power is smoothed by the super-capacitor stack.

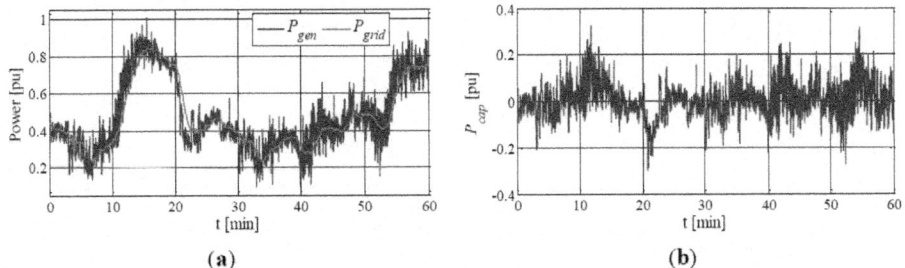

Figure 9. Simulation results under normal conditions (a) D-PMSG output power and the power transferred to the grid; (b) Power transferred to the super-capacitor stack.

4.3. Simulation Results under Unbalanced Grid Voltage Conditions

The wind speed is set at 11 m/s. The simulation results of unbalanced grid voltage sag are presented in Figure 10. Figure 10a presents the unbalanced grid voltage. During 10.0 to 10.2 s, phase A suffers an 80% voltage sag, phase B suffers an 85% voltage sag, and phase C suffers a 75% voltage sag. As the negative current is eliminated by the controller, the high voltage side converter currents remain balanced during all the simulation time, as shown in Figure 10b, but the active and reactive power of the high voltage side converter presented in Figure 10c, suffer 2ω oscillations during the voltage sag, as a result of the negative component of the unbalanced voltage. During the unbalanced grid voltage condition, the D-PMSG controller works to extract maximum amount of power from the wind, the generator speed pitch angle and output power of D-PMSG are presented in Figure 10d-f. But the power transferred to the grid is limited by the EPT high voltage side converter current, as shown in Figure 10b. This causes the EPT low voltage side DC voltage to increase, as shown in Figure 10g during 10 to 10.1 s the DC voltage increases from 1500 to 1650 V. When the EPT low voltage side DC voltage is more than 1650 V, the supplementary controller of the DC/DC controller starts to work to absorb the redundant power to prevent the DC capacitance from over voltage, and as shown in Figure 10g the DC voltage is less than 1800 V during unbalance grid voltage conditions. The power absorbed by the super-capacitor stack is presented in Figure 10h.

Another simulation of the system under unbalanced grid voltage conditions without super-capacitor is done to compare with the system with super-capacitor and the simulation results are presented in Figure 11. The unbalanced grid voltage is presented in Figure 10a. During the unbalanced grid voltage conditions, the D-PMSG controller works to extract maximum amount of power from the wind, but the power transferred to the grid, presented in Figure 11b, is limited by the EPT high voltage side converter current as shown in Figure 11a. This causes the EPT low voltage side DC voltage to increase, as shown in Figure 11c during 10 to 10.2 s the

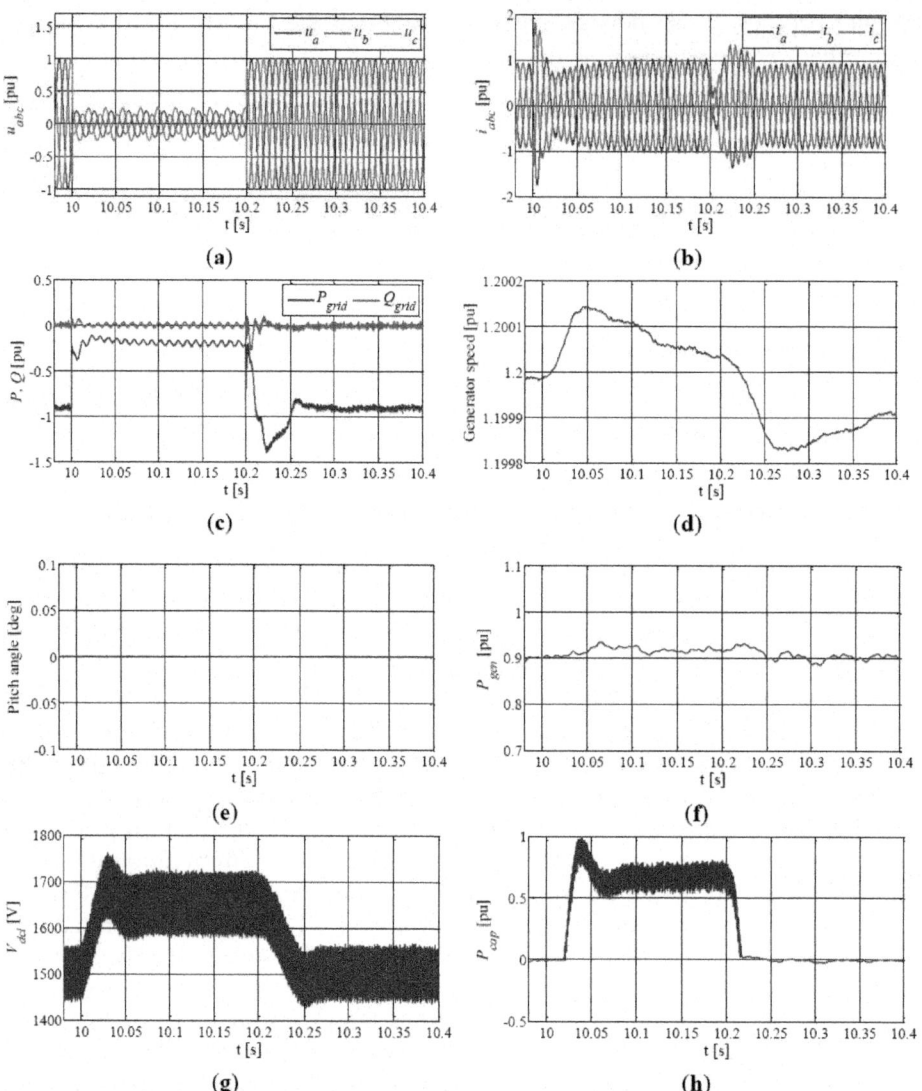

Figure 10. Simulation results under unbalanced grid voltage sag conditions. (a) Unbalanced grid voltage; (b) Grid side converter current; (c) Active and reactive power transferred to the grid; (d) Generator speed; (e) Pitch angle; (f) Output power of D-PMSG; (g) EPT low voltage side DC voltage; and (h) Active power absorbed by the super-capacitor stack.

DC voltage increases from 1500 to 2750 V and may damage the super-capacitor stack. From Figures 10 and 11, it can be seen that with super-capacitor the control scheme can improve the LVRT performance and prevent the DC capacitance from over voltage.

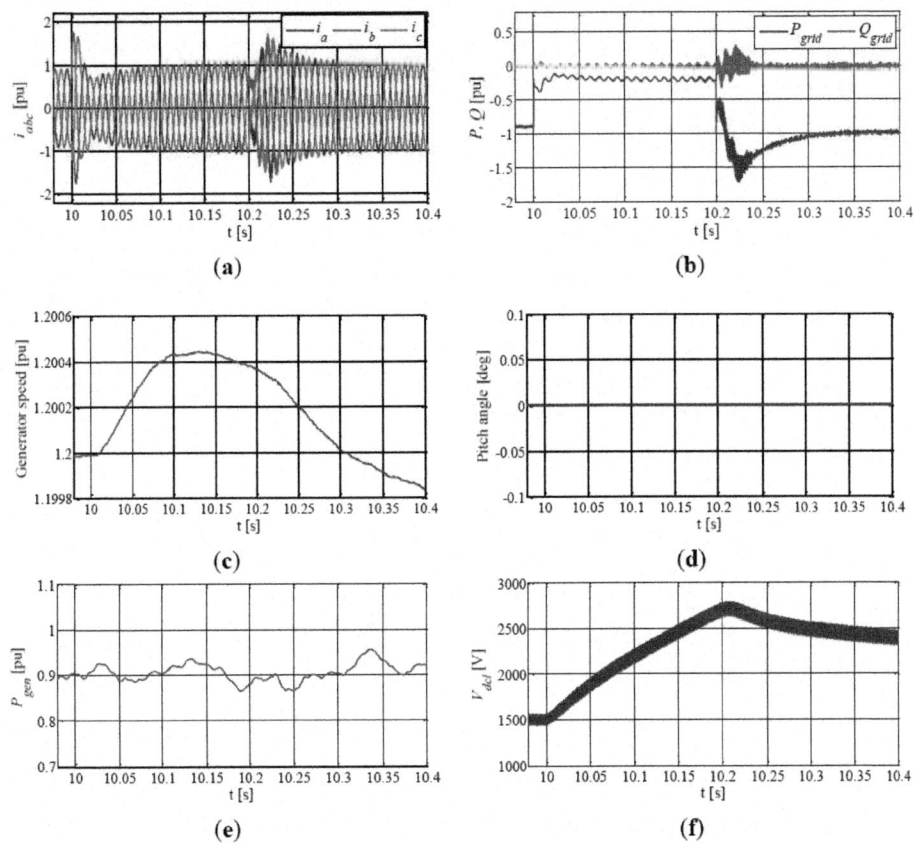

Figure 11. Simulation results under unbalanced grid voltage sag conditions without super-capacitor. (a) Grid side converter current; (b) Active and reactive power transferred to the grid; (c) Generator speed; (d) Pitch angle; (e) Output power of D-PMSG; and (f) EPT low voltage side DC voltage.

4.4. Simulation Results under Three Phase Ground Fault Conditions

The wind speed is set at 11 m/s. The three phase ground fault starts at 10.0 s and lasts for 0.2 s. The simulation results under three phase grid fault conditions are presented in Figure 12. Figure 12a presents the grid voltage. During 10.0 to 10.2 s the three phase voltages are 0 pu. The high voltage side converter currents are presented in Figure 12b. During the ground fault, as the voltage is 0 pu, the positive sequence current command $i^{P}_{d}{}^{*}$ and $i^{P}_{q}{}^{*}$ could not be calculated from Equation (8) and they are set to 0.65 and 0 pu, respectively. The active and reactive power transferred to the grid are presented in Figure 12c, which are 0 pu during the ground fault as the voltage is 0 pu. During the ground fault, the D-PMSG controller works to extract maximum amount of power from the wind which is presented in Figure 12d. However, the low voltage side DC voltage is limited to 1650 V as the redundant power is absorbed by the super-capacitor.

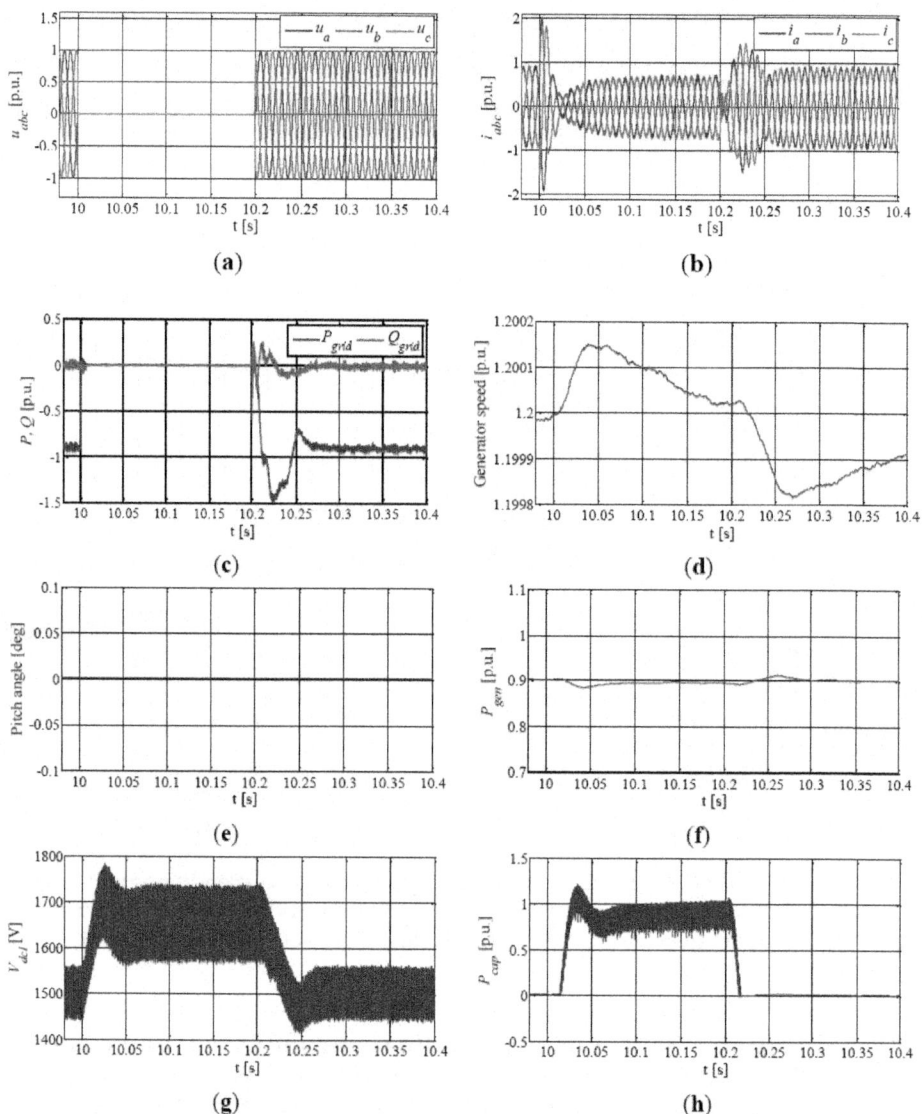

Figure 12. Simulation results under three phase ground fault conditions. (a) Unbalanced grid voltage; (b) Grid side converter current; (c) Active and reactive power transferred to the grid; (d) Generator speed; (e) Pitch angle; (f) Output power of D-PMSG; (g) EPT low voltage side DC voltage; and (h) Active power absorbed by the super-capacitor stack.

5. CONCLUSIONS

This paper has investigated how an Electronic Power Transformer incorporated with an energy storage system can be used to smooth the wind power fluctuations and enhance the LVRT capability of D-PMSGs in wind farms. The EPT system and D-PMSG system models and independent active and reactive power control strategies have been discussed The EPT control strategy for wind power fluctuation smoothing and LVRT capability enhancement of D-PMSGs has been proposed. Under normal operating conditions, the energy storage system absorbs the high frequency component of the D-PMSG output power to smooth the wind power fluctuations. Under grid fault conditions, the energy storage system absorbs the redundant power to help the D-PMSG ride through the low voltage conditions. This coordinated control strategy has been validated by simulation studies using MATLAB/Simulink, which show that the output wind power quality is improved and the D-PMSG can ride-through severe grid disturbances.

ACKNOWLEDGMENTS

This work was supported by the Project of the National Natural Science Foundation of China (51277083) and the Exquota Study Visit Funds China-UK. (51361130151).

Author Contributions

This paper is a result of the full collaboration of all the authors. Hui Huang and Wang Dan modelled the complete system and performed simulations. Manuscript preparation was completed by Hui Huang, Chengxiong Mao, Jiming Lu and Dan Wang. All author performed results analysis and contributed to the editing and reviewing of this document.

Conflicts of Interest

The authors declare no conflict of interest.

REFERENCES

1. Carrasco, J.M.; Franquelo, L.G.; Bialasiewicz, J.T.; Galvan, E.; Guisado, R.P.; Prats, A.M.; Leon, J.I.; Moreno-Alfonso, N. Power-electronic systems for the grid integration of renewable energy sources: A survey. *Ind. Electron. IEEE Trans.* **2006**, *53*, 1002–1016.
2. Fried, L. *Global Wind Statistics 2012*; Global Wind Energy Council (GWEC): Brussels, Belgium, 2013.
3. Li, H.; Chen, Z. Overview of different wind generator systems and their comparisons. *IET Renew. Power Gener.* **2008**, *2*, 123–138.
4. Zhang, K.; Duan, Y.; Wu, J.; Qiu, J.; Lu, J.; Fan, S.; Huang, H.; Mao, C. Low voltage ride through control strategy of directly driven wind turbine with energy storage system. In Proceedings of the 2011 IEEE Power and Energy Society General Meeting, San Diego, CA, USA, 24–29 July 2011; pp. 1–7
5. Mi, Z.; Chen, Y.; Liu, L.; Yu, Y. Dynamic performance improvement of wind farm with doubly fed induction generators using STATCOM. In Proceedings of the 2010 International Conference on Power System Technology (POWERCON), Hangzhou, China, 24–28 October 2010; pp. 1–6.
6. Obando-Montaño, A.F.; Carrillo, C.; Cidrás, J.; Díaz-Dorado, E. A STATCOM with supercapacitors for low-voltage ride-through in fixed-speed wind turbines. *Energies* **2014**, *7*, 5922–5952.

7. Han, C.; Huang, A.Q.; Baran, M.E.; Bhattacharya, S.; Litzenberger, W.; Anderson, L.; Johnson, A.L.; Edris, A.A. STATCOM impact study on the integration of a large wind farm into a weak loop power system. *Energy Convers. IEEE Trans.* **2008**, *23*, 226–233.
8. Conroy, J.F.; Watson, R. Low-voltage ride-through of a full converter wind turbine with permanent magnet generator. *IET Renew. Power Gener.* **2007**, *1*, 182–189.
9. Wang, D.; Mao, C.; Lu, J. Modelling of electronic power transformer and its application to power system. *IET Gener. Transm. Distrib.* **2007**, *1*, 887–895.
10. Manjrekar, M.D.; Kieferndorf, R.; Venkataramanan, G. Power electronic transformers for utility applications. In Proceedings of the Conference Record of the 2000 IEEE Industry Applications, Rome, Italy, 8–12 October 2000.
11. Ronan, E.R.; Sudhoff, S.D.; Glover, S.R.; Galloway, D.L. A power electronic-based distribution transformer. *Power Deliv. IEEE Trans.* **2002**, *17*, 537–543.
12. Van der Merwe, J.W. The solid-state transformer concept: A new era in power distribution. In Proceedings of the 2009 AFRICON, Nairobi, Kenya, 23–25 September 2009.
13. Wang, D.; Mao, C.; Lu, J.; Liu, H. Auto-balancing transformer based on power electronics. *Electr. Power Syst. Res.* **2010**, *80*, 28–36.
14. Wang, D.; Mao, C.; Lu, J.; Fan, S.; Peng, F. Theory and application of distribution electronic power transformer. *Electr. Power Syst. Res.* **2007**, *77*, 219–226.
15. Sang, Z.; Mao, C.; Lu, J.; Wang, D. Analysis and simulation of fault characteristics of power switch failures in distribution electronic power transformers. *Energies* **2013**, *6*, 4246–4268.
16. Liu, H.; Mao, C.; Lu, J.; Wang, D. Electronic power transformer with supercapacitors storage energy system. *Electr. Power Syst. Res.* **2009**, *79*, 1200–1208.
17. Liu, H.; Mao, C.; Lu, J.; Wang, D. Optimal regulator-based control of electronic power transformer for distribution systems. *Electr. Power Syst. Res.* **2009**, *79*, 863–870.
18. Brando, G.; Dannier, A.; Rizzo, R. Power electronic transformer application to grid connected photovoltaic systems. In Proceedings of the 2009 International Conference on Clean Electrical Power, Capri, Italy, 9–11 June 2009.
19. Gupta, R.K.; Castelino, G.F.; Mohapatra, K.K.; Mohan, N. A novel integrated three-phase, switched multi-winding power electronic transformer converter for wind power generation system. In Proceedings of the 35th Annual Conference of IEEE Industrial Electronics 2009, Porto, Portugal, 3–5 November 2009.
20. She, X.; Huang, A.Q.; Wang, F.; Burgos, R. Wind energy system with integrated functions of active power transfer, reactive power compensation, and voltage conversion. *Ind. Electron. IEEE Trans.* **2013**, *60*, 4512–4524.
21. Zhang, M.; Chen, J.; Wang, Z.; Wang, S.; Ouyang, L. A new permanent magnet synchronous wind-power generation grid-connected system. *Power Syst. Prot. Control* **2013**, *41*, 141–148.
22. Zhang, M.; Liu, J.; Jin, X. Research on the SVPWM solid state transformer applied in smart micro-grid. *Trans China Electrotech. Soc.* **2012**, *27*, 90–97. (In Chinese)
23. She, X.; Huang, A.Q.; Lukic, S.; Baran, M.E. On integration of solid-state transformer with zonal DC microgrid. *Smart Grid IEEE Trans.* **2012**, *3*, 975–985.
24. Muyeen, S.M.; Ali, M.H.; Takahashi, R.; Murata, T.; Tamura, J.; Tomaki, Y.; Sakahara, A.; Sasano, E. Comparative study on transient stability analysis of wind turbine generator system using different drive train models. *Renew. Power Gener. IET* **2007**, *1*, 131–141.
25. De Doncker, R.W.A.A.; Divan, D.M.; Kheraluwala, M.H. A three-phase soft-switched high-power-density DC/DC converter for high-power applications. *Ind. Appl. IEEE Trans.* **1991**, *27*, 63–73.
26. Song, H.; Nam, K. Dual current control scheme for PWM converter under unbalanced input voltage conditions. *Ind. Electron. IEEE Trans.* **1999**, *46*, 953–959.

27. Chinchilla, M.; Arnaltes, S.; Burgos, J.C. Control of permanent-magnet generators applied to variable-speed wind-energy systems connected to the grid. *Energy Convers. IEEE Trans.* **2006**, *21*, 130–135.
28. Miller, N.W.; Price, W.W.; Sanchez-Gasca, J.J. *GE-Power Systems Energy Consulting: Dynamic Modeling of GE 1.5 and 3.6 Wind Turbine-Generators*; General Electric International, Inc.: Schenectady, NY, USA, 2003.
29. Clark, K.; Miller, N.W.; Sanchez-Gasca, J.J. *General Electric International Technical Report: Modeling of GE Wind Turbine-Generators for Grid Studies*; General Electric International, Inc.: Schenectady, NY, USA, 2010.
30. Enjeti, P.N.; Choudhury, S.A. A new control strategy to improve the performance of a PWM AC to DC converter under unbalanced operating conditions. *Power Electron. IEEE Trans.* **1993**, *8*, 493–500.

This page left intentionally blank.